AF615218

RVSI

Acuity CiMatrix

Enjoy!

JcA

Acuity CiMatrix, 486 Amherst Street, Nashua, NH 03063

UNDERSTANDING AND APPLYING MACHINE VISION

MANUFACTURING ENGINEERING AND MATERIALS PROCESSING
A Series of Reference Books and Textbooks

FOUNDING EDITOR

Geoffrey Boothroyd
University of Rhode Island
Kingston, Rhode Island

1. Computers in Manufacturing, *U. Rembold, M. Seth, and J. S. Weinstein*
2. Cold Rolling of Steel, *William L. Roberts*
3. Strengthening of Ceramics: Treatments, Tests, and Design Applications, *Harry P. Kirchner*
4. Metal Forming: The Application of Limit Analysis, *Betzalel Avitzur*
5. Improving Productivity by Classification, Coding, and Data Base Standardization: The Key to Maximizing CAD/CAM and Group Technology, *William F. Hyde*
6. Automatic Assembly, *Geoffrey Boothroyd, Corrado Poli, and Laurence E. Murch*
7. Manufacturing Engineering Processes, *Leo Alting*
8. Modern Ceramic Engineering: Properties, Processing, and Use in Design, *David W. Richerson*
9. Interface Technology for Computer-Controlled Manufacturing Processes, *Ulrich Rembold, Karl Armbruster, and Wolfgang Ülzmann*
10. Hot Rolling of Steel, *William L. Roberts*
11. Adhesives in Manufacturing, *edited by Gerald L. Schneberger*
12. Understanding the Manufacturing Process: Key to Successful CAD/ CAM Implementation, *Joseph Harrington, Jr.*
13. Industrial Materials Science and Engineering, *edited by Lawrence E. Murr*
14. Lubricants and Lubrication in Metalworking Operations, *Elliot S. Nachtman and Serope Kalpakjian*
15. Manufacturing Engineering: An Introduction to the Basic Functions, *John P. Tanner*
16. Computer-Integrated Manufacturing Technology and Systems, *Ulrich Rembold, Christian Blume, and Ruediger Dillman*
17. Connections in Electronic Assemblies, *Anthony J. Bilotta*
18. Automation for Press Feed Operations: Applications and Economics, *Edward Walker*
19. Nontraditional Manufacturing Processes, *Gary F. Benedict*
20. Programmable Controllers for Factory Automation, *David G. Johnson*
21. Printed Circuit Assembly Manufacturing, *Fred W. Kear*
22. Manufacturing High Technology Handbook, *edited by Donatas Tijunelis and Keith E. McKee*
23. Factory Information Systems: Design and Implementation for CIM Management and Control, *John Gaylord*
24. Flat Processing of Steel, *William L. Roberts*
25. Soldering for Electronic Assemblies, *Leo P. Lambert*

26. Flexible Manufacturing Systems in Practice: Applications, Design, and Simulation, *Joseph Talavage and Roger G. Hannam*
27. Flexible Manufacturing Systems: Benefits for the Low Inventory Factory, *John E. Lenz*
28. Fundamentals of Machining and Machine Tools: Second Edition, *Geoffrey Boothroyd and Winston A. Knight*
29. Computer-Automated Process Planning for World-Class Manufacturing, *James Nolen*
30. Steel-Rolling Technology: Theory and Practice, *Vladimir B. Ginzburg*
31. Computer Integrated Electronics Manufacturing and Testing, *Jack Arabian*
32. In-Process Measurement and Control, *Stephan D. Murphy*
33. Assembly Line Design: Methodology and Applications, *We-Min Chow*
34. Robot Technology and Applications, *edited by Ulrich Rembold*
35. Mechanical Deburring and Surface Finishing Technology, *Alfred F. Scheider*
36. Manufacturing Engineering: An Introduction to the Basic Functions, Second Edition, Revised and Expanded, *John P. Tanner*
37. Assembly Automation and Product Design, *Geoffrey Boothroyd*
38. Hybrid Assemblies and Multichip Modules, *Fred W. Kear*
39. High-Quality Steel Rolling: Theory and Practice, *Vladimir B. Ginzburg*
40. Manufacturing Engineering Processes: Second Edition, Revised and Expanded, *Leo Alting*
41. Metalworking Fluids, *edited by Jerry P. Byers*
42. Coordinate Measuring Machines and Systems, *edited by John A. Bosch*
43. Arc Welding Automation, *Howard B. Cary*
44. Facilities Planning and Materials Handling: Methods and Requirements, *Vijay S. Sheth*
45. Continuous Flow Manufacturing: Quality in Design and Processes, *Pierre C. Guerindon*
46. Laser Materials Processing, *edited by Leonard Migliore*
47. Re-Engineering the Manufacturing System: Applying the Theory of Constraints, *Robert E. Stein*
48. Handbook of Manufacturing Engineering, *edited by Jack M. Walker*
49. Metal Cutting Theory and Practice, *David A. Stephenson and John S. Agapiou*
50. Manufacturing Process Design and Optimization, *Robert F. Rhyder*
51. Statistical Process Control in Manufacturing Practice, *Fred W. Kear*
52. Measurement of Geometric Tolerances in Manufacturing, *James D. Meadows*
53. Machining of Ceramics and Composites, *edited by Said Jahanmir, M. Ramulu, and Philip Koshy*
54. Introduction to Manufacturing Processes and Materials, *Robert C. Creese*
55. Computer-Aided Fixture Design, *Yiming (Kevin) Rong and Yaoxiang (Stephens) Zhu*
56. Understanding and Applying Machine Vision: Second Edition, Revised and Expanded, *Nello Zuech*

Additional Volumes in Preparation

UNDERSTANDING AND APPLYING MACHINE VISION

SECOND EDITION, REVISED AND EXPANDED

NELLO ZUECH

Vision Systems International
Yardley, Pennsylvania

MARCEL DEKKER, INC. NEW YORK • BASEL

Library of Congress Cataloging-in-Publication

Zuech, Nello.
Understanding and applying machine vision / Nello Zuech.— 2nd ed., rev. and expanded.
p. cm. — (Manufacturing engineering and materials processing)
Rev. ed. of: Applying machine vision, c1988.
ISBN 0-8247-8929-6
1. Computer vision. 2. Applying machine vision. I. Zuech, Nello. II. Title. III. Series.
TA1634.Z84 2000
006.3'7—dc21 99-051332

The first edition was published as *Applying Machine Vision* (John Wiley & Sons, Inc., 1988).

This book is printed on acid-free paper.

Headquarters
Marcel Dekker, Inc.
270 Madison Avenue, New York, NY 10016
tel: 212-696-9000; fax: 212-685-4540

Eastern Hemisphere Distribution
Marcel Dekker AG
Hutgasse 4, Postfach 812, CH-4001 Basel, Switzerland
tel: 41-61-261-8482; fax: 41-61-261-8896

World Wide Web
http://www.dekker.com

The publisher offers discounts on this book when ordered in bulk quantities. For more information, write to Special Sales/Professional Marketing at the headquarters address above.

Copyright © 2000 by Marcel Dekker, Inc. All Rights Reserved.

Neither this book nor any part may be reproduced or transmitted in any form or by any means, electronic or mechanical, including photocopying, microfilming, and recording, or by any information storage and retrieval system, without permission in writing from the publisher.

Current printing (last digit)
10 9 8 7 6 5 4 3 2 1

PRINTED IN THE UNITED STATES OF AMERICA

Preface

This book was written to inform prospective end users of machine vision technology, not designers of such systems. It provides the reader with sufficient background to separate machine vision promises from machine vision reality and to make intelligent decisions regarding machine vision applications.

The emphasis of the text is on understanding machine vision as it relates to potential applications and, conversely, understanding an application as it relates to machine vision. The book is designed to serve as a translator so a potential buyer can convey his requirements comprehensively. It will also provide the prospective buyer with basic understanding of the underlying technology embedded in machine vision systems.

The first chapter sets the tone for the book. It emphasizes that machine vision is a data collector, and thus the value of a machine vision system is in the data itself. Chapter 2 delves into the history of machine vision. Chapters 3 and 4 discuss principles in lighting, optics and cameras. The application engineering surrounding these elements of a machine vision system typically account for 50% of the design effort. Chapter 5 reviews the underlying image processing and analysis principles associated with machine vision. Chapters 6 and 7 discuss three-dimensional and color machine vision techniques.

The rest of the book is designed to provide a roadmap for successfully pursuing a machine vision application. Chapter 8 describes the various players that constitute the machine vision industry. Chapter 9 provides a means to perform a "back-of-the-envelope" estimate to determine the feasibility of a specific application. Chapter 10 reviews specific tactics to employ as one proceeds to deploy machine vision within a factory. Following the procedures given here will increase the probability of a successful installation.

Nello Zuech

Contents

UNDERSTANDING AND APPLYING MACHINE VISION

1

Machine Vision: A Data Acquisition System

Machine vision is not an end unto itself! It represents a piece of the manufacturing/quality control universe. That universe is driven by data related to the manufacturing process. That data is of paramount importance to upper management as it relates directly to bottom-line results. For competitiveness factors, top management can not delegate responsibility for quality control. Quality assurance must be built in - a function totally integrated into the whole of the design and manufacturing process. The computer is the means to realize this integration.

Sophisticated manufacturing systems require automated inspection and test methods to guarantee quality. Methods are available today, such as machine vision, that can be applied in all manufacturing processes: incoming receiving, forming, assembly, and warehousing and shipping. However, hardware alone should not be the main consideration. The data from such machine vision systems is the foundation for computer integrated manufacturing. It ties all of the resources of a company together - people, equipment and facilities.

It is the manufacturing data that impact quality, not quality data that impact manufacturing. The vast amount of manufacturing data requires examination of quality control beyond the traditional aspects of piece part inspection, into areas such as design, process planning, and production processes.

The quality of the manufacturing data is important. For it to have an impact on manufacturing, it must be timely as well as accurate. Machine vision systems when properly implemented can automate the data capture and can in a timely manner be instrumental in process control. By recording this data automatically from vision systems, laser micrometers, tool probes and machine controllers, input errors are significantly reduced and human interaction minimized.

Where data is treated as the integrator, the interdepartmental database is fed and used by all departments. Engineering loads drawing records. Purchasing orders and receives material through exercise of the same database which finance also uses to pay suppliers. Quality approves suppliers and stores results of incoming inspection and tests on these files. The materials function stocks and distributes parts and manufacturing schedules and controls the product flow. Test procedures stored drive the computer-aided test stations and monitor the production process.

The benefits of such an "holistic" manufacturing/quality assurance data management system include:

Increased productivity:

Reduced direct labor
Reduced indirect labor
Reduced burden rate
Increased equipment utilization
Increased flexibility
Reduced inventory
Reduced scrap
Reduced lead times
Reduced set-up times
Optimum balance of production
Reduced material handling cost and damage

Predictability of quality
Reduction of errors due to:

Operator judgement
Operator fatigue
Operator inattentiveness
Operator oversight

Increased level of customer satisfaction

Holistic manufacturing/quality assurance data management involves the collection (when and where) and analysis (how) of data that conveys results of the manufacturing process to upper management as part of a factory-wide information system. It merges the business applications of existing data processing with this new function.

It requires a partnership of technologies to maximize the production process to ensure efficient manufacturing of finished goods from an energy, raw material, and economic perspective. It implies a unified system architecture and information center software and database built together. This integrated manufacturing, design and business functions computer based system would permit access to data where needed as the manufacturing process moves from raw material to finished product.

Today such a data driven system is possible. By placing terminals, OCR readers, bar code readers (1D and/or 2D) and machine vision systems strategically throughout a facility it becomes virtually paperless. For example, at incoming receiving upon receipt of material, receiving personnel can query the purchasing file for open purchase order validation, item identification and quality requirements. Information required by finance on all material receipts is also captured and automatically directed to the accounts payable system.

The material can then flow to the mechanical and/or electrical inspection area where automatic test equipment, vision systems, etc. can perform inspection and automatically record results. Where such equipment is unavailable, inspection results can be entered via a data terminal by the inspector. Such terminals should be user-friendly. That is, designed with tailored keys for the specific functions of the data entry operation.

Actual implementation of such a data driven system will look different for different industries and even within the industry different companies will have different requirements because of their business bias. For example, a manufacturer of an assembled product who adds value with each step of the process might collect the following data:

Receiving inspection:
a. Total quantity received by part number
b. Quantity on the floor for inspection
c. Quantity forwarded to production stock
d. Calculation of yield

Inventory with audit (reconciliation) capability:
a. Ability to adjust, e.g., addition of rework
b. FIFO/LIFO
c. Part traceability provisions
d. Special parts

Production:
a. Record beginning/end of an operation
b. Ability to handle exceptions - slow moving or lost parts
c. Ability to handle rework
d. Ability to handle expedite provisions
e. Provide work in process by part number, operation
f. Provide process yield data by:
    Part number
    Process
    Machine
g. Current status reporting by:

Part number
Shop order number
Program operation
Rejection

h. Activity history of shop order in process including rework
i. Shop orders awaiting kitting
j. Shop orders held up because of component shortages
k. History file for last "X" months
l. Disc and terminal utilization

Quality:

Provide hard copy statistical reporting data (pie charts, bar diagrams, histograms, etc.)

Data input devices:

a. OCR
b. Bar code readers (1D and 2D)
c. Keyboards
d. Test equipment
e. Machine vision systems

Personnel:

a. Quality control inspectors
b. Production operators
c. Test technicians

With appropriate sensor technology, the results include unattended machining centers. Machine mounted probes, for example, can be used to set up work, part-alignment, and a variety of in-process gaging operations. Microprocessor-based adaptive control techniques are currently available which can provide data such as:

Tool wear
Tool wear rate greater or less than desired
Work piece hardness different from specification
Time spent
Percentage of milling vs. drilling time, etc.

Quality assurance can now use CAD/CAM systems for many purposes; for example, to prove numerical control machine programs, and provide inspection points for parts and tools.

After the first part is machined, inspection can be performed on an off-line machine vision system analogous to a coordinate measurement machine using

CAD developed data points. This verifies the NC program contains the correct geometry and can make the conforming part. At this point QA buys off the program software. While the program is a fixed entity and inspection of additional parts fabricated for shape conformance is not needed, inspection is required for elements subject to variables: machine controller malfunction, cutter size, wrong cutter, workmanship, improper part loading, omitted sequences and conventional machining operations. This may necessitate sample inspection of certain properties-dimensions, for example, and a 100% inspection for cosmetic properties - tool-chatter marks, for example.

The CAD/CAM system can be used to prepare the inspection instructions. Where automatic inspection is not possible, a terminal at the inspection station displays the view the inspector sees along with pertinent details. On the other hand, it may be possible in some instances to download those same details to a machine vision system for automatic conformance verification. CAD systems can also include details about the fixturing requirements at the inspection station. This level of automation eliminates the need for special vellum overlays and optical comparator charts. The machine vision's vellum or chart is internally generated as a referenced image in the computer memory.

While dimensional checks on smaller parts can be performed by fixturing parts on an X-Y table that moves features to be examined under the television camera, using a robot to move the camera to the features to be inspected or measured can similarly inspect larger objects. Again, these details can be delivered directly from CAD data.

Analysis programs for quality monitoring can include:

1. Histogram which provides a graphic display of data distribution. Algorithms generally included automatically test the data set for distribution, including skewness, kurtosis and normality.
2. Sequential plots, which analyze trends to tell, for example, when machine adjustments are required.
3. Feature analysis to determine how part data compares with tolerance boundaries.
4. Elementary statistics programs to help analyze data of workpiece characteristics-mean, standard deviation, etc.
5. X-bar and R control chart programs to analyze the data by plotting information about the averages and ranges of sequences of small samples taken from the data source.

A computer-aided quality system can eliminate paperwork, eliminate inspection bottlenecks and expedite manufacturing batch flow. The quality function is the driver that merges and integrates manufacturing into the factory of the future.

REFERENCES

Barker, Ronald D., “Managing Yields by Yielding Management to Computers,” Computer Design, June 1983, pp. 91-96.

Bellis, Stephen J., “Computerized Quality Assurance Information Systems,” Proceedings CAM-I Computer Aided Quality Control Conference, May, 1982, Baltimore, Md., pp. 107-112.

Bravo, P.F. and Kolozsvary, “A Materials Quality System in a Paperless Factory,” Proceedings CAM-I Computer Aided Quality Conference, May, 1982, Baltimore, Md., pp. 113-121.

Gehner, William S., “Computer Aided Inspection and Reporting CAIR,” Proceedings CAM-I Computer Aided Quality Conference, May, 1982, Baltimore, Md., pp. 175-182.

Koon, Troy and Jung, Dave, “Quality Control Essential for CIM Success,” Cadlinc, Inc., Elk Grove Village, II, Jan-Mar, 1984, pp. 2-6.

Kutcher, Mike and Gorin, Eli, “Moving Data, not paper, enhances productivity,” IEEE Spectrum, May, 1983, pp. 84-88.

Kutcher, Mike, “Automating it All,” IEEE Spectrum, May, 1983, pp. 40-43.

Papke, David, “Computer Aided Quality Assurance and CAD/CAM,” Proceedings CAM-I Computer Aided Quality Control Conference, May 1982, Baltimore, Md., pp. 23-28.

Schaeffer, George, “Sensors: the Eyes and Ears of CIM,” American Machinist, July, 1983, pp. 109-124.

2

Machine Vision: Historical Perspective

The concepts for machine vision are understood to have been evident as far back as the 1930's. A company - Electronic Sorting Machines (then located in New Jersey) - was offering food sorters based on using specific filters and photomultipliers as detectors. This company still exists today as ESM and is in Houston, TX. Satake, a Japanese company, has acquired them. To this day they still offer food sorters based on extensions of the same principles.

In the 1940's the United States was still using returnable/refillable bottles. RCA, Camden operation, designed and built an empty bottle inspection system to address the concern that the bottles be clean before being refilled. Again, the system used clever optics and a photomultiplier tube as the detector. As with the early food sorters, this technology was all analog-based. It is noted that over 3000 of these systems were installed. It is also understood that they adapted the basic principles to "check" detection - inspecting for cracks on the bottle lips.

The field of machine vision has evolved along with other evolutions involving the use of computers in manufacturing. The earliest related patents were issued in the early 1950s and concerned optical character recognition. Pattern recognition and analysis received a big push due to the research sponsored by the National Institute of Health (NIH) for chromosome analysis and various types of diagnostics based on blood and tissues associated with automatic tissue culture or petri culture analysis. These typically involved counting of cells designated by a common optical density.

The military supported a great deal of research and development (R&D) to provide operator assistance in interpreting and/or automating the interpretation of surveillance photos as well as for automated target recognition and tracking. In the late 1960's computer-vision-related research was being funded by the military at the AI Labs of MIT and Stanford University. NASA and the U.S. Geological Survey also supported R&D in this field. In an attempt to emulate the eye's performance, the military and the NIH have supported much research to provide an understanding of how the eye operates.

Along with the technology, government research support has spawned a cadre of people trained in computer-based vision. The National Science Foundation (NSF), the National Institute of Standards and Technology (formerly the Bureau of Standards), and the military have actually supported R&D specifically in the field of machine vision.

In 1962, optical character recognition was demonstrated using TV-based apparatus and computers. In the 1960's much R&D was being conducted driven by military objectives to enhance images for photoreconnaissance interpretation. Work also began during this time in imaging-related research supported by health/diagnostic objectives.

In the early 1960's IT&T delivered an image-dissector-based (an early TV camera) system to inspect reflectors, etc. to General Motors. At this same time, Proctor & Gamble was also experimenting with concepts to inspect Pampers.

In 1964, Jerome Lemelson was awarded a patent application for a generic concept: to capture electromagnetic radiation using a scanner device, digitize the signal and use a computer to compare the results to a reference stored in memory. A subsequent patent was awarded in 1971 with a similar specification and new claims. In the late 1970's he began filing more patents with essentially the same specification and new claims maintaining continuance to the original patent filing. This resulted in over a dozen machine-vision-related patents.

In 1965, a doctor intent on applying the technique to Pap smear diagnostics developed the concepts behind Object Recognition Systems' pattern recognition system. During this time there was other activity along these lines.

In 1965, Colorado Video was formed, providing a unit to digitize video images. It basically digitized one pixel per line for each scan. In other words, it required 500 scans to digitize an entire image. In 1970, driven by military objectives, GE demonstrated a 32 X 32 pixel CID camera and Bell Labs developed the concept of charge coupling, and created the CCD (charge coupled device). In 1971, Reticon developed its first sold state sensor.

In 1969, driven by a NASA project, EMR Photoelectric, a Schlumberger company, developed the Optical Data Digitizer. This was an all-digital camera, taking signals from a PDP-11 to expose, scan and digitize an image. It had features such as selective integration so one could restrict digitizing to just the areas of interest. Early versions of the camera were sold for: TV interferometric-based measurements, optical computing applications, TV-based spectrometers, X-Ray digitizing, and motion analysis for prosthesis evaluation/biomechanics. By 1974 the system had been used to read characters on rifles and inspect fuses. The OCR system actually used a 256 X 256 X 4- bit frame grabber.

In the late 1960s, minicomputer-based image analysis equipment became available for biological and metallurgical pattern recognition examinations. In general, commercially available products had limited performance envelopes and were very expensive. In the early 1970s, the NSF began to support applied research focused on specific advanced manufacturing technology issues. Several of these included projects on machine vision. In virtual synchronization with the flow of the results of these research efforts, microcomputer performance was improving, and costs were decreasing. Similarly, advances were being made in fiber optics, lasers, and solid-state cameras.

By the early 1970's several companies had been formed to commercialize TV-based inspection systems. For the most part these commercialized products were analog-based. Autogage was a company started by Jay Harvey in Connecticut. Ball Corporation and Emhart introduced systems to inspect the sidewall of bottles. Intec was established in 1971 offering a flying spot scanner approach to inspecting products produced in web form.

In 1971, Solid Photography (now Robot Vision Systems Inc.) was established to commercialize 3D techniques of capturing data from a person to be the basis of creating a bust of the person. The Navy funded research to extend the techniques to inspecting ship propellers. Around this same time, Diffracto was established in Canada to commercialize 3D sensing techniques.

By the mid-1970s, exploratory steps were being taken to apply the technology in manufacturing. Several companies began to offer products that resembled what now appear to be machine vision systems. Virtually every installation had a significant custom content so that the system was really dedicated to performing one task and one task only - controlling the quality of French fries, for example (Figure 2.1).

Some systems also became available that performed tasks with potential for widespread adoption: Pattern recognition systems (Figure 2.2) that automate the wire-bonding process in the manufacture of semiconductor circuits is one such example. Computer-based electro-optical products also entered the marketplace to automatically inspect web products (Figure 2.3), silicon wafers, gage diameters (Figure 2.4), and so on. Toward the end of the 1970s, products became available to perform off-line dimensional measurements essentially automating optical comparator and coordinate measuring-machine-type products (Figure 2.5).

By 1973, several companies had commercialized solid-state cameras. Until this development, all the systems that were based on conventional analog vidicon cameras suffered from the need to continuously "tweak" them to compensate for drift that was generally unpredictable. Fairchild introduced their first commercial

Figure 2.1 - Key technology system designed to detect blemishes in french fries as well as other vegetables.

Figure 2.2 - Early View Engineering wire-bonding pattern recognition system shown on Kulicke and Soffa wire bonder.

Figure 2.3 – Honeywell Measurex scanner verifying the integrity of paper.

Figure 2.4 - Early laser gauging system from Z-Mike (formerly Zygo and now part of LaserMike) used to monitor the results of a grinding process.

Figure 2.5 - Version of an off-line dimensional measuring system "Voyager" offered by GSI Lumonics.

CCD sensor; a 500 X 1 linear array ($4000). GE sold its first 32 X 32 pixel CID for $6500. Reticon also introduced 32 X 32 pixel and 50 X 50 pixel cameras. In 1973 Reticon also introduced their 64 Serial Analog Memory and the "SAD m100" for processing image data. They also established a system integrator group.

Two other events cited in 1973 were: GM simulated automobile wheel mounting using vision-guided robotics and ORS introduced a microprocessor-based pattern recognition system based on the 8080 microprocessor. The development of the microprocessor and the development of solid state cameras are what really made the applications of machine vision possible and cost effective.

Around 1973, the National Science Foundation began funding research in machine vision at the University of Rhode Island, Stanford Research Institute and Carnegie Mellon. These three schools all formed industrial affiliate groups as an advisory panel and a means of effecting technology transfer. This led to a number of pioneering demonstration projects related to manufacturing applications. For example, the University of Rhode Island demonstrated a vision-guided bin-picking robot.

Also in the early 1970's research in the field of computer vision was initiated at many other universities including: University of Missouri, Columbia, Case Western Reserve, several University of California schools, University of Maryland, and University of Michigan.

In 1974, GE introduced a 100 X 100 pixel CID camera for $6500 and later in the year a 244 X 128 version also for $6500. By 1976 they had cut the price to $2800.

In 1975, EMR introduced a TV-based off-line dimension-measuring system. It used a PDP-8 computer and essentially the principles of their Optical Data Digitizer. Shortly after, View Engineering also introduced their system.

In 1976, the following are understood to have occurred: GM first published work on its automatic IC chip inspection system; Fairchild introduced the 1024 X 1 linear array; Reticon introduced its CCD transfer device technology.

In 1977: Quantex introduced the first real-time image processor in a single box; GE introduced their first commercial vision inspection system for sale; SRI introduced a vision module - a laboratory system with camera, analog preprocessor and DEC computer intended for prototyping applications; ORS systems were available for sale; Hughes researchers demonstrated a real-time 3 X 3 Sobel (edge segmenting) operator using a CCD chip for storage and outboard hardware for processing. This development pioneered dedicated hardware and eventually application-specific integrated circuits to speed up software image processing functions.

In 1977, Leon Chassen received a patent applying the principles of structured light to log profiling as the scanner input to an optimization program to increase the yield cut from a log. Shortly thereafter he established Applied Scanning to commercialize the technique.

By 1978, ORS had established a relationship with Hajime in Japan that led to the commercialization of the technology there. In 1978 Octek, another early machine vision company, was founded. By the late 1970s companies, such as Imanco in the UK and Bausch & Lomb in the States, had introduced TV-based computer workstations for metallographic analysis as well as microscopic biomedical analysis.

In 1979, View commercialized pattern recognition techniques out of Hughes for alignment applications for wire bonders and other production equipment for the semiconductor industry. Around this time, KLA was also formed and announced the development of a photomask inspection system. By the early 1980s, Texas Instruments had an in-house group developing machine vision solutions for their own manufacturing requirements. These included pattern-recognition systems for alignment, wafer inspection, and even an off-line TV-based dimensional measuring system.

In 1980, Machine Intelligence Corporation (MIC) was formed to commercialize the SRI machine vision technology. In 1981, Cognex was formed and introduced their first product performing a binary correlation running on a DEC LSI 11/23. Also in 1981, MIC introduced their VS-110, a system intended to perform high speed inspection on precisely indexed parts, by referencing the part's image to a master image stored in memory. The first industrial application of binary thresholded template matching was aimed at highly registered and controlled parts in the electronic industry.

In 1981, Perceptron was formed by principals that came out of General Motors. Also in 1981, Machine Vision International (originally called Cyto Systems) was formed to commercialize the parallel-pipeline cytocomputer coming out of the Environmental Research Institute of Michigan (ERIM) and morphological principles out of University of Michigan and the Ecole des Mines in France. In 1982, Applied Intelligent Systems, Inc. was founded by another group coming out of ERIM to commercialize another version of their developments.

In 1983, Itran shipped their first factory-oriented system to AC spark plugs to perform speedometer calibration. Their system used normalized gray scale correlation for the "find" or locator function. By this time, the industry also witnessed the beginning of the establishment of an infrastructure to support the application of machine vision. Merchant system integrators began to emerge as well as independent consultants. Around this time, GM announced the conclusion of their in-house analysis that suggested that they alone would require 44,000 machine vision systems.

In 1984, Allen Bradley acquired the French company Robotronics and became a machine vision supplier.

By 1984, the Machine Vision Association (MVA) was spun out of Robotics International, a professional interest group within the Society of Manufacturing Engineering. Also in 1984, the Robotics Industries Association spawned the Automated Vision Association, a trade association for the machine vision industry.

It has since been renamed the Automated Imaging Association. The term "machine vision" was defined by this group and became the accepted designation to describe the technology. Also, around this time, GM took a 20% position in four machine vision companies: Applied Intelligent Systems, Inc. (for cosmetic inspection applications), View Engineering (for metrology), Robot Vision Systems Inc. (for 3D robot guidance systems) and Diffracto (for 3D systems to measure gap and flushness on assemblies). (They have since disengaged themselves from these companies.)

By this time, one could say that the machine vision industry was well on its way to evolving into a mature industry. With advances in microprocessor technology and solid state cameras and the cost declining of these basic components, things were in place for the industry to grow. Some other noteworthy events:

1984 - NCR and Martin Marietta introduce the GAPP, a single chip systolic array for use in parallel processing for pattern and target recognition.

1985 - VIDEK, a Kodak subsidiary at the time (now Danfoss-Videk), shipped its first unit with dedicated hardware boards to perform edge segmentation, histogramming and matrix convolutions. EG&G introduced a strobe specifically for machine vision - high speed and power.

1987 - Hitachi introduces the IP series using the first dedicated image processing VLSI chip.

1988 - Cognex introduced their VC-1, the first VLSI chip dedicated to image analysis for machine vision. It performs measurements such as blob locations, normalized correlation, feature vectors, image projection at any angle, spatial averaging and histograms.

1988 - Videk introduced the Megaplus camera - 1024 X 1024 resolution.

1988 - Sharp introduces 3" X 3" complete image processing function "core board."

1988 - LSI Logic introduces the RGB - HSI converter CMOS chip. It was first applied to color-based machine vision/image processing boards by Data Translation. In 1988, Imaging Technology, Inc. introduced their RTP-150 using LSI Logic's real time Sobel processor and Rank Value Filter chips.

1989 - LSI Logic introduces a contour tracing chip and Plessey Semiconductors introduced ASICs to perform 2D convolutions at 10 MHz rates.

1991 - Dickerson Vision Technologies is one of the first to offer a "smart camera" - a camera with embedded microprocessor that results in a general purpose machine vision flavor

1992 - Cognex introduces the VC-2, a complement to the VC-1.

1994 - Cognex introduces the VC-3, an improved VC-1 with higher throughput. They also introduced their Checkpoint system.

1994 - Itran and Automatix merged to form Acuity Imaging.

1995 - Acuity Imaging became a subsidiary of RVSI; Cognex acquired Acumen. Some suggested that consolidation began.

1995 - KLA sales in machine vision based products exceed $300M. Cognex became the first supplier of general-purpose machine vision to have sales exceed $100M.

From the mid to late 1980's application-specific machine vision systems were being developed to address the needs of virtually every manufacturing industry: Key Technology and Simco Ramic (now Advanced Machine Vision Corporation) for food grading and sorting applications at food processors; Design Systems a 3D system for water jet cutters and portioning of fish, poultry, and meat; Kanebo and Fuji for tablet inspection in the pharmaceutical industry; Eisai for particulate in solution detection; Ball and Inex Vision Systems a family of products for glass container inspection; Pressco and Ball, products for the metal container industry; Control Automation (now Teradyne) and IRI/Machine Vision Products populated printed circuit board inspection systems; Orbotech and others, systems to inspect bare boards; and so forth.

Since 1980, about 100 companies have been spawned that have introduced more flexible machine vision products as well as several hundred companies that now offer similar equipment with a similar complement of components dedicated to specific tasks: trace verification on printed circuit boards (Figure 2. 6) or thick-film hybrid circuits (Figure 2. 7), LED/LCD verification and evaluation systems, character readers (Figure 2.8), photomask/reticle inspection, wet- and dry-product color sorters, drop analyzers, seam tracking (Figure 2.9), particulate-in-solution analysis, tablet/capsule inspection, integrated circuit (IC) mask blank inspection (Figure 2.10), and so on. These have been designed to satisfy specific needs in specific industries.

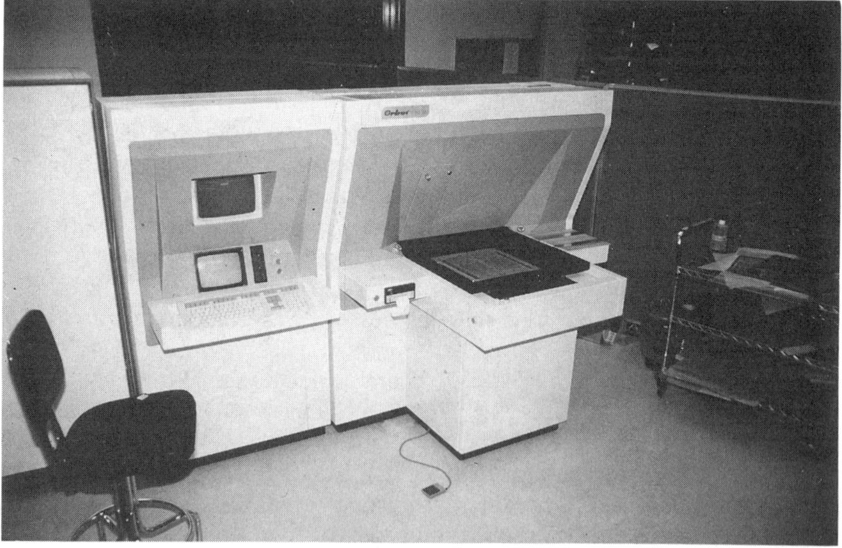

Figure 2.6 - Early version of an Orbot (now part of Orbotech) printed circuit board trace inspection system.

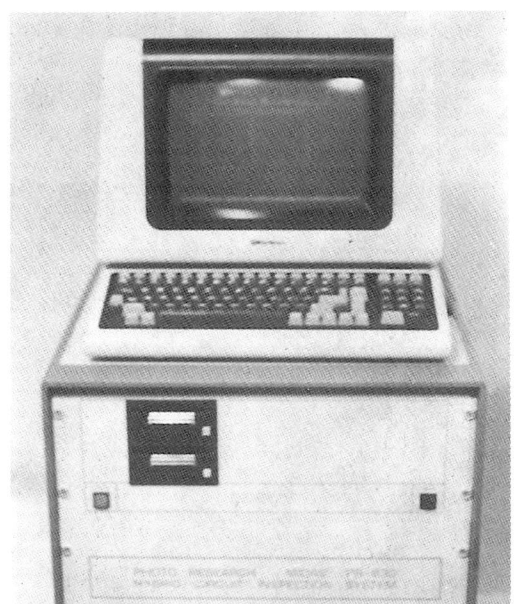

Figure 2.7 - Early version of a Midas Vision System (formerly Vanzetti Vision System) for thick film verification.

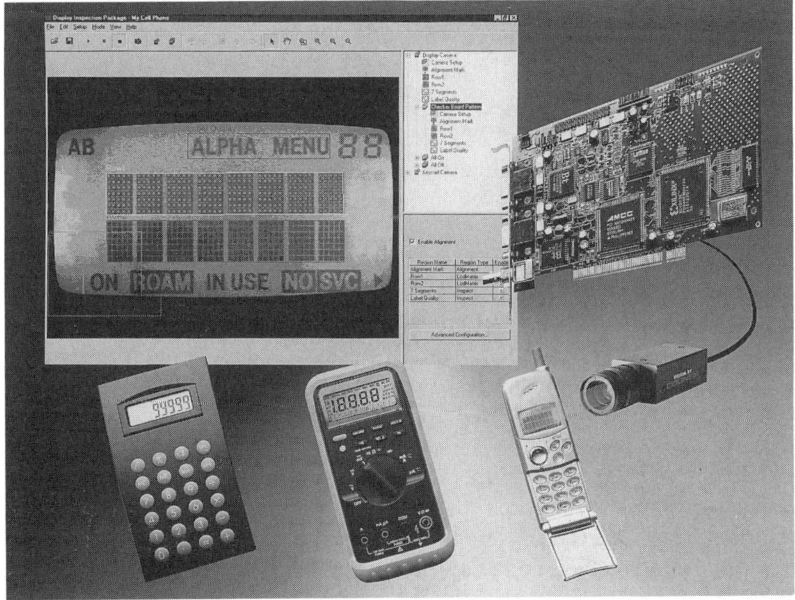

Figure 2.8 - Cognex's "Display Inspect" system for inspecting displays.

Figure 2.9 - Servo-robot machine vision based seam tracking and adaptive control system applied to aluminum welding.

Of the companies offering products that are more configurable, one can characterize these products as "solutions looking for problems." Classifying these product offerings is difficult. Several approach it on the basis of simple versus complex; binary versus gray scale; hardware/firmware versus software processing; image processing computers versus computers for image processing; backlighted, front lighted, or structured lighting; and so on. Many permutations exist within this framework.

Alternatively, one can classify them by classes of applications addressed by the systems: gauging, robot guidance, cosmetic inspection, verification, contouring, identification, and pattern recognition. The data sheets of most machine vision companies would have you believe their products can address all of these tasks.

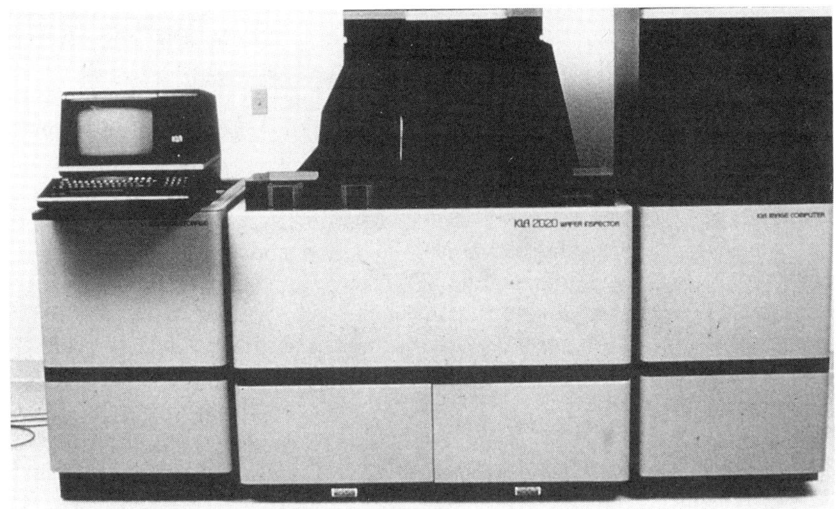

Figure 2.10 - Early version of a KLA-Tencor photomask and reticle automatic optical inspection system.

In fact, most machine vision systems have evolved based on a single idea about how human vision works. Consequently, while well suited for some specific tasks, the performance of other tasks will only be successful under the most ideal conditions. In many cases, machine vision vendors have not come to grips with the limitations of the performance envelopes of their products. Consequently, virtually every application involves experimenting with samples to assess whether the job can be performed.

Not only are the vendors unable to quantify the application parameters related to their products, but most applications are further complicated because the application parameters themselves are qualitative rather than quantitative.

The significance in understanding the difference between products can be the difference between a successful installation and a white elephant. Successful deployment of machine vision involves a good fit between the technology and the application. In addition, it requires empathy for the market and the specific application on the part of the vendor.

Since 1995, many companies have merged and companies outside of the machine vision industry have acquired machine vision companies: Electroscientific Industries acquired Intelledex and Applied Intelligent Systems, Inc. and General Scanning acquired View Engineering - a sign of the maturing of the market.

Along the way a fair number of companies have disappeared (International Robotmation Intelligence (IRI), Octek, Machine Intelligence, Inc. (MIC), Machine Vision International (MVI), to list some of the more noteworthy; however, a fair number of new companies have also been established. The cost of entry remains perceptively low to enter the machine vision market. The challenge, however, has been to establish efficient and effective distribution channels.

As a company planning to implement a machine vision system, it is important to work with suppliers that have the resources to support an installation and the appearance of staying power.

Advances in electronics have made it technically feasible to consider applying machine vision to many situations. Microelectronics has resulted in improved solid-state sensors, higher density memory, and faster microcomputers. Advances in the personal computer and MMX technology now make it possible to perform many of the compute-intensive machine vision algorithms fast enough for many applications. The adoption of Windows™-based graphic user interfaces has resulted in more user-friendly systems, especially for shop floor personnel.

Simultaneously, the costs associated with these improved products have decreased, making it possible to cost-effectively apply the technology. The good news is that all these factors continue to improve so that ever more applications can be addressed cost-effectively.

3

Description of the Machine Vision Industry

The machine vision industry consists of establishments that supply technology used in manufacturing industries as a substitute for the human vision function. It is made up of suppliers of systems that embody techniques leading to decisions based on the equivalent functionality of vision, but without operator intervention.

Characteristics of these techniques include non-contact sensing of electromagnetic radiation; direct or indirect operation on an image; the use of a computer to process the sensor data and analyze that data to reach a decision with regard to the scene being examined; and an understanding that the ultimate function of the system's performance is control (process control, quality control, machine control or robot control). What follows is meant to provide an understanding of what machine vision and is and what it is not.

The machine vision industry is a segment of the larger industry characterized as electronic imaging. Within electronic imaging there are basically two major components: one deals with the application of computers to generate images such as in CAD, visualization, animation systems, etc., and the second deals with the application of computers to acquired images. The common technology that serves as the infrastructure to a number of these distinctive markets includes cameras, frame grabbers, computers, and image processing and analysis software (firmware).

Within this second segment, there are a number of distinct classifications. These include images acquired as a result of remote sensing techniques such as those from NASA and military satellites or aircraft reconnaissance. In addition there are those systems that acquire larger area formats such as engineering drawings as input to a computer. In the area of documents, there are small area document scanners - page readers.

There are also those systems that typically use more conventional television techniques to acquire an image. These include systems that are used in conjunction with medical diagnostics and scientific research. Finally, in the last class of products associated with the segment of electronic imaging dealing with the use of computers to operate on acquired images, is the machine vision class.

In these cases, the images are related to a manufacturing process and involve operating on those images for the purposes of production control, process control, quality control, machine control, or robot control. Machine vision represents a very small segment of the electronic imaging market that involves the use of computers operating on acquired images and an even smaller segment of the entire electronic imaging market.

3.1 CLARIFICATION OF WHICH APPLICATIONS ARE AND WHICH ARE NOT INCLUDED AS MACHINE VISION

Machine vision techniques are being adopted in other fringe applications, such as biometrics/access control, traffic control, and in the automotive after-market. In the latter case, they are being used for such applications as inspecting for gap and flushness measurements on cars being repaired after crashes, inspecting for wheel alignment, verifying headlight aiming, and verifying color match properties.

In a number of cases, companies involved in these types of applications are system integrators who are integrating conventional machine vision products offered by machine vision vendors or image processing board suppliers. Because of the non-manufacturing nature of the applications, these systems are not typically included as part of the machine vision market.

Today machine vision technology is found embedded in bar-code scanners. With traditional bar codes that are one-dimensional, linear array-based and area array-based machine vision techniques have emerged in products offered alongside those that use laser scanner techniques. Currently there is a growing interest in two-dimensional codes whose advantages include savings in the space needed to encode a given amount of information and an ability to store and read data at any angle.

Conventional machine vision pattern recognition techniques are being adapted to this application which generally involves binary data processing. These products, when delivered by machine vision companies, are included as part of the machine vision market.

Motion analysis is another fringe application of machine vision. Today motion analysis systems are being deployed which use television and computers to interpret television images for analyzing machinery and robots for accuracy, as well as humans for prosthesis fitting, rehabilitative purposes, and athletic conditioning such as golf swings and diving.

Another class of applications that uses image processing techniques typically found in the laboratory setting which now has migrated onto production floors are those involving the analysis of interferometric images. Often these are images reflecting the surface conditions of manufactured parts. In some cases these techniques have been adapted to on-line applications in a manufacturing environment.

There is also a growing interest in the use of a variety of range sensing and laser radar techniques in conjunction with vehicle guidance systems specifically designed for intelligent automobile navigation. For the most part, this activity is still confined to research laboratories. Because of the non-manufacturing nature of these applications, any systems sold for such purposes are not typically considered as part of the machine vision market.

Two other computer vision markets not considered machine vision are the postal service and trash-separation markets. In the postal service, in addition to optical character recognition (OCR), machine vision techniques are being used for package handling and for finding address blocks before reading. In the case of trash separation, there is an apparent market potential to use machine vision techniques to separate classes of containers and within classes to separate by color.

Computer vision techniques are also finding their way into the security market. In retail, systems exist that monitor the items being rung up to assure the integrity of the cashier or to verify that the carriage is empty.

3.2 MACHINE VISION INCLUDES A BROAD RANGE OF TECHNICAL APPROACHES

Machine vision technology is not a single technical approach. Rather, there are many techniques and implementations. Machine vision involves three distinct activities and each has multiple approaches to accomplish the result desired for an application: image formation/acquisition, image processing, and image analysis/decision making action.

Image formation/acquisition involves the stage of the process that couples object space to image space and results in the quantitized and sampled data formatted in a way that makes it possible for a computer or dedicated computer-like circuitry to operate on the data. This stage itself typically includes: lighting, optics, sensor, analog-to-digital converter, and frame buffer memory to store the image data, as well as application-specific material handling and complementary sensors such as presence triggers.

Each of these can be executed in any number of ways, sometimes dictated by the application specifics. Lighting can be active or passive. Passive lighting refers to those lighting techniques designed to provide illumination (ideally as uniform as possible) of the object so that the sensor receives a sufficient amount of light to yield a reliable electron image by the photon to electron imaging transducer. Again, there are specific passive lighting techniques - fluorescent, incandescent, fiber optics, and projected.

Active lighting involves those arrangements that operate on the image of the light itself. "Structured" light is one such popular arrangement. Typically, a line of light is projected across the object and the image of the line of light, as deformed by the geometry of the object, is acquired and analyzed. There are many forms of active lighting arrangements.

Similarly, there are different sensors that can be used to acquire an image. In a flying spot scanning arrangement, the image is acquired from a single element point detector as a time-dependent signal representative of the image data. Linear arrays are also used and are especially beneficial if an object is consistently moving past the scanner station (Figure 3.1). The image data is acquired one line at a time.

The sensor most frequently identified with machine vision is the area array; however, it is important to understand that the machine vision industry is not restricted to only those systems that acquire images using area arrays.

Beyond differences in sensing or converting the optical image into an electronic image, there are differences in how the analog signal is handled, converted to a digital signal, and stored. Some executions do some signal enhancement processing in the analog domain. Some operate on the digitized image in real time without storing a frame.

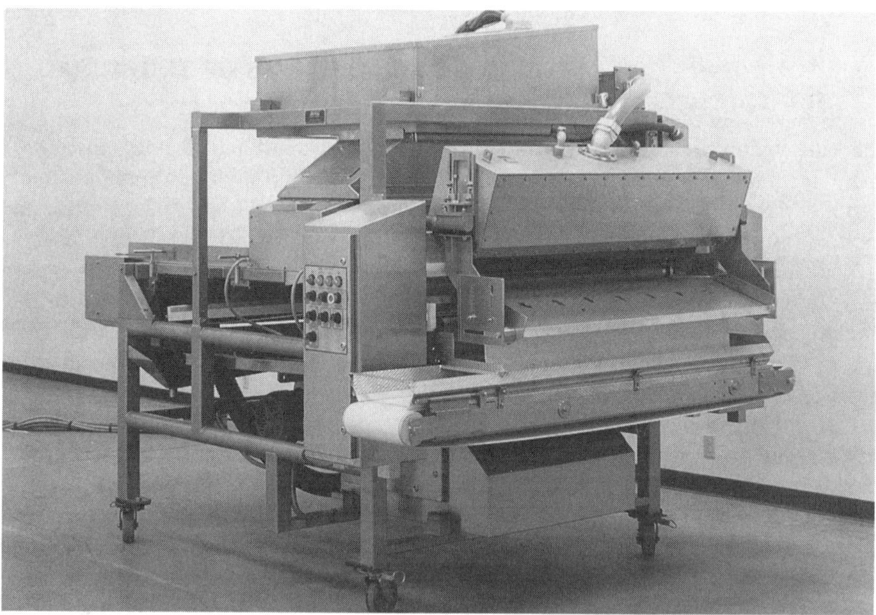

Figure 3.1 - Kroma-Sort System 480 from SRC Vision/AMVC suitable to separate unwanted conditions from many different types of produce and fruit.

Processing on the image also varies with each execution. Some vision platforms have extensive compute capacity to enhance and segment images, some less so. Different executions base analysis and decisions on different routines.

This discussion on differences is meant solely to emphasize that they exist. True general-purpose machine vision does not exist. Rather, there are many embodiments of the complement of techniques together representing machine vision.

3.3 WHAT TECHNICAL APPROACHES ARE INCLUDED AS MACHINE VISION

All techniques performing a machine vision application are considered as machine vision. For example, techniques for inspecting webs or products that are produced or treated in a continuous flat form for purpose of both quality and process control are in widespread use in many industries.

The paper industry looks for tears and other anomalies. Certain materials are coated and the coating is inspected for "holidays" (missing coating), bubbles, runs, streaks, etc. (Figure 3.2). Early techniques for this inspection function were typically based on flying spot scanners, usually laser scanners. A number of companies have products to perform the same function, i.e. going after the same market, using high-speed linear arrays, and applying image processing instead of signal processing.

Another major issue has to do with whether the automatic optical inspection techniques in widespread use in the semiconductor and electronics industries are machine vision systems. For the most part, these systems satisfy the definition of machine vision, although these products generally involve a person who makes a final judgment on all reject conditions automatically detected because of the incidence of false rejects experienced.

As the technology becomes more robust, it is reasonable to believe that these systems will ultimately experience fewer false reject rates - lowered to an acceptable level which would further reduce labor content. Consequently, these systems are also included as part of the machine vision market.

3.4 MACHINE VISION INDUSTRY BUSINESS STRATEGIES

In addition to recognizing that there are variations in machine vision technology, one must also recognize that various companies participating in the machine vision market have made certain business decisions that also dictate the segment of the market in which they compete. Consequently, at least four business strategies have emerged, influencing product design implementations.

Figure 3.2 – Honeywell Measurex's web imaging system scanning paper for defective conditions.

Some companies are offering multi-purpose machine vision. These are still "idiot savants," but they generally have some robustness, making them sufficiently flexible to handle a variety of applications under certain constraints. Companies such as CiMatrix/Acuity Imaging/RVSI, Electro Scientific Industries/Applied Intelligent Systems Inc., Cognex, PPT Vision, and DVT offer such products.

The second segment of the market includes those companies that have adapted the multi-purpose vision platform for use in a specific application niche. Generally, complementary material handling is what adapts these systems to the dedicated application. Should the application disappear, the company that bought the product could unbundle the vision platform and, with appropriate application engineering, configure the system for another application. These companies include CiMatrix/Acuity Imaging/RVSI and Parish Automation/GSMA who offer systems dedicated to consumer package inspection applications, and PPT Vision and Cognex who offer systems specifically for ball grid array inspection. Typically these really represent "canned" applications of the multi-purpose platform, not a turnkey system.

The third segment includes those companies that offer application-specific machine vision systems or machine vision for dedicated tasks. For example, Inex

Vision Systems/BWI offers bottle inspection systems (Figure 3.3), and MVT Technology, Ltd. is a system dedicated to in-line automatic inspection and measurement of solder paste deposits on SMT PCBs (Figure 3.4). While there may be some flexibility in the fundamental platform used, it has been optimized for a specific generic application - flaw detection, for example.

The fourth segment includes those companies that offer customized machine vision systems generally built around a common technology base. Companies like Perceptron offer systems designed for gap and flushness measurements in sheet metal assemblies based on a family of lighting/sensor probes (Figure 3.5), and Diffracto offers a system to inspect metal panel surfaces.

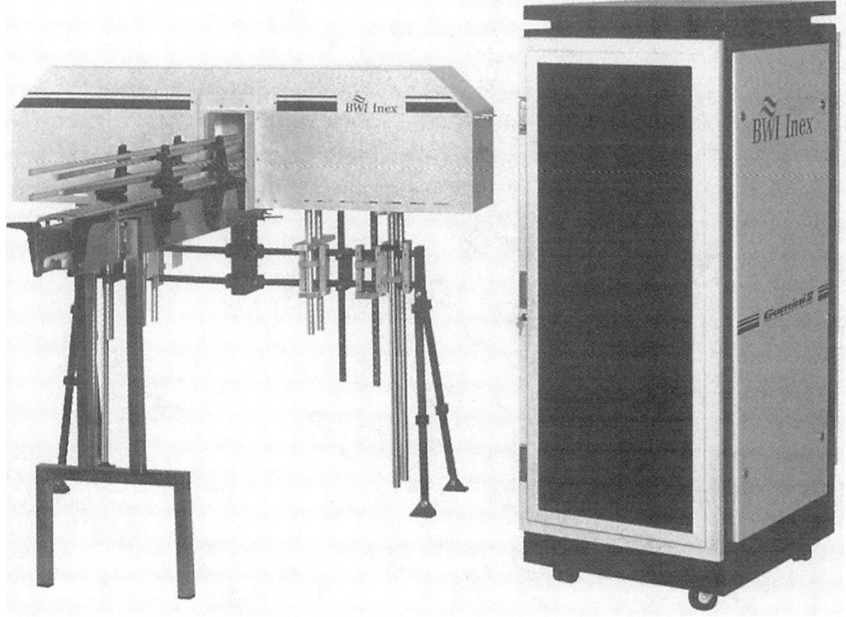

Figure 3.3 - Inex Vision Systems/BWI bottle inspection system.

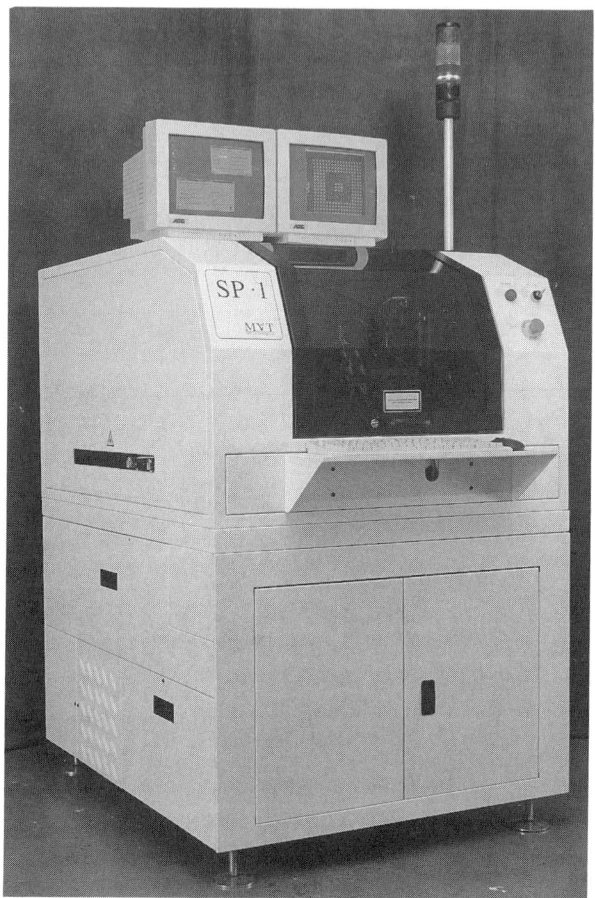

Figure 3.4 - SP - 1 from MV Technology system for in-line automatic inspection of solder paste deposits on SMT PCBS.

The vast majority of machine vision vendors are players in niche applications in specific manufacturing industries. While generic machine vision platforms have been applied in many industries, no single company has emerged within the machine vision industry as a dominant player with a product(s) that has been applied across a significant number of manufacturing industries.

Several companies offer general-purpose vision platforms that have sufficient functionality permitting them to be configured for a variety of applications. Some of these same companies are suppliers of products that address a specific set of applications such as optical character recognition (OCR) and optical character verification (OCV). Some companies are suppliers of image processing board sets that also offer the functionality of a vision platform and can be utilized to address many applications like the general-purpose vision platforms.

Figure 3.5 - Turnkey system from Perceptron performing 3D dimensional analysis on "body-in-white" in automotive assembly plant for critical dimensions, gap and flushness.

The vast majority of the suppliers that make up the machine vision industry are suppliers of industry-specific niche application products. There is often as much value associated with peripheral equipment necessary to provide a turnkey solution, as there is value of the machine vision content in the system.

It is becoming increasingly more difficult to classify companies in the machine vision market. Suppliers of general-purpose systems are extending their lines to include products that might have earlier been classified as application-specific machine vision systems. Similarly, suppliers of image processing boards are offering boards with software that makes their products appear to be a general-purpose machine vision system. There are a couple of board suppliers that today actually offer turnkey, application-specific machine vision systems. There are several suppliers of application-specific machine vision systems with turnkey systems that address specific applications in different markets (e.g., unpatterned and patterned/print web inspection (Figure 3.6), or 3D systems for semiconductor and electronic applications).

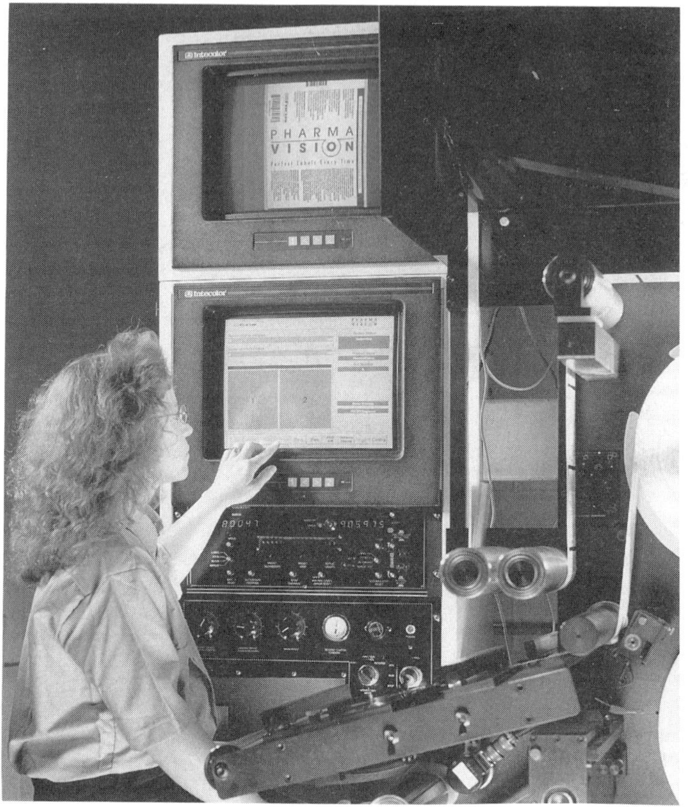

Figure 3.6 - PharmaVision system from Focus Automation inspecting a roll of pharmaceutical labels on a rewinder for print defects.

3.5 MACHINE VISION INDUSTRY-RELATED DEFINITIONS

The following are definitions associated with the different segments of the machine vision industry:

Merchant machine vision vendor - a company that either offers a general-purpose, configurable machine vision system or a turnkey application-specific machine vision system. In either case, without the proprietary machine vision functionality, there would be no purchase by a customer. The proprietary machine vision hardware could be based either on commercially available image board level products or proprietary vision computer products.

Image processing board set suppliers (IPBS) - A company offering one or more products, such as a frame grabber, that often incorporates an A/D, frame storage, and output look-up tables to display memorized or processed images. These boards can operate with either digital or analog cameras. In some cases, they merely condition the image data out of a camera making it compatible with processing by a personal computer.

Often these boards will be more "intelligent," incorporating firmware that performs certain image-processing algorithmic primitives at real time rates, and off-loading the computer requirements to the firmware from the computer itself. The interface supplied generally requires a familiarity with image processing and analysis, since one will generally start at the algorithm level to develop an application. IPBS can be sold to GPMV builders, ASMV, builders, merchant system integrators, OEMs, or end-users.

General-purpose machine vision system vendor (GPMV) - A company offering products that can be configured or adapted to many different applications. The vision hardware design can be either based on commercially available image board level products or proprietary vision computer products. The graphic user interface is such that little or no reference is made to image processing and analysis. Rather, the interface refers to generic machine vision applications (flaw inspection, gaging, assembly verification, find/locate, OCR, OCV, etc.) and walks the user through an application set-up via menus or icons.

These systems may or may not have the ability to get into refining specific algorithms for the more sophisticated user. GPMV systems are sold to application-specific machine vision system builders, merchant system integrators, OEMs, or end-users.

A GPMV supplier can use some combination of:

Proprietary software
Proprietary frame grabber + proprietary software
Commercial frame grabber + proprietary software
Proprietary IPBS + proprietary software
Commercial IPBS + proprietary software
Proprietary hardware + proprietary software.

Application-specific machine vision vendor (ASMV) - A company supplying a turnkey system that addresses a single specific application that one can find widely throughout industry or within an industry. Interface refers specifically to the application itself, not to generic machine vision applications or imaging functions. In other words, machine vision technology is virtually transparent to the user.

The vision hardware can be either based on commercially available image board level products, general-purpose machine vision systems, or proprietary vision computer products. ASMV systems are typically sold directly to end-users.

An ASMV supplier can use some combination of:

Proprietary frame grabber + proprietary software
Commercial frame grabber + proprietary software
Proprietary IPBS + proprietary software
Commercial IPBS + proprietary software
Proprietary hardware + proprietary software
Commercial GPMV + proprietary software.

Machine vision software supplier (MVSW) - A company supplying software that adapts image processing and analysis hardware to generic machine vision applications (flaw inspection, gauging, locate/find, OCR, OCV, etc.). Usually the software is designed to adapt a commercially available image processing board for use in machine vision applications. Alternatively, it may adapt a personal computer to a machine vision application. MVSW can be sold to GPMV builders, ASMV, builders, merchant system integrators, OEMs, or end-users.

Web scanner supplier - A company providing a turnkey system to inspect unpatterned products produced in webs (paper, steel, plastic, textile, etc.). These systems can capture image data using area cameras, linear array cameras, or laser scanners. The vision hardware used in the system can be based on commercially available image board level products, general-purpose machine vision systems or proprietary vision computers. Web scanners are typically sold to end-users.

3D-machine vision or laser triangulation supplier - A company providing a system that offers 3D measurements based on the calculation of range using triangulation measurement techniques. The system can use any number of detection schemes (lateral effect photodiode, quadrant photodetector, matrix array camera, linear array camera) to achieve the measurement. The lighting could be a point source, line source, or specific pattern.

The simpler versions collect data one point at a time. Some use a flying spot scanner approach to reduce the amount of motion required to make measurements over a large area. Others use camera arrangements to collect both 2D and 3D data simultaneously. Laser triangulation-based machine vision systems can be sold to GPMV builders, ASMV, builders, merchant system integrators, OEMs, or end users.

Merchant system integrator - A company providing a machine vision system with integration services and adapting the vision system to a specific customer's requirements. A system integrator is project-oriented. Merchant system integrators typically sell to an end user.

A merchant system integrator provides:

1. Turnkey system based on:
   Commercial frame grabber + proprietary software or commercial software
   Commercial IPBS + proprietary software of commercial software
   Commercial GPMV + proprietary software or commercial software

2. Plus value added: application engineering, GUI, material handling, etc.

Captive system integrator - A company purchasing a machine vision product for its own use and employing its own people to provide the integration services. The machine vision product will typically be either a general-purpose machine vision system or an image board set.

Original equipment manufacturer (OEM) - A company offering a product with a machine vision value adder as an option. An OEM includes machine vision in its product, but without machine vision, the system would still have functionality for a customer.

Absent from this list of supplier types "value adder remarketer (VAR)." This term is so general that it loses its meaning. Virtually every other type of company associated with applying machine vision is essentially a value adder. In other words, a company that manufactures application-specific machine vision systems based on a commercial general-purpose machine vision product or image processing board set is a value adder to those products.

An OEM is a company adding a whole lot of value - generally the functionality required by the user of its piece of equipment. A merchant system integrator adds value to either a general-purpose machine vision system or image processing boards -- the value being project-specific software and hardware application engineering.

The distinctions between an ASMV, OEM, and merchant system integrator are:

ASMV - turnkey system provider; functionality purchased includes entire system; any single element of system has no value to customer alone; sells many of the same system

OEM - machine vision is an optional value adder to existing functionality

Merchant system integrator - project-based business.

3.6 SUMMARY

This discussion is meant to clarify the vendor community for the prospective buyer of a machine vision system. It is important to understand that there are different players with different business goals as well as expertise. Successful deployment depends on matching the supplier's product and skill mix to the application.

4

The "What" and "Why" of Machine Vision

Machine vision, or the application of computer-based image analysis and interpretation, is a technology that has demonstrated it can contribute significantly to improving the productivity and quality of manufacturing operations in virtually every industry. In some industries (semiconductors, electronics, automotives), many products can not be produced without machine vision as an integral technology on production lines.

Successful techniques in manufacturing tend to be very specific and often capitalize on clever "tricks" associated with manipulating the manufacturing environment. Nevertheless, many useful applications are possible with existing technology. These include finding flaws (Figure 4. 1), identifying parts (Figure 4.2), gauging (Figure 4.3), determining *X, Y,* and Z coordinates to locate parts in three-dimensional space for robot guidance (Figure 4.4), and collecting statistical data for process control and record keeping (Figure 4.5) and high speed sorting of rejects (Figure 4.6).

Machine vision is a term associated with the merger of one or more sensing techniques and computer technologies. Fundamentally, a sensor (typically a television-type camera) acquires electromagnetic energy (typically in the visible spectrum; i.e., light) from a scene and converts the energy to an image the computer can use. The computer extracts data from the image (often first enhancing or otherwise processing the data), compares the data with previously developed standards, and outputs the results usually in the form of a response.

It is important to realize in what stage of the innovation cycle machine vision finds itself today. Researchers who study such cycles generally classify the stages as (1) research, (2) early commercialization, (3) niche-specific products, and (4) widespread proliferation. In the research stage, people that are experts in the field add new knowledge to the field. In the early commercialization phase, entrepreneurial researchers develop products that are more like "solutions looking for problems." It requires a good deal of expertise to use these products. The individuals applying stage 2 technology are generally techies who thrive on pioneering.

Stage 3 sees the emergence of niche-specific products. Some suggest this is the stage machine vision finds itself in today. Machine vision systems embedded in a piece of production equipment are generally totally transparent to the equipment operator. Application-specific machine vision systems generally have a graphic user interface that an operator can easily identify with as it speaks only in terms with which he is familiar.

Nevertheless, while the fact that a machine vision system is being used may be disguised, it still requires an understanding of the application to use it successfully.

Figure 4.1 - Early version of a paint inspection system that looks for cosmetic defects on auto body immediately after paint spray booth.

Figure 4.2 - Cognex Vision system verifying and sorting foreign tires based on tread pattern identification.

Stage 4 is characterized by technology transparency - the user does not know anything about it, other than that it is useful. Most car drivers understand little about how a car operates, other than what it does when you turn the key. Interestingly, when the car was a "stage 2" technology, a driver also had to be able to service it because of frequent breakdowns experienced. Since then an infrastructure of service stations and highways has emerged to support the technology. In stage 2 there were over 1100 car manufacturers in the United States alone! The industry consolidated as it moved from stage 2 to stage 4.

Clearly, while some consolidation has taken place in the machine vision industry, there are still hundreds of players. This is an indicator of more of a Stage 3 technology. This means that one should have some level of understanding of the technology to apply it successfully. Machine vision is far from a commodity item. The first step is to become informed - the very purpose of this book.

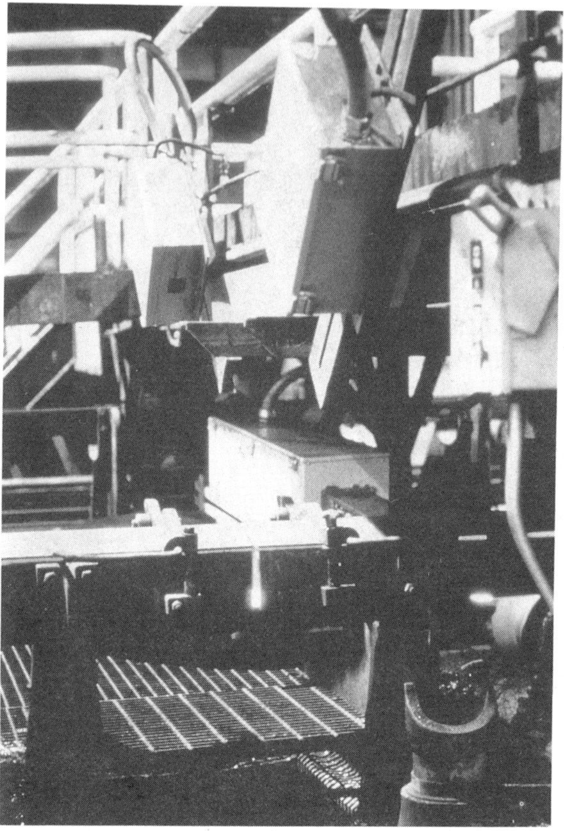

Figure 4.3 - Early system installed on a steel line by Opcon designed to measure cylindrical property of billet.

It is not clear that machine vision as we have defined it will ever become transparently pervasive in our lives or truly a stage 4 technology. The reality is that the underlying technology will definitely become stage 4 technology. The area of biometrics that often uses the same computer vision technology is expected to become a major tool in accessing automated teller machines, cashing checks, accessing computers, etc. There is no doubt there will be other markets in which the underlying technology will become pervasive. For example, if the automobile is to ever achieve autonomous vehicle status, computer vision in some form will make it possible.

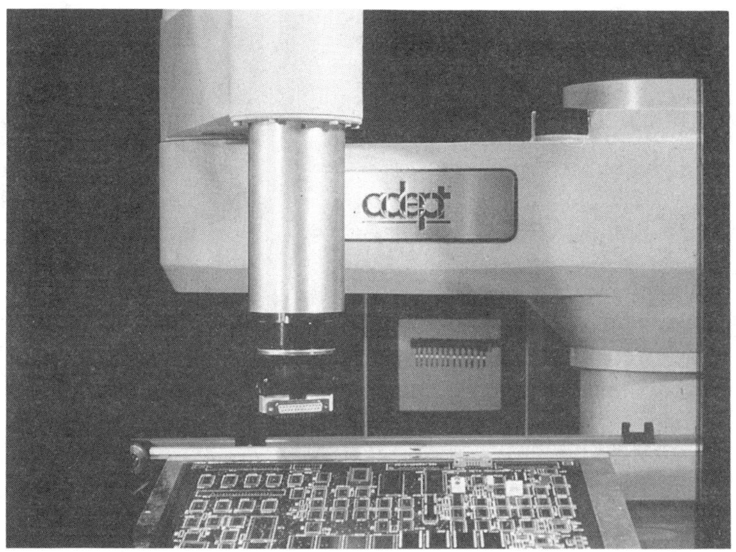

Figure 4.4 - Adept vision-guided robot shown placing components on printed circuit board.

4.1 HUMAN VISION VERSUS MACHINE VISION

Significantly, machine vision performance today is not equal to the performance one might expect from an artificially intelligent eye. One "tongue-in-cheek" analysis by Richard Morley and William Taylor of Gould's Industrial Automation Section quoted in several newspaper articles in the mid-1980's suggests that the optic nerve in each eye dissects each picture into about one million spatial data points (picture elements). Retinas act like 1000 layers of image processors. Each

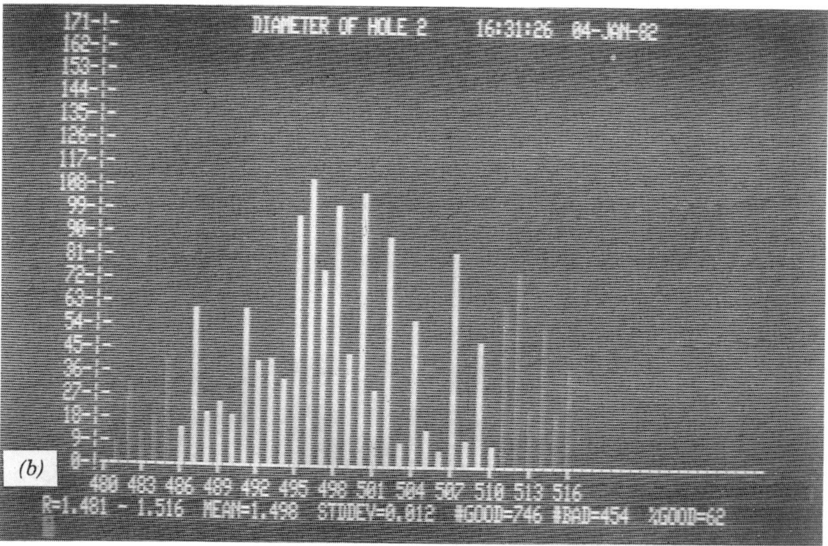

Figure 4.5 - Early RVSI (Automatic) system at end of stamping line examining hole presence and dimensions to monitor punch wear (a) and example of data (b).

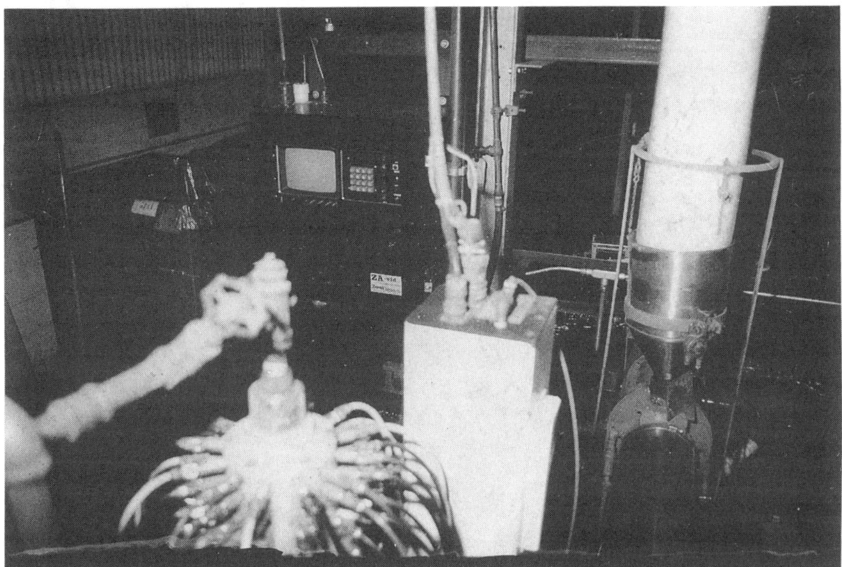

Figure 4.6 - Zapata system inspecting bottle caps to verify presence and integrity of liners at rates of 2600 per minute.

layer does something to the image (a process step) and passes it on. Since the eye can process about 10 images per second, it processes 10,000 million spatial data points per second per eye.

While today there are machine vision systems that operate at several billion operations per second, these still do not have anywhere near the generic vision capacity of humans. Significantly, the specification of MIPS, MOPS, and so on, generally has little relevance to actual system performance. Both hardware and software architectures affect a system's performance, and collectively these dictate the time needed to perform a complete imaging task.

Based on our eye-brain capacity, current machine vision systems are primitive. The range of objects that can be handled, the speed of interpretation, and the susceptibility to lighting problems and minor variations in texture and reflectance of objects are examples of limitations with current technology. On the other hand, machine vision has clear advantages when it comes to capacity to keep up with high line speeds (Figure 4.6). Similarly, machine vision systems can conduct multiple tasks or inspection functions in a virtually simultaneous manner on the same object or on different objects (Figure 4.7). With multiple sensor inputs it can even handle these tasks on different lines.

Some comparisons that can be made between human and machine vision are as follows:

Human vision is a parallel processing activity. We take in all the content of a scene simultaneously. Machine vision is a serial processor. Because of sensor

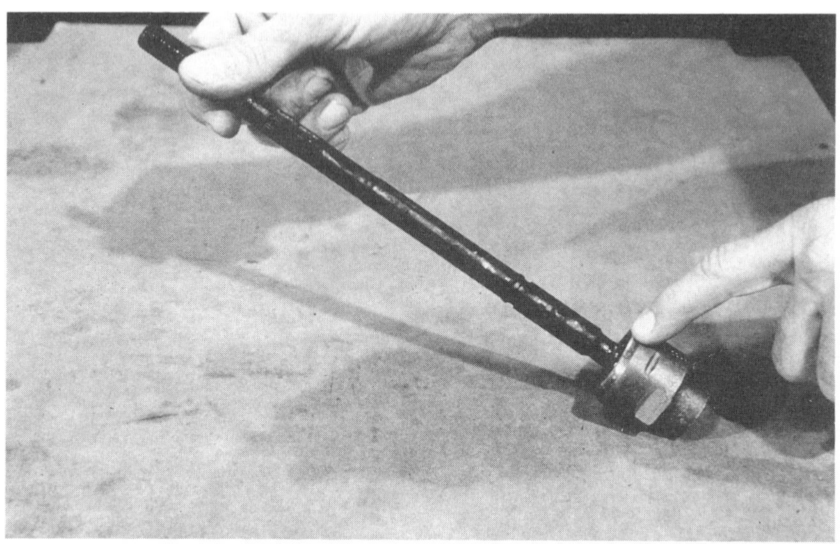

Figure 4.7 - (a) Early RVSI (Automatix) system with multiple cameras inspects tie rod to verify presence of thread, assembly, completeness and swage angle; (b) with multiple cameras inspects tie rod to verify presence of thread, assembly, completeness, and swage angle; (c) with multiple cameras to inspect tie rods to verify presence of thread, assembly, completeness, and swage angle; and (d) with multiple cameras to inspect tie rods to verify presence of thread, assembly, completeness, and swage angle.

(b)

(c)

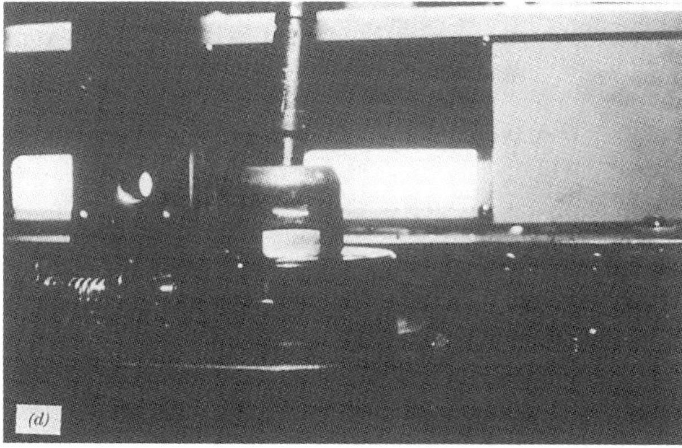
(d)

technology, information about a scene is derived serially, one spatial data point at a time.

Human vision is naturally three-dimensional by virtue of our stereovision system. Machine vision generally works on two-dimensional data.

Human vision interprets color based on the spectral response of our photoreceptors. Machine vision is generally a gray scale interpreter regardless of hue, based on the spectral response of the sensor world. Significantly, sensors exist that permit viewing phenomenon beyond the range of the eyes (Figure 4.8).

Human vision is based on the interaction of light reflected from an image. In machine vision any number of illumination methods are possible, and the specific one used is a function of the application.

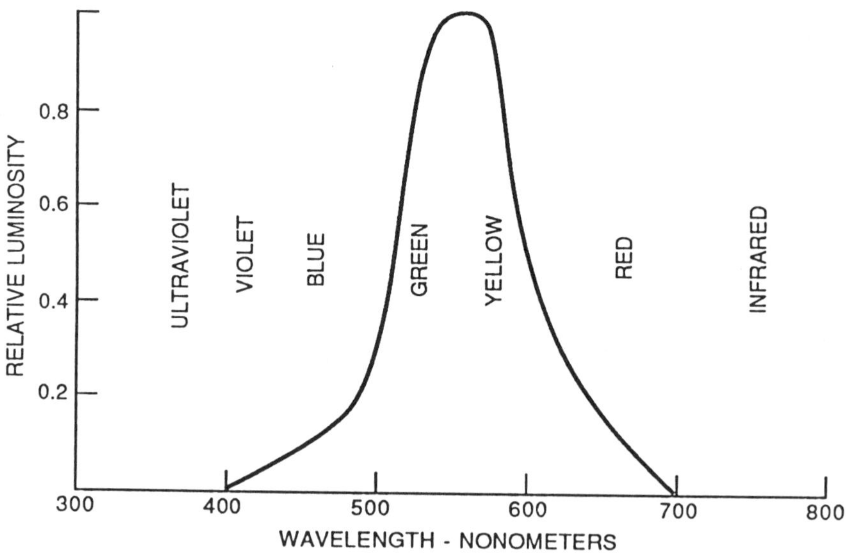

Figure 4.8 - Light spectrum.

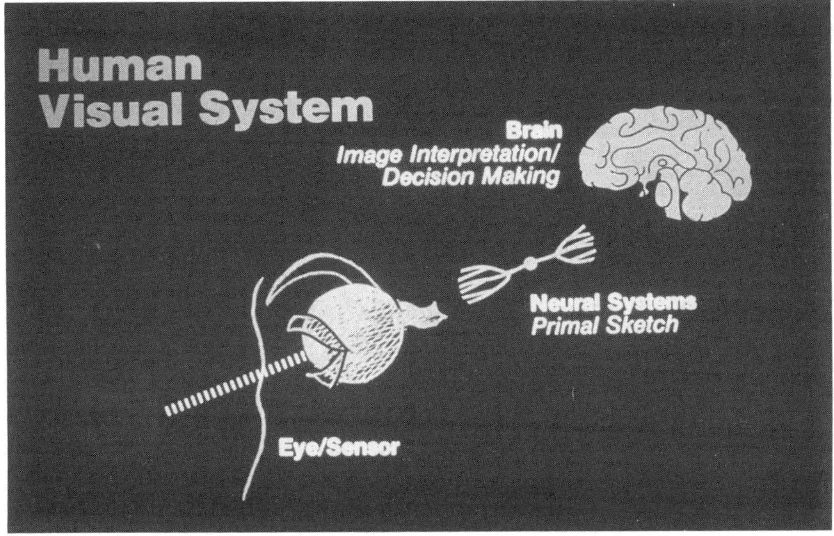

Figure 4.9 - Rendering of eye (courtesy of RVSI/Itran).

Tables 4.1 and 4.2 summarize the comparison between machine vision and human vision. A key difference is that machine vision can be quantitative while human vision is qualitative and subjective.

The process of human vision begins when light from some source is reflected from an object. The lens (Figure 4.9) in the eye focuses the light onto the retina. The light strikes pigments in the rods and cones, where a photochemical reaction generates signals to the attached neurons. The neural network modifies these signals in a complex manner before they even reach the optic nerve and are passed onto the occipital nerve, where cognitive processing of the image starts. Generally speaking, early on we establish models of our surroundings and interpret what we observe based on a priori known relationships stemming from learned models. Machine vision has a long way to go.

Table 4.1 Machine Vision versus Human Vision: Evaluation of Capabilities

CAPABILITIES	MACHINE VISION	HUMAN VISION
Distance	Limited capabilities	Good qualitative capabilities
Orientation	Good for two dimensions	Good qualitative capabilities
Motion	Limited, sensitive to image blurring	Good qualitative capabilities
Edges/regions	High contrast image required	Highly developed
Image shapes	Good quantitative measurements	Qualitative only
Image organization	Special software needed: limited capability	Highly developed
Surface shading	Limited capability with gray scale	Highly developed
Two-dimensional interpretation	Excellent for well-defined features	Highly developed
Three-dimensional interpretation	Very limited capabilities	Highly developed
Overall	Best for quantitative measurement of structured scene	Best for qualitative interpretation of complex, unstructured scene

4.2 MACHINE VISION DEFINITION

What do we mean by machine vision? Distinctions are made between image analysis, image processing, and machine vision. Image analysis generally refers to equipment that makes quantitative assessments on patterns associated with biological and metallurgical phenomena. Image processing refers generally to equipment designed to process and enhance images for ultimate interpretation by people. The instruments used to interpret meteorological and earth resources data are examples.

Machine vision has been defined by the Machine Vision Association of the Society of Manufacturing Engineers and the Automated Imaging Association as the use of devices for optical, noncontact sensing to automatically receive and interpret an image of a real scene in order to obtain information and/or control machines or process.

Significantly, machine vision involves automatic image interpretation for the purpose of control: process control, quality control, machine control, and robot control.

Table 4.2 Machine Vision versus Human Vision: Evaluation of Performance

PERFORMANCE CRITERIA	MACHINE VISION	HUMAN VISION
Resolution	Limited by pixel array size	High resolution capability
Processing speed	Fraction of second per image	Real-time processing
Discrimination	Limited to high-contrast images	Very sensitive discrimination
Accuracy	Accurate for part discrimination based upon quantitative differences; accuracy remains consistent at high production volumes	Accurate at distinguishing qualitative differences; may decrease at high volumes
Operating cost	High for low volume; lower than human vision at high volume	Lower than machine at low volume
Overall	Best at high production volume	Best at low or moderate production volume

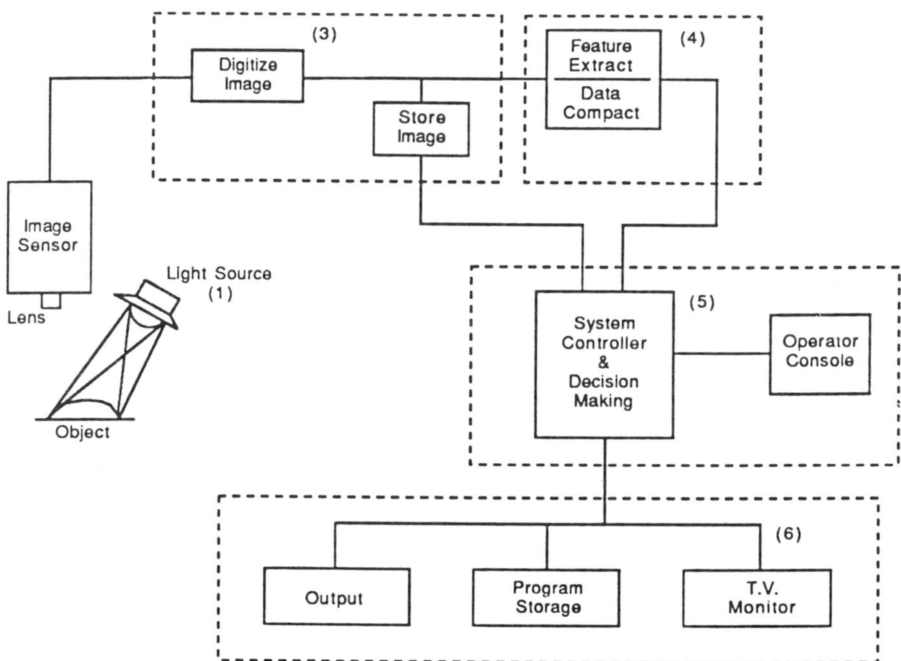

Figure 4.10 - Functional block diagram of basic machine vision system.

A fundamental machine vision system (Figure 4.10) will generally include the following functions:

Lighting. Dedicated illumination.

Optics. To couple the image to a sensor.

Sensor. To convert optical image to analog electronic signal.

Analog-to-Digital (AID) Converter. To sample and quantize the analog signal. (Note: some cameras have digital outputs so a separate A/D function is not required.)

Image Processor/vision engine. Includes software or hardware to reduce noise and enhance, process, and analyze image.

Computer. Decision-maker and controller.

Operator Interface. Terminal, light pen, touch panel display and so on, used by operator to interface with system.

Input-Output. Communication channels to system and to process.

Display. Television or computer monitor to make visual observations.

The fundamental machine vision functional block diagram of virtually all machine vision suppliers looks the same (Figure 4.10). Significantly, each of the discrete functions described in this figure may have different form factors. For example, the A/D converter could be a function on a frame grabber or image processing board, a part of the proprietary design of a vision engine or integrated into the sensor/camera head. Similarly, the display may be a unit separate and independent from the operator interface display or integrated with that display. The image processor/vision engine could in fact be software that operates within the computer or an image processing board or a proprietary hardware design. In other words, depending on the system and/or the applications one might observe different implementations of the functionality depicted in Figure 4.10.

What happens in machine vision? It all starts with converting the optical picture to a digital picture. In general, the systems operate on the projected image of a three-dimensional scene into a two-dimensional plane in a manner analogous to what takes place in a photographic camera. Instead of film, a sensor acts as the transducer and when coupled with an A/D converter, the system characterizes the scene into a grid of digital numbers (Figure 4.11). The image information content at discrete spatial locations in the scene is derived in this manner.

One analogy is to consider the image as on a piece of graph paper (Figure 4.12) with each location mapped onto the corresponding grid. This array has a finite number of discrete elements called picture elements, or pixels (also sometimes called pels). The number of *X and Y* elements into which the image can be discretely segmented are called resolvable elements. One definition of the resolution of a system is therefore the number of *X* and *Y* pixels. A pixel is correspondingly the smallest distinguishable area in an image.

Figure 4.11 - Camera with analog-to-digital converter results in digital representation of image.

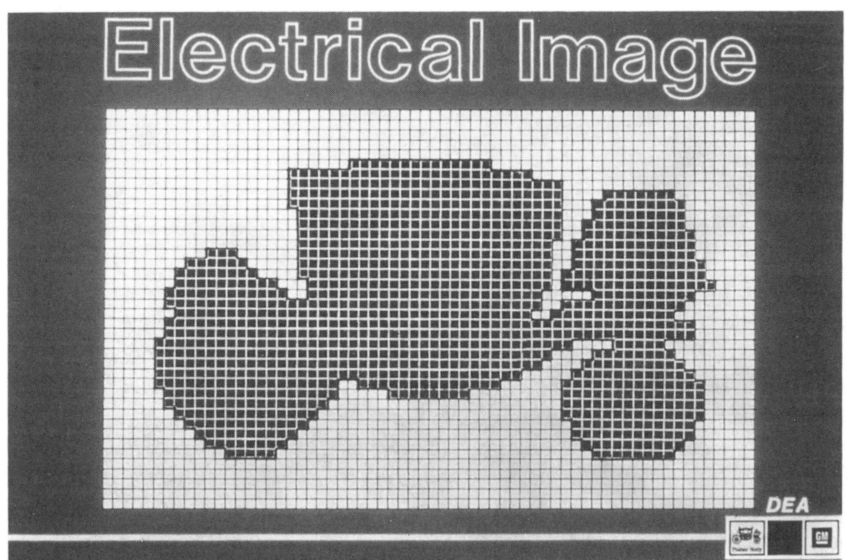

Figure 4.12 - Mapping of three-dimensional scene into two-dimensional plane.

The quantized information content in each pixel corresponds to intensity. This information is defined as a "bit" and relates to image brightness when digitized into a number of quantized elements:

$$2^B = bits$$

For example, $2^4 = 16$. In other words, the 4 bits corresponds to interpreting the scene as 16 shades of gray; 6 bits, 64 shades; 8 bits, 256 shades. In terms of shades of gray, a person is supposed to have an ability to distinguish a single hue (color) into 60 or so shades. However, unlike people, who can interpret hues and therefore characterize as many as 4000 shades by hues, machine vision systems available today generally only interpret all hues into the shades of gray defined by the specific system. In other words, they generally cannot distinguish an object's hue and can become confused if two hues have the same gray value.

TABLE 4.3 Object Properties in Pixel Gray Value

Object properties
Color
Hue
Saturation
Brightness
Specular properties
Reflectance
Texture
Shading
Shadows
Nonuniformities
Lighting
Optics/vignetting

Table 4.3 depicts the properties of an object that contribute to the value of the shade of gray at a specific pixel site. In addition, this property can be influenced by the medium between object and illumination and object and sensor, by filters between object and illumination and object and sensor, by optical properties such as vignetting and dirt on the optics, and by sensor pixel sensitivity variations. Figure 4.13 reflects the digital representation of a scene, and Figure 4.14 depicts the digitally encoded values of the gray shades that are being fed to the computer, reflecting the properties in one small section of the scene. In terms of resolution, the greater the resolving power of the system, the truer the fidelity of the image the system receives as the basis on which to make decisions.

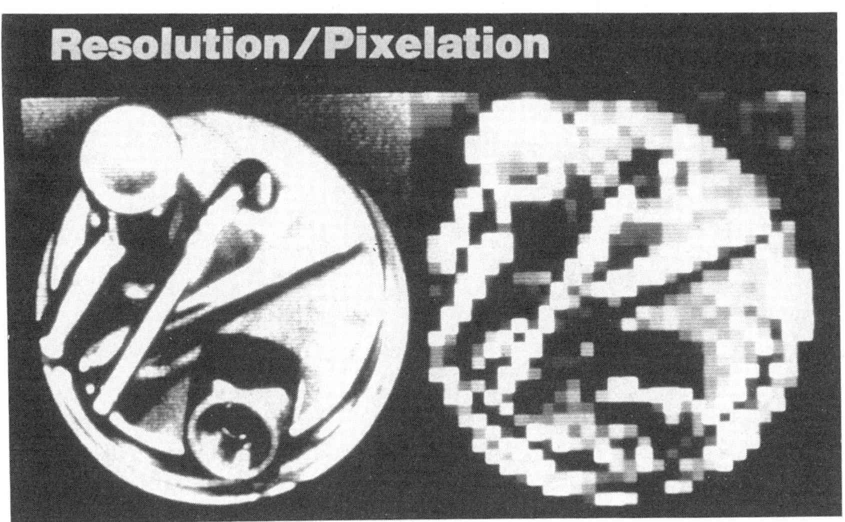

Figure 4.13 - Depiction of resolution/pixelation; digitally encoded values of shades of gray (courtesy of RVSI/Itran).

14	17	14	19	14	17	14	21
8	17	8	17	8	17	21	30
24	19	8	14	17	21	29	28
30	17	19	19	21	27	32	29
29	26	25	27	29	30	28	27
24	27	24	24	28	24	26	24
19	24	21	24	24	24	27	21
22	24	19	24	24	21	24	24

Figure 4.14 - Reflects encoded gray values of small piece of picture (RVSI/Itran).

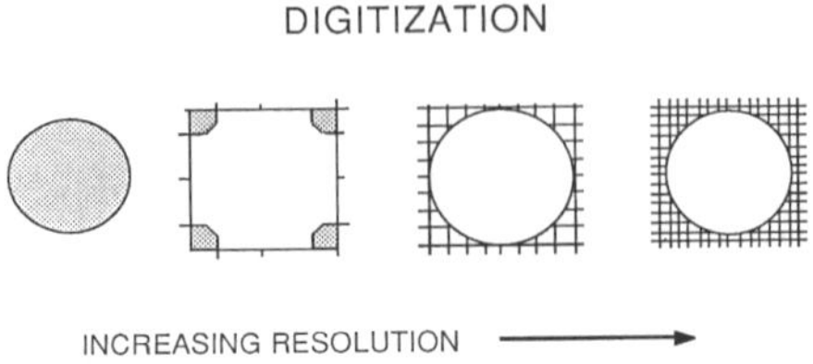

Figure 4.15 - Resolution and image fidelity (courtesy of General Scanning/SVS).

Figure 4.15 shows the impact of higher resolution to more faithfully reproduce the image for computer interpretation. In practice the sensor and the time available on which to make a decision limit resolution. The application dictates the complexity of the processing required and this in combination with the amount of time available and the resolution dictates the computational power required.

In other words, compromises may be required (stemming from the amount of data generated by a sensor as resolution increases) in computing power and time it takes to perform all the computations. In principle, however, the larger the resolution of the sensor (Figure 4.16), the smaller the detail one can observe in the scene. Correspondingly, keeping detail size the same, the field of view on which one can operate increases.

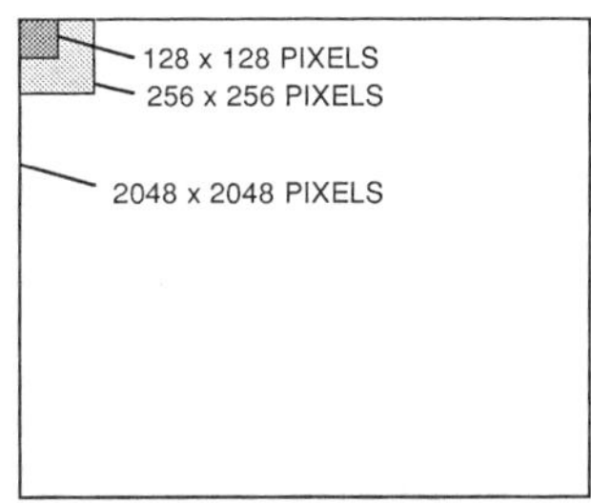

Figure 4.16 - Resolution versus field-of-view (courtesy of General Scanning/SVS).

Edge Detection

For pixel E, the Sobel Value is computed as:

GRAD X = (A+2B+C) - (G+2H+1)

GRAD Y = (C+2F+1) - (A+2D+G)

And the Sobel Value is:

$$\text{Sobel} = \sqrt{\text{GRAD X}^2 + \text{GRAD Y}^2}$$

A	B	C
D	E	F
G	H	I

Figure 4.17 - Neighborhood processing (courtesy of RVSI/Itran).

The challenge in machine vision is the computational power required to handle the amount of image data generated:

256 X 256 X 30 ~ 2 MHz
512 X 512 X 30 ~ 8 MHz

These are 8-bit bytes if processing 256 shades of gray images. Data arrives at the rate of one pixel in every 100 or so nanoseconds. This is why in the many machine vision systems, resolution is only nominally 512 X 512, and each picture element in the image is assigned to either black or white. This significantly reduces the amount of data that has to be handled.

Gray scale systems require far more computer power and algorithms for processing data. Conventional data-processing computer architectures require 20 or more instructions for gray scale image acquisitions and a "nearest neighbor" processing on one pixel (Figure 4.17). This refers to an operation in which a pixel's value is changed in some way based on replacing that pixel with an altered value, where the basis for the alteration is derived from the values associated with neighboring pixels.

If a machine can perform two hundred million instructions per second, it will be able to perform this type of operation at a rate of 10,000,000 pixels per second - a 512 X 512 image will take 20-30 milliseconds. The actual computational requirements of an application are a function of image size, response times, number of processing steps required, complexity of processing, and number of cameras. Actual processing can require 100-10,000 operations per pixel depending on the requirements. Image preprocessing can be minimized by optimizing staging to eliminate positioning uncertainty or other uncertainties stemming from shadows, highlights occlusions, and so on.

Systems for processing color images are another order of magnitude more complex. To minimize complexity, machine vision systems generally operate on two-dimensional information. With certain lighting techniques, the three-dimensional properties of an object can be inferred from a two-dimensional scene. For example, by examining how a stripe of light bends over a three-dimensional object, a machine vision system can infer dimensional data and the distance of an object. An alternate approach to obtain three-dimensional detail has been to employ two cameras and use stereo correspondence analysis based on triangulation principles.

4.3 MACHINE VISION APPLICATIONS

Table 4.4 depicts the type of information that can be extracted and analyzed from an image of an object: spectral, spatial, and temporal. The actual data operated on and the type of analysis that a machine vision system must perform is a function of application, which includes task, object, and related application issues (material handling, staging, environment, etc.). The task refers to:

- Inspection
  - Gauging
  - Cosmetic (flaw detection)
- Verification
- Recognition
- Identification
- Location analysis
  - Position
  - Guidance

Tables 4.5 and 4.6 depict taxonomies of generic machine vision applications outlined in a study conducted at SRI International by Charles Rosen in the late 1970s.

Gaging deals with quantitative correlation to design data, seeing that measurements conform to designs (Figure 4.18). Cosmetic inspection (flaw detection) is a qualitative analysis involving detection of unwanted defects, unwanted artifacts with an unknown shape at an unknown position (Figure 4.19).

Figure 4.18 - Perceptron on-line gauging system checking sheet metal assemblies for gap and flushness.

Table 4.4 Hierarchy of Types of Visual Information Extractable from Image of Single Object

Spectral
Frequency: color
Intensity: gray tones
Spatial
Shape and position (one, two and three dimensions)
Geometric: shape, dimensions
Topological: holes
Spatial coordinates: position orientation
Depth and range
Distance
Three-dimensional profile
Temporal
Stationary presence and/or absence
Time dependent: events, motions, processes

Table 4.5 Machine Vision Applications: Inspection

Highly quantitative mensuration, critical dimensions: critical exterior and interior dimensions of key features of workpieces

A. Qualitative-semiquantitative mensuration
   1. Label reading and registration
   2. Sorting
   3. Integrity and completeness
      a. All parts and features present; right parts
      b. Burrs; cracks; warping; defects, approximate size and location of key features
4. Cosmetic and surface finish properties: stains and smears; colors, blemishes, surface discontinuities
5. Safety and monitoring

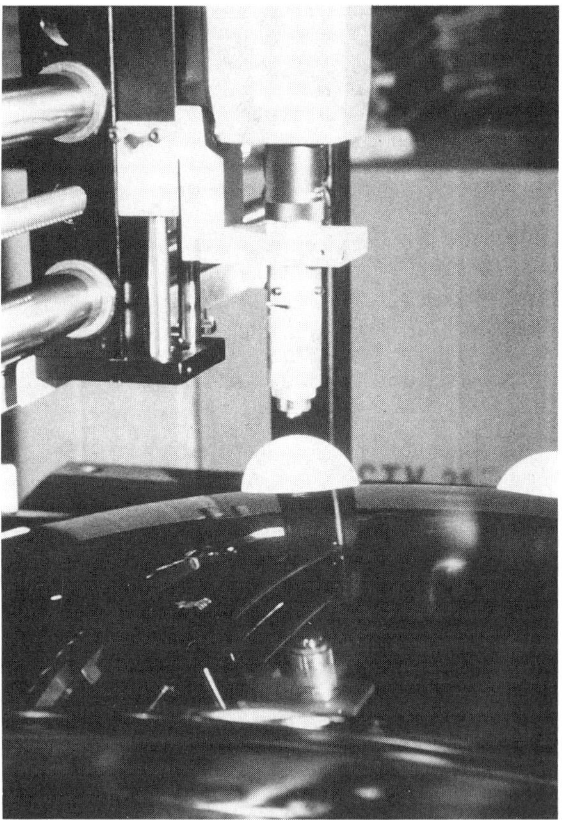

Figure 4.19 - ORS Automation machine vision system inspecting faceplate of cathode ray tubes for imperfections.

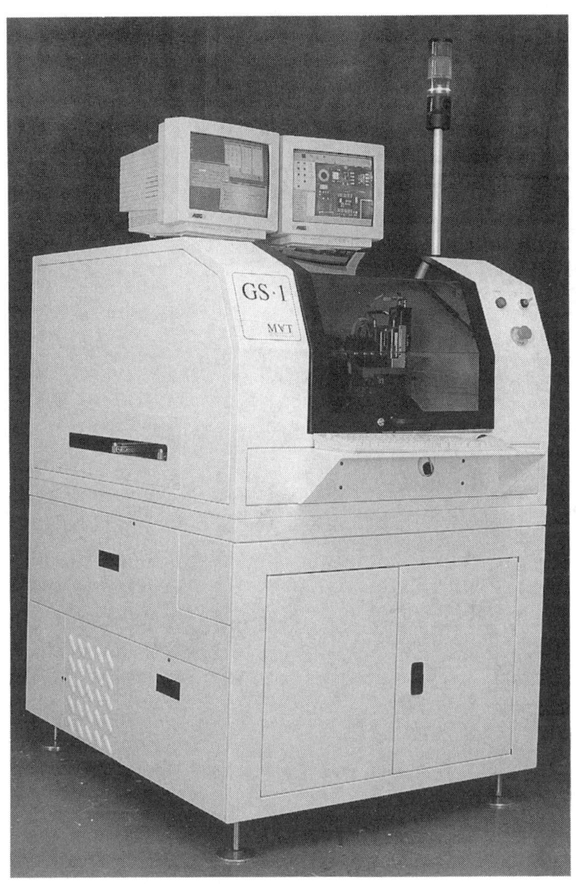

Figure 4.20 - GS-1 system from MV Technology Ltd for in-line automatic inspection and measurement of populated SMT PCBS.

Verification is the qualitative assurance that a fabrication assembly has been conducted correctly (Figure 4.20). Recognition involves the identification of an object based on descriptors with the object (Figure 4.21). Identification is the process of identifying an object by use of symbols on - an object (Figure 4.22). Location analysis is the assessing of the position of an object (Figure 4.23). Guidance means providing adaptively positional information for feedback to control motion (Figure 24).

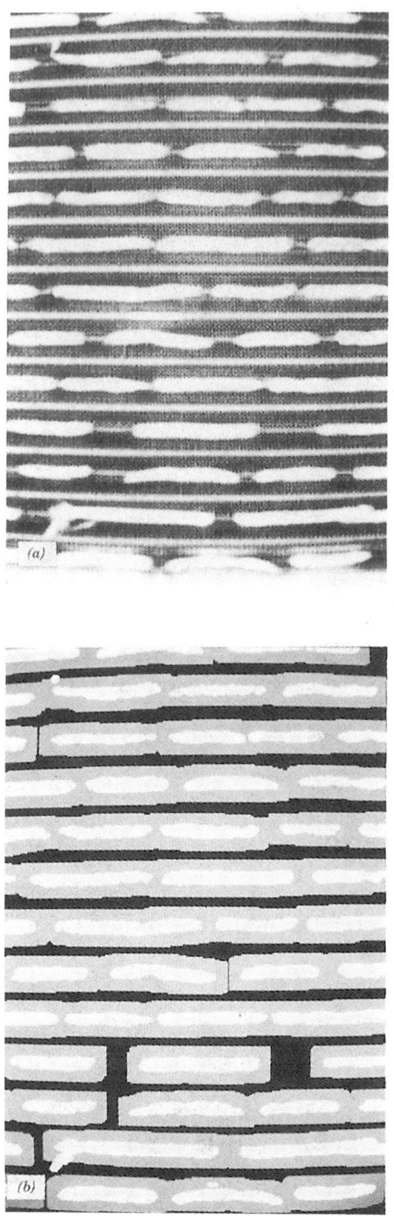

Figure 4.21 - System that can recognize green beans and distinguish them from foreign objects such as stems.

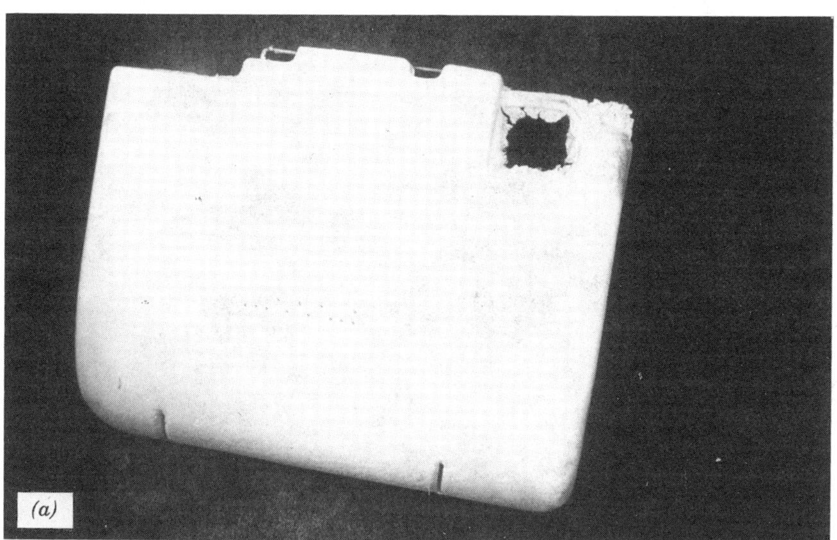

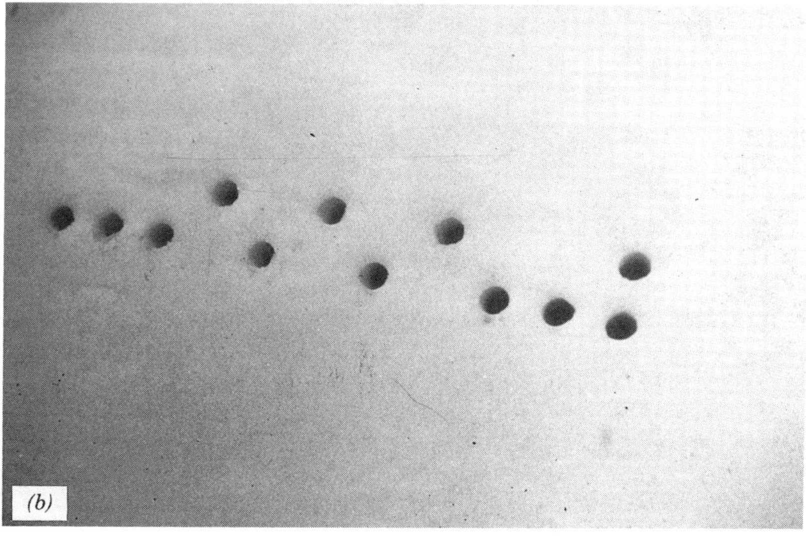

Figure 4.22 - Early system from Penn Video used to identify different foam auto seats based on dot matrix pattern.

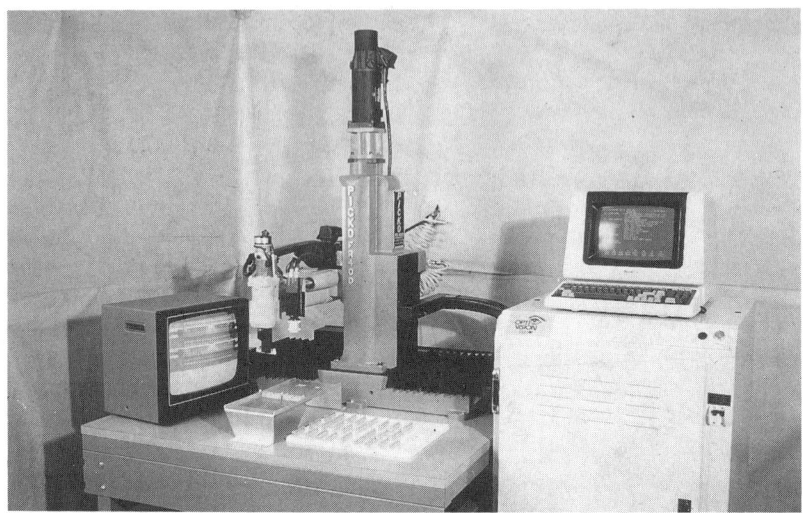

Figure 4.23 - Early system from Gould Electronics providing two-dimensional vision to guide robot for assembly operation.

Figure 4.24 - Early system from Machine Vision International provides correction to six degrees of freedom associated with position of car as it installs windshield.

TABLE 4.6 Machine Vision, Robotic Related

Manipulation of Separated Workplaces on Conveyors	Bin Picking	Manipulation of Manufacturing Process	Assembly
Workplaces lying Stably on belt	Workpieces completely random spatial organization	Finishing, sealing, deburring, cutting, process control, flash removal, liquid gasketing	In-process Inspection
Workplaces hung on hooks partially			Fastening, spot welding, riveting, arc
constrained	Workplaces highly organized and separated		Welding, bolting, screwing, nailing, gluing, stapling
	Workplaces partially organized spatially and unseparated		Fitting, parts presentation
			Mating of parts

Several analyses of applications that have been conducted suggest that approximately 42% of the applications relate to inspection (gaging, cosmetic, and verification), 45% to visual servoing location analysis of which robot guidance is only one application, and 13% to part identification and recognition. Significantly, robot vision applications often require systems capable of inspection in addition to guidance.

4.4 OVERVIEW OF GENERIC MACHINE VISION BENEFITS AND JUSTIFICATION

The opportunities for machine vision are largely in inspection and assembly operations. Even in the latter case, many of the applications will involve inspection (e.g., of tasks), verification, flaw detection, and so on. In conjunction with such tasks, people are only 70-85% effective, especially when dealing with repetitive tasks. According to researchers at the University of Iowa, people asked to perform the visual sorting task of picking out a minority of black Ping-Pong balls from a production line of white ones allowed 15% of the black balls to escape. Even two operators together were only about 95% effective.

People have a limited attention span, which makes them susceptible to distractions. People are also inconsistent. Individuals themselves often exhibit different sensitivities during the course of a day or from day to day. Similarly, there are inconsistencies from person to person, from shift to shift, and so on. The eye's response may also be a performance limiter.

However, people offer some advantages over machine vision. People are more flexible and can be trained for many tasks. People can make adjustments to compensate for certain conditions that should be ignored. For example, a label inspection system would have to be tolerant of the range of blue saturation that may be permissible. A person can accept anything between pastel yellow and virtually orange if that much of a variance is acceptable. On the other hand, to be tolerant of such a variance, a machine vision system may require its threshold sensitivity be set such that it then accepts labels that are torn. People are also quite capable of interpreting the true nature of a condition and, when trained, can take routine action to correct for a pending process failure.

The justification for machine vision need not be based solely on labor displacement. A 1984 Booz-Allen Hamilton study (Duncan and Bowen) cited two elements in the cost of quality: the cost of control and the cost of failure. The essence of the study suggests that one must consider the savings stemming from the cost of failure in any justification equation. The cost of control is generally easy to quantify and includes the prevention and appraisal measures employed in a factory to find defects before products are shipped to customer-inspection and quality control labor costs and inspection equipment.

The cost of failure is much more difficult to quantify and includes internal failures resulting in materials scrap and rework and external failures that result in warranty claims, liability, and recall orders as well as the hidden costs (e.g., the loss of customers).

Machine vision should be considered wherever the prevention of failure or the reduction of the cost of failure is a priority, which should be throughout manufacturing industries. Machine vision can be the primary means to avoid internal and external failures.

For example, use of a machine vision system in a manufacturing process can avoid the production of scrap. Unlike a human inspector who will only detect a reject condition, a machine vision system can spot trends, - trends indicative of incipient conditions that will lead to the production of scrap.

Laser gauges, as well as linear array sensors, are available that can make measurements right on or immediately after a machine tool. The dimensional or surface finish data gathered by such systems are used as a guide for readjusting the machine tool or replacing the cutting tool before the machine produces scrap.

Many industries have jumped on the statistical process control (SPC) philosophy bandwagon. Trend analysis, frequency distribution, and histogram formats for each of the sensors in a system are used to interpret data and report changes in production quality levels. In many such cases, this kind of data is available only the first time from systems that perform 100% inspection. Assessment of the data and its interpretation in the light of corrective action to take to prevent out-of-specification conditions is a process made possible because of the machine vision equipment. Both process control and quality control are possible with machine vision systems.

Significantly, avoiding deviations in quality can impact on downstream operations such as assembly. By guaranteeing that every piece is in an acceptable condition, one can avoid schedule upsets or the need to reschedule an operation because only defective parts are available. Among others, the result of process monitoring and trend analysis could be increased machine uptime or improved capital productivity, that is, increased production capacity without additional equipment and associated floor space.

Despite the amount of data now available for processing, an ancillary benefit is reduced paperwork since record keeping **is** automated. Data transfer between a hierarchy of controllers and computers is easily possible.

In those cases where rejects are not prevented, machine vision system can possibly be used to detect conditions before value is added. A good example of this can be found in the electronics industry (Figure 4.25). It has been estimated that a fault found on a bare printed circuit board immediately after fabrication only costs $0.25 to repair. Once the board is fully loaded with components, the cost to repair that same bare-board reject condition is estimated at $40 before installation in a piece of equipment or shipment. As can be appreciated, the costs become commensurately higher to effect that same repair with each value-adding step in manufacturing.

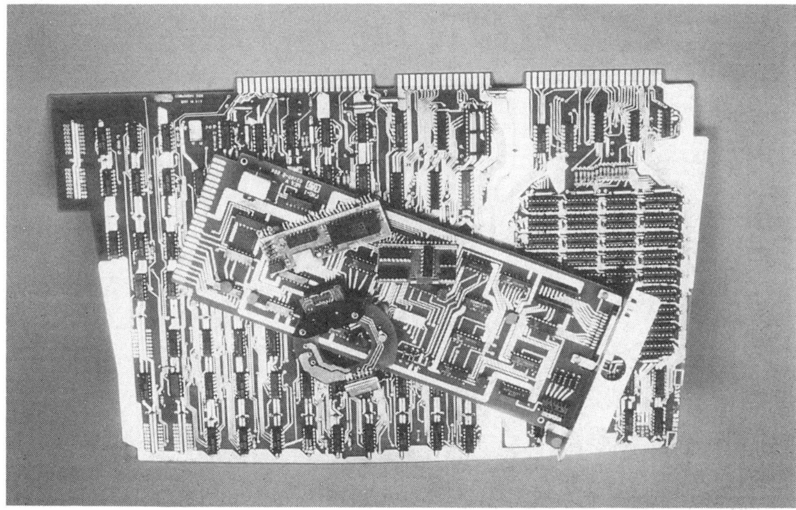

Figure 4.25 - Printed circuit board inspection in electronics industry offers many opportunities for detection of reject conditions before adding value (courtesy of Teradyne/Control Automation).

Similarly, where rejects are not preventable, separating scrap into that which can be reclaimed from that which cannot is possible with machine vision. In the case of thick-film circuits, for example, the detection of the reject before firing permits the reuse of the substrate. In the case of machined parts, parts that have dimensions that exceed the maximum tolerance limit can generally be reworked, while those that exceed the minimum tolerances cannot. Machine vision systems designed to make measurements on parts can be used to make the distinction, both on-line, as with the previously mentioned laser gage types, and off-line, with television optical comparator analogs.

Real-time machine vision techniques can flag conditions and indicate the need for corrective action before a process goes out of specification or at the very least after only a few rejects are experienced. Significant reductions in scrap and rework costs can be achieved from the consistency of flexible automation such as machine vision (Figure 4.26).

Figure 4.26 - Inex Vision Systems/BWI monitors positioning of labels at rate of 1200 per minute.

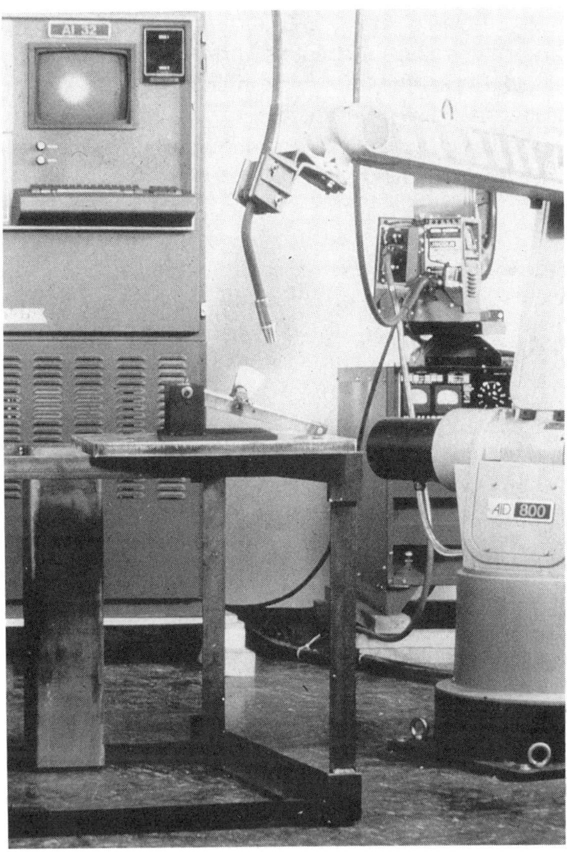

Figure 4.27 - Early RVSI/Automatic vision system for weld seam guidance to relax requirements of fixturing.

Clearly, machine vision has value although the tangible costs of scrap and rework are often hidden in manufacturing overheads, thus making it difficult to expose the true cost associated with producing a bad product; for example, a percentage of work-in-process inventory might be held pending scrap or rework decisions. Rework, similar to inventory, is subject to shrinkage and to annual carrying costs. Unfortunately, it is difficult to quantify the savings that result from making the product right the first time.

Another area to investigate that represents an opportunity for machine vision is one where expensive hard tooling is required to hold a part for an operation (Figure 4.27). This may be avoidable totally or is at least a cheaper flexible fixturing substituted if a machine vision system is used. In this case the system can provide location analysis, that is, so-called software fixturing. A key to this requirement is where setup time is lengthy and the amount of time a part is actually being operated on is very small relative to the total cycle time associated with an operation. Significantly, machine vision may offer increased flexibility, especially in assembly operations, that is, flexibility to produce a wider variety of parts with shorter lead times, better response times to changes in designs, and so on.

Another area for the use of machine vision is where a high incidence of equipment breakdown (Figure 4.28) is experienced because of such problems as over- and under-size or misshapened and/or warped parts or misoriented parts. A machine vision system upstream of the feeder mechanism can reduce or even eliminate downtime.

A situation that definitely warrants a machine vision system is one that involves the inventory of parts because inspection may result in the rejection of a complete batch based on statistical sampling techniques. A 100% inspection assures every part is good so "just-in-time" inventory can be a by-product, with a corresponding reduction in the material handling and damage experienced by handling. Similarly, machine vision opportunities exist where inspection is a production bottleneck.

Figure 4.28 - Early Opcon system verifies that label is properly positioned. Labels applied inadvertently to area where flash exists gum up blades of deflashing unit, resulting in equipment downtime.

As with the justification for robots, one can look for applications related to unhealthy or hazardous environments. The Occupational Safety and Health Administration (OSHA) has expressed concerns about the operator's well being: the noise level is too high, the temperature is too hot, the products are too heavy, and so on. It may be that the environment includes contaminants (metal dust or vapors) that can be injurious to a person. The converse may also be a justification. People may bring contaminants into the environment that can damage the product, for example, dust, causing damage to polished surfaces.

Where operation experiences errors due to operator judgment, fatigue, inattentiveness, or oversight brought about because of the dullness of the job, machine vision opportunities exist (Figures 4.29, 4.30, 4.31 and 4.32). Certainly, when an operation is experiencing a capital expansion mode, machine vision should be considered in lieu of alternative, less effective, more costly methods. Where automation is contemplated as a substitute for people, it should be understood that as people are removed, so are their senses, especially sight. When contemplating automation, an analysis is necessary to assure that loss of sight will not affect the production process.

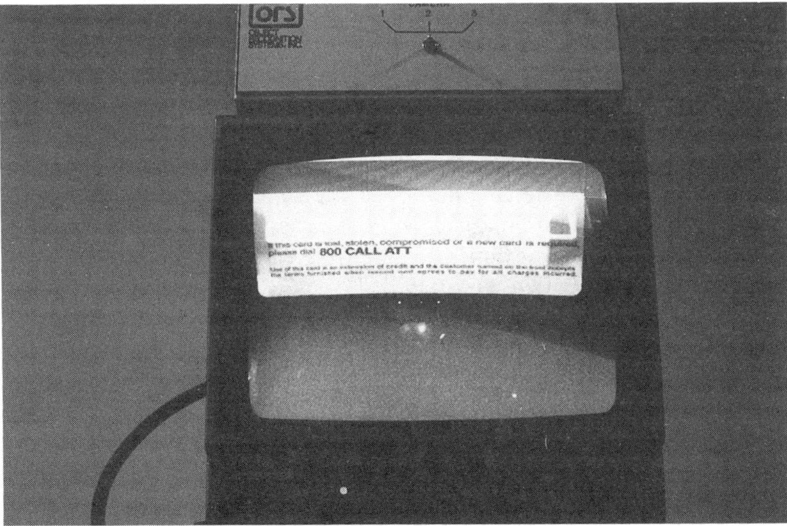

Figure 4.29 - ORS Automation system inspects magnetic and signature strips of credit cards for blemishes at rate of 400 per minute.

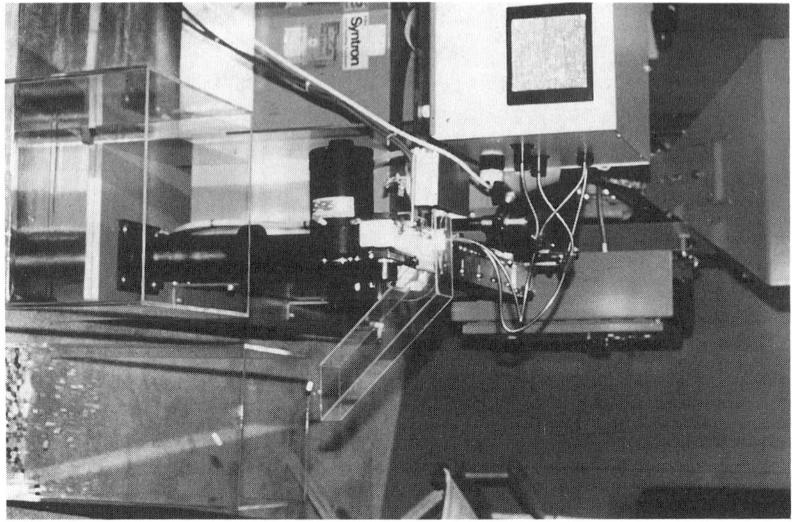

Figure 4.30 - Early system from Vanzetti Vision Systems performs "stranger elimination" function to guarantee all capsules are right ones based on color at rates up to 3600 per minute.

Table 4.7 summarizes how to identify potential applications for machine vision. Unquestionably, any operation can identify opportunities for machine vision by performing an introspective examination of its operations. Table 4.8 summarizes the benefits that can accrue; these benefits can be the basis for justifying the purchase of machine vision. The adoption of this technology, with the result of the objective 100% inspection of products, will cut costs, improve quality, reduce warranty repairs, reduce liability claims, and improve consumer satisfaction - all components in an improved profit picture.

Table 4.7 Identifying Applications

1. Lowest value added
2. Process control
3. Separate scrap that can be reworked
4. Avoid expensive hard tooling
5. Avoid equipment breakdown
6. Avoid excess inventory
7. Hazardous environment
8. Operator limitations essential

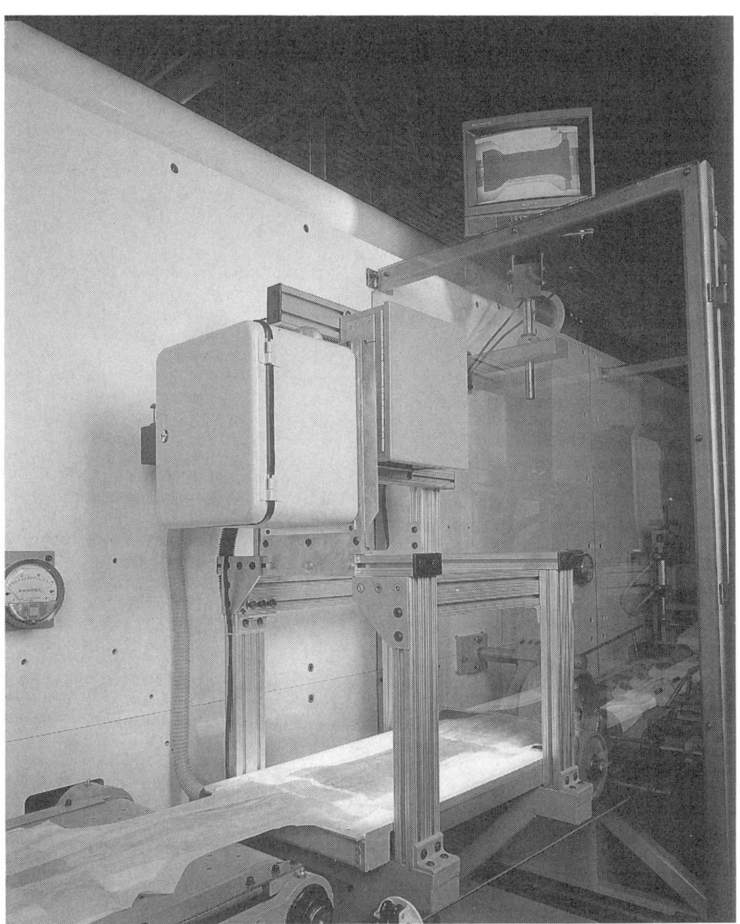

Figure 4.31 - Vision system from Systronic performs on-line inspection of diapers to verify presence of all features.

Figure 4.32 – Vision system from Avalon Imaging verifying empty state of plastic injection molding.

Table 4.8 Machine Vision Benefits and Justification Summary

Economic Motivations

To reduce costs of goods manufactured by:

(a) Detecting reject state at point of lowest value added
(b) Automating to reduce work-in-process inventory
(c) Saving on tooling and fixturing costs
(d) Being able to separate scrap than can be reclaimed from that which cannot
(e) Providing early warning to detect incipient reject state to reduce scrap
(f) Reducing scrap and reworking inventory costs
(g) Reducing in warranty repairs, both in the field and returned goods
(h) Reducing service parts distribution costs
(i) Reducing liability costs
(j) Reducing liability insurance
(k) Improving production yield
(l) Reducing direct and indirect labor and burden rate
(m) Increasing equipment utilization
(n) Reducing setup time
(o) Reducing material handling cost and damage
(p) Reducing inventory
(q) Reducing paper
(r) Eliminating schedule upsets

Quality Motivations

To improve quality by:

(a) Conducting 100% inspection versus sample inspection
(b) Improving effectiveness of quality check to improve goods shipped and thereby improving customer satisfaction
(c) Providing predictability of quality

People Motivations

(a) Satisfy OSHA
(b) Remove from hazardous, unhealthy environment
(c) Avoid contaminants in clean room
(d) Avoid strenuous task
(e) Avoid labor turnover and training costs
(f) Avoid need to hire for seasonal work
(g) Eliminate monotonous and repetitive job
(h) Expedite inspection task that is production bottleneck
(i) Reduce need for skilled people
(j) Avoid errors due to operator judgment, operator fatigue, operator inattentiveness, and operator oversight
(k) Improve skill levels of workers

Miscellaneous Motivations

(a) Substitute capital for labor in expansion mode
(b) Automate record keeping and capture statistics quicker
(c) Feedback signals based on trend analysis to control manufacturing process
(d) Function as "eyes" for automation
(e) Enhance reputation as quality leader
(f) Accelerate response to design changes
(g) Get new technology into business

REFERENCES

Birnbaum, J., "Toward the Domestication of Microelectronics," *Computer,* November 1985.

Duncan, L. S., and Bowen, G. L., "Boosting Product Quality for Profit Improvement," *Manufacturing Engineering,* Society of Mechanical Engineers, April 1984.

Gevarter, W. B., "Machine Vision: A Report on the State of the Art," *Computers in Mechanical Engineering,* April 1983.

Kanade, T., "Visual Sensing and Interpretation: The Image Understanding Point of View," *Computers in Mechanical Engineering,* April 1983.

Lerner, E. J., "Computer Vision Research Looks to the Brain," *High Technology,* May 1980.

Lowe, D. G., "Perceptual Organization and Visual Recognition," National Technical Instrumentation Service Document AD A-150826.

5

Machine Vision: Introductory Concepts

Machine vision all begin with an image - a picture. In many ways the issues associated with a quality image in machine vision are similar to the issues associated with obtaining a quality image in a photograph. In the first place, quality lighting is required in order to obtain a bright enough reflected image of the object. Lighting should be uniformly distributed over the object. Non-uniform lighting will affect the distribution of brightness values that will be picked up by the television camera.

As is the case in photography, lighting tricks can be used in order to exaggerate certain conditions in the scene being viewed. For example, it is possible that shadows can, in effect, include high contrast information that can be used to make a decision about the scene being viewed.

The types of lamps that are used to provide illumination may also influence the quality of the image. For example, fluorescent lamps have a higher blue spectral output than incandescent lamps. While the blue spectral output is more consistent with the spectral sensitivity of the eye, higher infrared output is typically more compatible with the spectral sensitivity of solid state sensors that are used in machine vision.

It has been found that the sensitivity of human inspectors can be enhanced as a consequence of using softer lighting or fluorescent lamps with gases that provide more red spectral output; so too it may also be the case in machine vision. That is, that the lamp's spectral output may influence the contrast associated with the specific feature one is attempting to analyze. This has been demonstrated in the case of many organic products.

As in photography, machine vision uses a lens to capture a picture of the object and focus it onto a sensor plane. The quality of the lens will influence the quality of the image. Distortions and aberrations could effect the size of features in image space. Vignetting in a lens can affect the distribution of light across the image plane. Magnification of the lens has to be appropriate for the application. As much as possible the image of the object should fill the image plane of the sensor.

Allowances have to be made for any registration errors associated with the position of the object and the repeatability of that positioning. The focal length and aperture have to be optimized in order to handle the depth of field associated with the object.

The imaging sensor that is used in the machine vision system will basically dictate the limit of discrimination of detail that will be experienced with the system. Imaging sensors have a finite number of discrete detectors and this number limits the number of spatial data elements that can be processed or into which the image will be dissected. In a typical television-based machine vision system today the number of spatial data points is on the order of 400 to 500 horizontal X 400 to 500 vertical.

Basically, what this means basically is that the smallest piece of information that can be discriminated is going to be a function of the field of view. Just like in photography, one can use panoramic optics to take a view of a mountain range, and although a family might be in the picture in the foothills of the mountains, it is unlikely that you would be able to discriminate the family in the picture. On the other hand, using a different lens and moving closer to the family, one would be able to capture the facial expressions of each member, but the resulting picture would not include the peaks of the mountains.

So, for example, given that an application requires a one-inch field of view, and a sensor with the equivalent of 500 spatial data points is used, one would have a spatial data point that would be approximately .002 inches on the side. Significantly, the ability of machine vision today to discriminate details in a scene is generally better than the size of a spatial data point.

In a manner basically analogous to how an eye can see stars in a night sky because of the contrast associated with the star light, so too in machine vision techniques exist which allow systems to be able to discriminate details smaller than a spatial data element. Again, contrast is critical. The claims for subpixel sensitivity vary from vendor to vendor and depend very much on their execution and the application.

In all machine vision systems up until this point in our discussion, the information or the image has been in an analog format. For a computer to operate on the picture the analog image must be digitized. This operation basically consists of sampling at discrete locations along the analog signal that corresponds to a plot of time vs. brightness, and quantizing the brightness at that sample point.

The actual brightness value is dependent on: the lighting, the reflective property of the object, conditions in the atmosphere between the lighting and the object and between the object and the camera, and the specific detector sensitivity in the imaging sensor. Most vision systems today characterize the brightness into a value of between 0 and 255. The brightness so characterized is generally referred to as a shade of gray.

For the most part today, machine vision systems are monochromatic. Consequently, the color may also be a factor in the brightness value. That is, it is possible to have a shade of red and a shade of green (and so on) all of which would have the same brightness value. In many cases where color issues are a concern, filters are used in order to eliminate all colors that are not of interest to the particular application. In this way the gray shades are an indicator of the saturation level associated with a specific color. Color cameras can also be used to acquire the data and segmentation based on the specific color enabled.

At last we have a picture that has been prepared for a computer. In most machine vision systems today, the digitized image is stored in memory that is separated from the computer memory. This dedicated memory is referred to as a frame store - where frame is synonymous with the term used in television to describe a single picture. In some cases the dedicated hardware that includes the frame store also includes the analog-to-digital converter as well as other electronics to permit one to view images after processing steps have been conducted on the image to view the effects of these processing procedures.

Now the computer can operate on the image. The operation of the computer on the image is generally referred to as image processing. In addition to operating on the image, the computer is also used to analyze the image and make a decision on the basis of the analyzed image and perform an operation accordingly. What is typically referred to as the vision engine part of the machine vision system is the combination of image processing, analysis and decision-making techniques that are embodied in the computer.

A good analogy can be made to a toolbox. Virtually all machine vision systems today include certain fundamental tools, much like a hammer, screwdriver or pliers. Beyond these, different suppliers have developed additional tools, more often than not driven by a specific class of applications. Consequently the description frequently given for machine vision as being an "idiot savant" is quite apropos. That is, most of the platforms are brilliant on one set of applications but "idiots" or truly not the optimal for other applications.

It is important, therefore, to select the vision platform or toolbox with the most appropriate tools for an application. Significantly, no machine vision systems exist today that come anywhere near simulating the comprehensive image understanding capabilities that people have. It is noted that for many applications many different tools will actually do the job and in many cases without sacrificing performance. On the other hand, in some cases while the tools appear to do the job, performance might be marginal, in a manner analogous to when we attempt to use a flat head screwdriver in order to turn a screw with a Phillips head.

Image processing is generally performed on most images for basically two reasons: to improve or enhance the image and, therefore, make the decision associated with the image more reliable, and to segment the image or to separate the features of importance from those that are unimportant. Enhancement might be performed, for example, to correct for the non-uniformity in sensitivity from photo site to photo site in the imaging sensor, correct for distortion, correct for non-uniformity of illumination, to enhance the contrast in the scene, correct for perspective, etc.

These enhancement steps could be as simple as adding or subtracting a specific value to each shade of gray or can involve a variety of logical operations on the picture. There are many such routines. One routine that is commonly found as a tool for image processing in most vision platforms today is a histogram routine. This involves developing a frequency distribution associated with the number of times a given gray shade is determined.

One use of histograms is to improve contrast. This involves mathematically redistributing the histogram so that pixels are assigned to gray shades covering 0 to 255, for example. In an image with this type of contrast enhancement, it could be easier to establish boundaries or easier to establish a specific gray shade level or threshold to use to binarize the image. Binarizing an image, or segmenting an image based on a threshold above which all pixels are turned on and below which all pixels are turned off, is a conventional segmentation tool included in most vision platforms and can be effective where high contrast exists.

Where contrast in a scene is not substantial, segmentation based on edges may be more appropriate. Edges can be characterized as locations where gradients or gray shade changes take place. Both the gradient as well as the direction of change can be used as properties to characterize an edge. Significantly, edges can be caused by shadows as well as reflectance changes on the surface in addition to the boundaries of the object itself. Artifacts in the image may also contribute to edges. For example, unwanted porosity may also be characterized by increased edges.

There are many different ways edges are characterized. One of the simplest is just using the fact that there are sharp gray scale changes at an edge. Significantly, however, edges in fact appear across several neighboring pixels and what one has is in fact a profile of an edge across the pixels. Because of this, there are ways to mathematically discriminate the physical position of an edge to a value less than the size of the pixel. Again, there are many ways that these subpixel calculations have been made and the results are very application dependent. Consequently, although claims are made of one part in ten or better subpixelling capability, it is important to understand that the properties of a given application can reduce the effectiveness of subpixelling techniques.

Having performed image-processing routines to enhance and segment an image, the computer is now used to analyze the image. The specific analysis conducted is again going to be very application-dependent. In the case of a robot guidance application, for example, a geometric analysis would typically be conducted on the segmented image. Looking at the thresholded segmented image or edge segmented image one would be able to calculate the centroid property and furnish this as a coordinate in space for the robot to pick up an object, for example.

In the case of using vision systems to perform inspections of one type or another, there are literally hundreds of different types of analysis techniques that have emerged. The number of pixels associated with the binarized or thresholded picture, for example, could be counted. This could be a relatively simple measure of the completeness of an object. The number of transitions or times that one goes from black to white can be counted. The distance between transitions can be counted and can serve as a measurement between boundaries of an object. The number of pixels that are associated with an edge can be counted. Vectors associated with the direction of the gradient at an edge can be used as the analysis features. A model based on the edges can be derived where the edges can be characterized as vectors of a certain length and angle. Geometric features can be extracted from the enhanced image and used as the basis of decisions.

These same techniques can be used in conjunction with pattern recognition applications. In each case, one or more of the above-mentioned features extracted from the image can define a pattern. For example, maybe a combination of the transition counts and edge pixels would be sufficient to make a judgement about patterns where that combination is sufficient to distinguish between the patterns. Another approach might be to use geometric properties to distinguish patterns. These might include length and width ratios, perimeter, etc.

The computer, having reduced the image to a set of features used as the basis of analysis, would typically then use a deterministic or probabilistic approach to analyze the features. A probabilistic approach is one that basically suggests that given a certain property associated with a feature, there is a high probability that the object is in fact good. So, for example, using the total number of pixels as an indication of the completeness of an object one would be able to suggest that if the total number of pixels exceeded, say, 10,000 there is a high probability that the object is complete. If less than 10,000 the object should be rejected because it would be characterized as incomplete. Some refer to this as goodness-of-fit criteria. It is also possible to set a boundary around these criteria. That is, it should fall between 10,000 and 10,500. An indication of a pixel count greater than 10,500 could be an indication, for example, of excess flashing.

A deterministic approach is one that will use physical feature properties as the criteria. For example, the distance between two boundaries has to be one inch +/- .005". The perimeter of the object must fall between 12 inches +/- .020". The pattern must match the following criteria in order to be considered a match:

length/width ratio of a certain value, perimeter of a certain value, centroid of a given calculated value, etc.

In a deterministic mode each of the features can be associated with a vector in decision space. In a pattern recognition application, the combined feature vector or the shortest distance to the known feature set for each of the patterns is the one that would be selected. This type of evaluation is referred to as decision theoretic. Another type of analysis is one based on syntactic techniques. In these cases, primitives associated with pieces of the image are extracted and the relationship between them is compared to a known database associated with the image.

In other words, the primitives and their relationship to each other have to abide to a set of rules. Using syntactic techniques one may be able to infer certain primitives and their position knowing something about other primitives in the image and their position with respect to each other. This could be a technique to handle parts that might be overlapping and still be able to make certain decisions associated with those parts even though one can not see them entirely.

As you can see, there are many vision tools that are available and the specific tools that one requires are application-dependent.

5.1 WHO IS USING MACHINE VISION

Today one can find machine-vision-type technology in virtually every manufacturing industry. The largest adopter by far is the electronics industry. In microelectronics, machine vision techniques are used to automatically perform inspections throughout the integrated circuit manufacturing process: photomask fabrication, post die slicing inspection, pre-cap inspection and final package inspection for mark integrity.

Throughout the manufacturing process, machine vision is also used to provide feedback for position correction in conjunction with a variety of manufacturing processes such as die slicing and bonding and wire bonding. In the macroelectronic industry machine vision is being used to inspect printed circuit boards for conductor width spacing, populated printed circuit boards for completeness, post solder inspection for solder integrity.

As in microelectronics, it is also being used to perform positional feedback in conjunction with component placement. It has become an integral part of the manufacturing process associated with the placement of chip carriers with relatively high-density pin counts.

In industries that produce products on a continuous web, such as the paper, plastic, and textile industries, machine vision techniques are being used to perform an inspection of the integrity of the product being produced. Where coatings are applied to such products, machine vision is also being used to guarantee the coverage and quality of coverage. In the printing industry, one finds machine vision being used in conjunction with registration.

The food industry finds machine vision being used in the process end to inspect products for sorting purposes, that is, sorting out defective conditions or misshapen product or undersize/oversize product, etc. At the packaging end, it is being used to verify the size and shape of contents, such as candy bars and cookies, to make sure they will fit in their respective packages.

Throughout the consumer manufacturing industries one will find machine vision in various applications. These include label verification, that is, verifying the position, quality and correctness of the label. In the pharmaceutical industry one finds it being used to perform character verification, that is, verifying the correctness as well as the integrity of the character sets corresponding to date and lot code.

The automotive industry finds itself using machine vision for many applications. These include looking at the flushness and fit of sheet metal assemblies, including the final car assembly; looking at paint qualities, such as gloss; inspecting for flaws on sheet metal stampings; verifying the completeness of a variety of assemblies from ball bearings to transmissions; etc.; used in conjunction with robots to provide visual feedback for: sealant applications, windshield insertion applications, robotic hydropiercing operations, robotic seam tracking operations, etc.

Virtually every industry has seen the adoption of machine vision in some way or another. The toothbrush industry, for example, has vision systems that are used to verify the integrity of the toothbrush. The plastics industry looks at empty mold cavities to make sure that they are empty before filling them again. The container industry is using machine vision techniques widely. In metal cans they look at the quality of the can ends for cosmetic flaws, presence of compound, score depth on converted ends, etc. The can itself is examined to inspect it for defective conditions internally.

The glass container industry uses machine vision widely to inspect for sidewall defects, mouth defects and empty bottle states as well as dimensions and shapes. In these cases, vision techniques have proven to be able to handle 1800 to 2000 objects per minute.

5.2 DEPLOYING MACHINE VISION

How do I know what machine vision techniques are most suitable for my application? A studied approach is usually required unless the application is one that has a system that has been widely deployed throughout an industry. In that case the pioneering work has already been done. Adaptation to one's own situation, while not trivial, may have little risk. To find out if your application has been solved, ask around. Today, most machine vision companies are quite helpful. If they do not offer the specific solution, but know of any other company that does, they will generally respond with candor and advise accordingly. Consultants may also be able to identify sources of specific solutions.

Having identified those sources they should be contacted to identify their referenceable accounts and these in turn should be contacted to determine: why they were selected, what has been the experience, service, etc., and would they purchase the same product? This should help to narrow down the number of companies to be solicited for the project. In this case, the ultimate selection will no doubt be largely based on price, though policies such as training, warranty, service, spare parts, etc., should also be considered, as they will impact the life cycle cost.

What do you do if you find your application is not a proliferation of someone else's success? In this case a detailed application description and functional specification should be prepared. This means really getting to know the application - what are all the exceptions and variables? The most critical ones are position and appearance. These must be understood and described comprehensively.

What are the specific requirements of the application? Will the system first have to find the object - even minor translation due to vibration can be a problem for some machine vision executions. In addition to translation, will the part be presented in different rotations? Are different colors, shades, specular properties, finishes, etc. anticipated? Does the application require recognition? Is it gaging? What are the part tolerances? What percent of the tolerance band would it be acceptable to discriminate? If flaw detection, what size flaw is a flaw? Is the flaw characterized by reflectance change, by geometric change, etc.?

Having prepared the spec, at least a preliminary acceptance test for system buy-off should be prepared and solicitations should be forwarded to potential suppliers. How do you identify those suppliers? A telephone survey of the 150 or so companies is one approach. Again, use of a consultant can greatly accelerate the search. In any event, the leading question should be whether or not they have successfully delivered systems that address similar requirements.

Since we have already established that the application does not represent the proliferation of an existing system solution, the best one can expect is to find a number of companies that have been successful in delivering systems that address needs similar to yours and seemed to have been able to handle similar complexities. So, for example, if the application is flaw detection - are the type and size flaws similar? Are the part, size and geometric complexity and material similar? Is part positioning similar, etc.?

This survey should narrow the number of companies to be solicited to four to six. The solicitation package should demand a certain proposal response. It is important to get a response that reflects that the application has truly been thought about. It is not sufficient to get a quotation and cover letter that basically says "trust me" and "when I get the order I will think about how I'm going to handle it." The proposal should give system details. What lighting will be used and why was that arrangement selected? How about the camera properties, have they been thought through? How about timing, resolution, sampling considerations, and so forth?

Most importantly, does the proposal reflect an understanding of how the properties of the vision platform will be applied and can it defend that those properties are appropriate for the application? How will location analysis be handled? What image processing routines will be enabled specifically to address the application? A litany of the image processing routines inherent in the platform is not the issue. Rather, what preprocessing is being recommended, if any? What analysis routines, etc.? Along with this, an estimate should be prepared of the timing associated with the execution from snapping a picture through to signal availability reflecting the results of a decision. This should be consistent with your throughput requirements.

When a vendor has thought through the application this way and conducted a rather comprehensive analysis, he is in a good position to provide both a schedule of project development tasks and a good estimate of the project cost. An excellent paper that describes this systematic approach as applied to a "Frobus Assembly" was written by Dr. Joseph Wilder and can be found in SPIE Volume 849 "Automated Inspection and High Speed Vision Architectures."

By insisting on this type analysis in the proposal, both vendor and buyer should avoid surprises. Among other things, it will give the buyer a sense that the application is understood. Those proposals responsive in this manner should be further evaluated using a systematic procedure such as Kepner Tregoe decision-making techniques. These involve establishing criteria to use as the basis of the evaluation, applying a weighting factor to the criterion and then evaluating each of the responses against each weighted criterion to come up with a value.

This value represents a measure of how a company satisfies the criterion along with the relative importance of that criterion to the project. In some cases, the score given should be 0 if the approach fails to satisfy one of the absolute requirements of the application. A good paper describing the application of these techniques to evaluating machine vision proposals was written by Ed Abbott and delivered at the SME sponsored Vision 85 Conference, March 25-28, 1985.

Having made a decision on a vendor, justifying the project may be the next issue. Significantly, justification based solely on labor displacement is unlikely to satisfy the ROI requirements. Quantifying additional savings is more difficult but in reality may yield an even greater impact than labor savings. Product returns and warranty cost should be evaluated to assess how much they will be reduced by the machine vision system. The cost of rework should be a matter of record. In addition, however, a value can be calculated for the space associated with rework inventory as well as the rework inventory itself. The cost of rejects and related material costs, the cost of waste disposal associated with rejects, and the cost of freight costs on returns are all very tangible quantifiable costs.

There are other savings, through less tangible, which should be estimated and quantified. These include items such as:

1. The cost of overruns to compensate for yield.
2. The avoidance of inspection bottlenecks and impacton inventory income and inventory turnover.
3. The elimination of adding value to scrap conditions.

4. The potential for increased machine uptime and productivity accordingly.
5. The elimination of schedule upsets due to the production of items that require rework.

Another observation is that when considering the savings due to labor displacement, it is important to include all the savings. These include:

1. Recruiting
2. Training
3. Scrap rework created while learning a new job
4. Average workers compensation paid for injuries
5. Average educational grant per employee
6. Personnel/payroll department costs per employee

Overall the deployment of machine vision will result in improved and predictable quality. This in turn will yield improved customer satisfaction and an opportunity to increase market share - the biggest payback of all.

6

Image Acquisition

6.1 INTRODUCTORY CONCEPTS

The application of machine vision technology involves dealing with many visual variables. In most cases one must be forgiving of some variables and sensitive to others, the system must detect. Even simple applications have to contend with variables that a person performing the same task can easily dismiss. Label presence and/or absence detection is often cited as a machine vision application. One variable a system must be able to contend with is hue saturation. For example, a label may be perfectly acceptable as long as its "color" is yellow, from pastel to virtually orange. A person can easily accept this range and still detect a missing or skewed label. A machine vision system may be tricked into thinking the label is missing when actually the acceptable hue saturation change is what was experienced.

Lighting and optics in many applications can be optimized to enhance the contrast associated with the variable for which a system is purchased or, conversely, can assist in diminishing the influence of the variable on the data required to make a reliable "vision/decision."

In a given machine vision installation, dedicated lighting is recommended. However, secondary sources of illumination may come from ambient lighting and the reflection of the primary source of light off of other equipment or even a floor after polishing! The net result is a complex pattern of light in which uniformity is a compromise. This can affect the nature of shadows and shading on the surface, which can affect the interpretation of the object.

Gray Levels encoded by a machine vision system are physical measurements are a function of illumination, viewpoint, surface reflectance, and orientation. In general, variations in illumination, surface reflectance, and surface orientation give rise to shaded images as seen from a viewpoint; surface reflectance, and surface orientation give rise to shaded images as seen from a given viewpoint.

The reflectivity property of an object is determined by such surface characteristics as texture or color. An object of uniform color and smoothness reflects light uniformly from all points on its surface. Real objects do not behave this way, experiencing varying surface properties, and consequently, variations in brightness are observed in images.

Human perception of brightness or the color of a part of an image is strongly influenced by the brightness or color of the neighboring parts. The relative constancy of the perceived lightness and color of a surface under different illumination conditions also affect human perception. A gray piece of paper taken from a poorly lit room into the bright outside sunshine still appears gray, even though the amount of light reflected is several orders of magnitude greater than from a white piece of paper indoors. Similarly, a surface of a certain color generally retains its perceived hue even though the color of illumination changes.

A sensor, whose response is proportional to the average light intensity in a small neighborhood, makes local measurement of the brightness of a small area of an image, while to the eye the perceived brightness is dependent on the intensities of the surrounding pixels. The human perception of color involves differentiation based on three independent properties: intensity, hue, and saturation (Figure 6. 1).

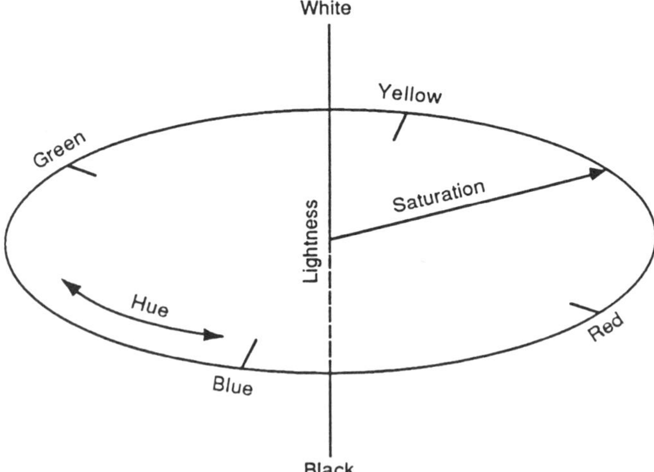

Figure 6.1 - Three dimensional color system (courtesy of Hunter Lab).

Hue corresponds to color (blue, red, etc.), intensity is the lightness value, and saturation is the distance from lightness per hue. For a monochromatic source of light, our perception of hue is directly dependent on wavelength. However, in normal environments, objects of a certain hue, say, red, do not reflect just the light of a single wavelength but rather a broad spectrum that has a distribution different from that of the white light. Different hues are also obtained by mixing colors, for example, green from blue and yellow, even though the mixed light may contain no green wavelength.

Colors are synthesized from a mixture of three primary colors. Commonly used primary colors are red, green, and blue. These need not be monochromatic but may have a wide spectrum distribution of light intensity. Television cameras, as with the eye, do not measure the perceived attributes of intensity, hue, and saturation directly. Instead, the measured attributes correspond to the intensity of the three primary-color components in the viewed surface.

Surface orientation stemming from the geometry of the surface of an object can result in light intensity variations. A surface perpendicular to the light appears brighter than if aligned at some other angle. Complex geometric shapes will result in complex variations in surface brightness in the image. Significantly, in some applications the data of value about the scene can be inferred from shadows.

The integration of the four variables (illumination, viewpoint, surface reflectance, and surface orientation) along with the nonobject or system factors enumerated in Chapter 1 results in the encoded value at a pixel site. The resulting distribution of light intensities forms an image, a two-dimensional representation of a typically three-dimensional scene.

6.1.1 Application Features

Features that characterize an image of a scene include position (location and orientation in space), geometry or shape, and light intensity distribution. The location of an object refers to its coordinates in space. Orientation refers to the direction of a specific axis of the object. The geometric features of an object can be used to distinguish it from other objects or to verify its dimensional properties. The light intensity distribution of an object may also be a means of recognizing or distinguishing one object from another. It can also be used to tell something about the surface or cosmetic properties of an object as well as to verify that certain conditions are satisfied.

As indicated, the requirements addressed by machine vision technology are varied. While the front end always consists in some form of image acquisition to be further processed and analyzed, the intended outcome of the analysis can be the analysis of one or several quite different attributes of the object based on the requirements of the application:

1. The simple presence or absence of an object, or a part of an assembly.

2. The general shape and/or profile of an object or one of its parts, and its distribution in class groups.

3. The particular location or orientation of a part in an assembly.

4. The determination of the color and/or shade of an object and/or of some of its parts.

5. The determination of surface conditions of an object, such as finish, polish, texture, dust. These are usually unwanted attributes in unpredictable and random locations.

6. The optical density at specified colors, or integrated color bands (for example, tinted eyeglasses).

7. The determination of a dimensional property, such as length, thickness, depth, azimuth, angle, depth of thread, and their distribution in class groups.

8. Combinations of 1-7.

9. The use of motion analysis to obtain 3-D shape information as well as direction information (e.g., for an autonomous land vehicle).

10. Object recognition as distinct from checking the simple presence or absence of an object (e.g., in distinguishing between a square and a circle).

Some of these parameters relate to the cosmetic appearance of a product. The need for machine vision inspection arises in those cases from the psychological expectation that the good appearance of a product will result in its better acceptance in the marketplace.

Other parameters relate to the integrity of a product. An engine block, for example, should have its crankcase fully bolted, before being placed in the chassis. A pharmaceutical blister package must contain a specified number of tablets; they should be completely formed and of the right color. Still other parameters relate to the reproducibility or constancy of tolerances. The human eye, though sensitive and discriminating, cannot make quantitative judgements that are reproducible from event to event, even less from individual to individual observer. The color of a fabric, for example, could slowly change from hour to hour in one direction, without a human inspector perceiving it. A closure must fit the mating part of a container within specified tolerances. A slight change in the bevel of a hypodermic needle would pass unnoticed by the inspector.

Still other conditions may prevail when, because of their small size, color, physical inaccessibility, or other limitation, some of the "visual" parameters cannot be seen by the unaided human eye, but only by an appropriate sensor.

6.2 LIGHT AND LIGHTING

Whether it involves human or machine hardware, visual data acquisition proceeds essentially in three steps. In human vision, for example, the object should be properly lighted so as to make it "visible" to the human eye; second, the human eye, itself a lensing system, is needed to image the object on the target of the sensor, or the retina; finally, the retina should somehow "read" and a signal to be further processed conveyed to the brain by the optical nerve.

Similarly in machine vision, the first step is to properly light the object to render it detectable by the sensor. The second step consists of imaging the object on the target of the sensor.

6.2.1 Contrast and Resolution

Image-capturing typically involves acquiring the two dimensional projections of a three-dimensional object. The two most important qualities of the image are contrast and resolution. That is, the attributes of the image that will become the basis of an action or decision must be distinguishable and resolvable (or measurable).

Contrast is the range of differences between the light and dark portions of an image. Normally, contrast is measured between the feature(s) containing the needed information and the immediate objects surrounding these features (referred to as background).

The ideal image has ultimate contrast, with the desired information having intensity values of absolute white, and background (everything else) intensity values of absolute black.

Resolution (or the ability to see two closely spaced objects) is a distance measurement associated with the smallest detectable object. The resolution required depends on the task of the machine vision system. If a system is needed to locate a part in X and Y to within one inch, the system resolution needs to be less than one inch. Unlike contrast, infinite resolution is not always desired. An example is measuring the gap of a spark plug. Too much resolution would result in an image of the electrode surface appearing mountainous, uneven and pitted, not smooth and flat.

Lighting and optics can have an effect on both contrast and resolution in many applications. Lighting and optics can be optimized to enhance the delectability associated with the variables for which a system is purchased, or conversely, can assist to diminish the influence of the variable on the data required to make a reliable "vision decision."

6.2.2 Lighting

In a given machine vision installation, dedicated lighting is strongly recommended because secondary sources of illumination may come from ambient lighting and the reflection of the primary source of light off other equipment, objects, or the floor and windows. The net result is a complex pattern of light in which uniformity is a compromise. This can affect the nature of shadows and shading on the surface which can affect the recognition of the object.

The objectives of lighting are:

1. Optimize the contrast (gray scale difference) associated with the condition one seeks to detect versus the normal state.
2. Normalize any variances due to ambient conditions.
3. Simplify image processing and, therefore, computing power required.

Lighting in a machine vision application can make the difference between a successful and unsuccessful application. Illumination can either enhance features to be detected or obscure them. Poorly designed lighting can produce glare which may saturate the camera, create shadows which can include the data to be detected or obscure them, and generate low contrast or nonuniformity making the inspection difficult. Sufficient illumination is also required because sensors have designated minimum levels - the minimum amount of light required to produce a voltage video signal.

Lighting considerations in a given application include:

1. The *type* of light. Incandescent, fluorescent, quartz halogen, sodium, lasers, light emitting diodes, etc., all emit different wavelengths (colors) of light. The type used should illuminate the part's surface and also be measurable by the sensor type used.
2. The lighting *technique* is the geometric setup between the part, any lights and the sensor. This depends on what information is desired; maybe a silhouette is needed, or the bottom of a deep bore needs to be illuminated. Different lighting techniques would be used for each case.
3. The control of the illumination may include the passive blocking of light with covers or shades.
4. *Geometry* of propagation. There are three types of sources: point, diffuse and collimated. A single point of light will create shadows that accent edges. If shadows will block areas where information is needed, a diffuse light source will eliminate shadows.
5. At times the *shape* of the light "beam" can be used as a method to gain information from the scene. Shapes can include points, lines, crosses, gratings, etc. These are called "structured lighting techniques."

The specific lighting technique used for a given application depends on:

1. The object's geometric properties (specularity, texture, etc.)
2. The object's color
3. The background
4. The data to be extracted from the object (based on the application requirement).

As an introductory comment, it is noted that quite contrary to common thinking, the human eye or a sensor does not really "see" an object; rather they see the reaction of an object to incident light. This is the reason an object is seen differently when illuminated by different types of light and is not seen at all in the absence of exciting light. Hence, it is important to recall the properties of light that determine its interaction with objects.

6.2.3 Light

Light is electromagnetic energy generated by random atomic processes, in turn caused by heat, collisional, or electrical excitation. This energy statistically builds up to an oscillating electric field propagating in space as a traveling wave with a velocity of 3 x 10^{10} cm/sec in a vacuum (Figure 6.2).

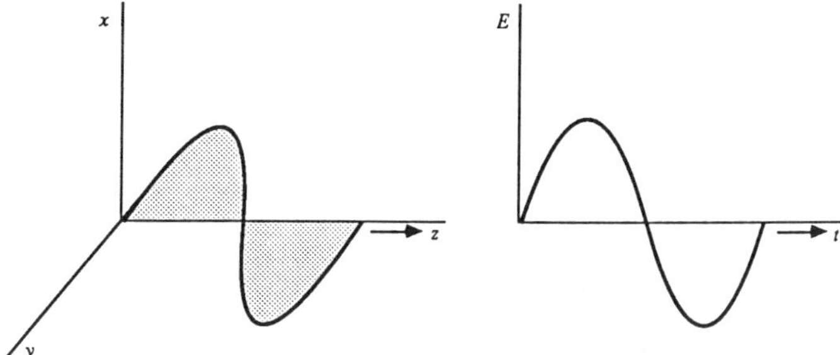

Figure 6.2 - Light as traveling wave.

6.2.3.1 Wavelength. The frequency of a field or its reciprocal, the wavenumber, or the wavelength, which is the wavenumber multiplied by the velocity, covers an extremely wide range, extending from the gamma rays at the short end, of 10^{-10} cm, to the radiowaves at the very long end, of several kilometers (Figure 6.3). Wavelength determines whether and how much light interacts with matter and how much of this interaction is detected by the human eye or by a sensor.

6.2.3.2 Polarization. The wave generated by an elementary source, being transverse, is essentially polarized (Figure 6.2). Most practical sources, however, are made up of many elementary sources that are all random. Hence, most practical sources have polarization distributed uniformly within the 360 degrees of the wavefront.

6.2.3.3 Geometry of Propagation. The direction of illumination or its propagation in space depends on the geometry of the integrated source. In that respect, there are essentially three types of sources: point, diffuse, and collimated (Figure 6.4).

1. Point source essentially originates from a geometric or very small point - a spherical source with infinitesimal diameter. Its propagation is uniform within the 4 Π solid angle. A point source illuminates an object from a single direction and hence causes reflections and shadows. Shadows are sometimes desired and include the data upon which a decision is based. Point sources of light cast very strong shadows. This is sometimes desired if the presence or absence of shadow can reveal information about the presence or absence of depth. Point sources of light are also very good for revealing surface defects.

Good point source illumination can be obtained from:

Incandescent spot lights (lights emitted by a hot object)

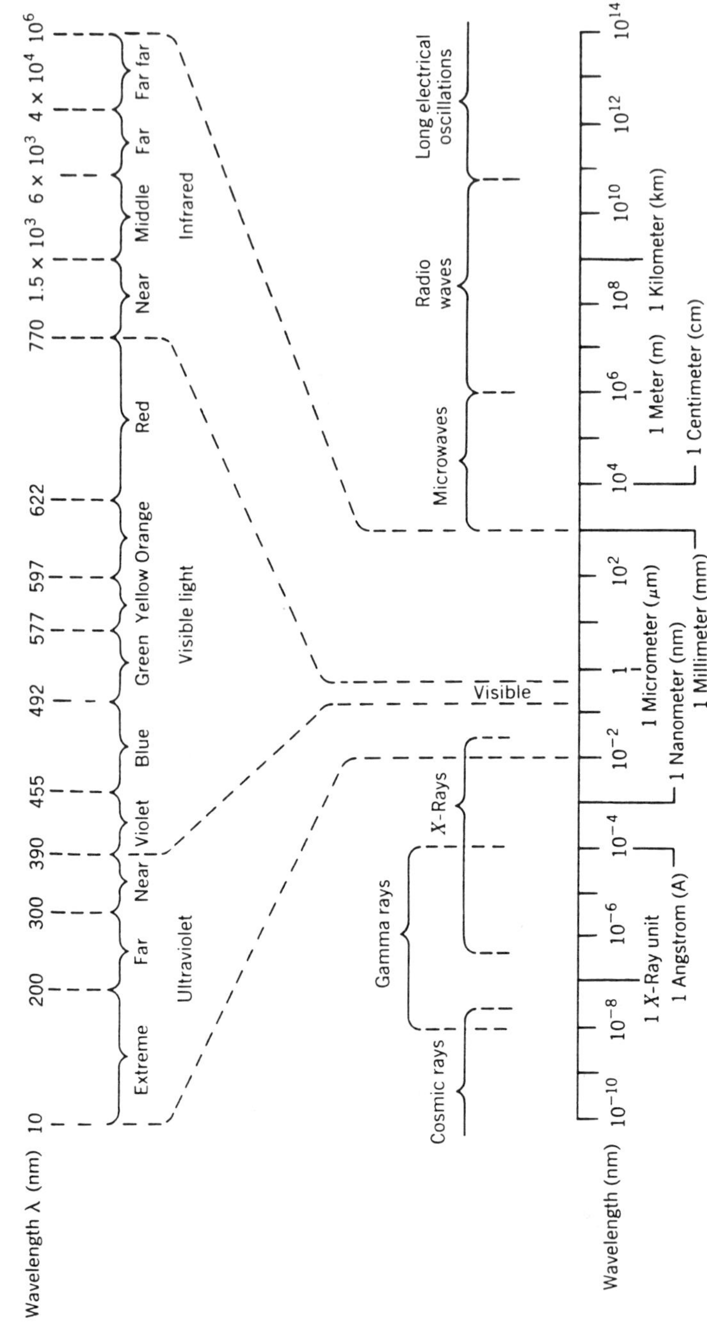

Figure 6.3 - Electromagnetic spectrum.

Fiber optic sources, slightly overlapping, provide good uniform point source illumination. (This is the best choice, but it is an intense light having a short lifetime and a high price tag.)

Unfrosted incandescent light bulbs (e.g., Tensor, or conventional room light bulbs).

2. Diffused light originates from an "extended source," consisting of many point sources distributed uniformly on a plane in space. It is incident on an object from all directions, creates little or no reflection, and no shadows. Most machine vision applications work best when illuminated with diffuse light. Good sources of diffuse lighting include:

Fluorescent ring lights

Long fluorescent bulbs with or without light diffusers

Fiber optic ring illuminators (very intense light)

Banks of light-emitting diodes

3. A collimated beam consists of a single unidirectional and generally quasi-parallel beam. It originates from a single source optically located at an infinite distance. It creates reflection and sharp shadows.

6.2.4 Practical Sources

Sources of light for machine vision vary from the common incandescent lamp to sophisticated lasers. One source of light almost never used is ambient light. The vagaries of ambient light are typically beyond the capability of state-of-the-art machine vision systems to handle.

6.2.4.1 Incandescent Light Bulb. Light (and heat) is obtained from an incandescent tungsten metal filament heated to 2000-2500 K by passing an electrical current. Its spectral response is that of the "black body" (as defined in physics text books), whose emitted wavelength distribution depends on the temperature (Figure 6.5). At 2000 K, it peaks at 1500 nm and dies out at around 200 nm. For certain vision applications, such as those requiring gray scale rendition, it may be best to regulate the current to the bulb.

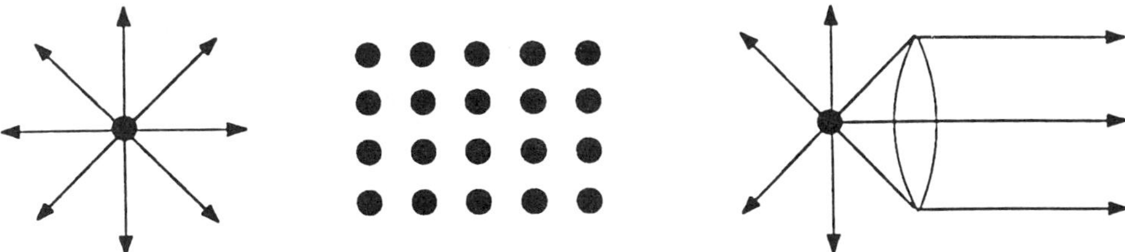

Figure 6.4 - Three types of light sources: point, diffuse and collimated.

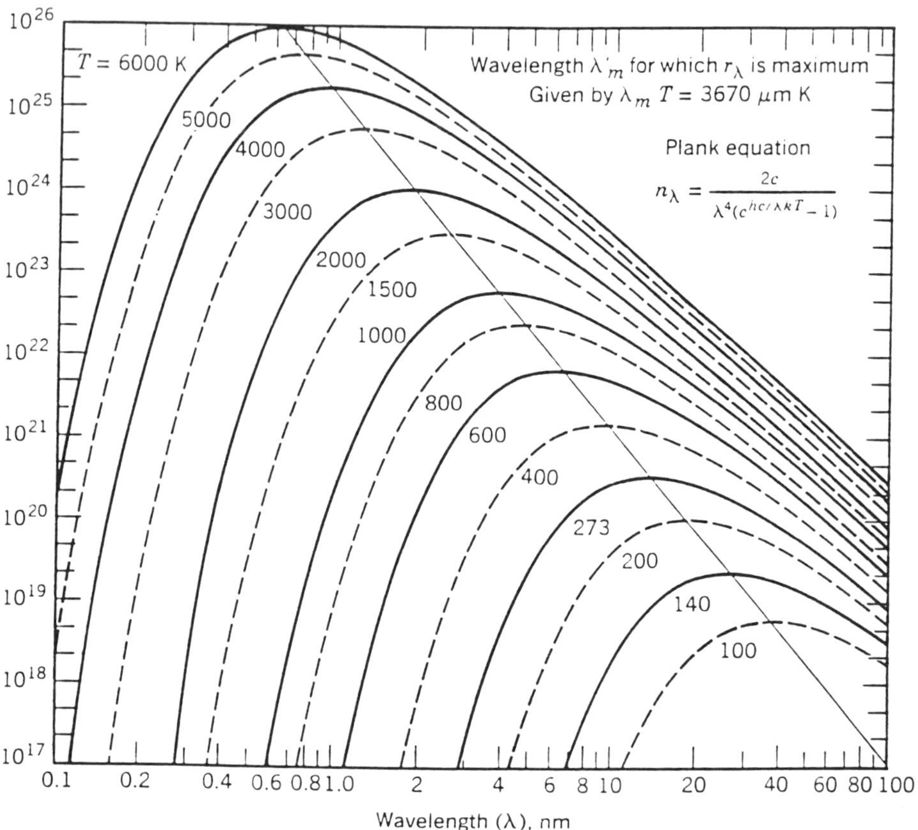

Figure 6.5 Wavelength versus number of photons.

The light bulb can be used as a point source, provided the tungsten filament is coiled and very small, as in projection lamps and beams for cars. Under similar conditions and when provided with proper optics, it can be used as a quasi-collimated source. It can also be made into a moderately good diffused source with the addition of a diffuser such as frosted and opal glass.

At conservatively low operating temperatures, its life is very long, but its efficiency is very poor. Raising the temperature boosts its efficiency but also drastically lowers its life expectancy because of the fast evaporation of the tungsten metal and its condensation on the cool walls of the bulb.

The main advantages of incandescent lamps are their availability and low cost. However, a large performance disparity exists between lamps of the same wattage, which may be a factor in machine vision performance.

Their disadvantages include short lifetimes, much of their energy is converted to heat, output that declines with time due to evaporation of the tungsten filament (the impact of which can be minimized by using a camera with an automatic light or gain control), and high infrared spectral content. Since solid-state cameras have a high infrared sensitivity, it may be necessary to use filters to optimize the contrast and resolution of the image.

Incandescent lamps typically require standard 60-Hz ac power. The cyclical light output that results might be a problem in some applications. This can be avoided using dc power supplies. When operated with dc power suppliers the lifetime will be degraded. The lifetime of incandescent lamps can be extended somewhat by operating at below voltage ratings. Significantly, the spectral content of the light will be different than if operated at rated voltages.

6.2.4.2 Quartz Halogen Bulbs. Quartz halogen bulbs contain a small amount of halogen gas, generally iodine. The iodine combines with the cool tungsten on the inside of the wall, and the tungsten-iodine diffuses back to the filament, where it disassociates and recycles. This allows the tungsten to be operated at a higher temperature, resulting in a more efficient source with more white-light emission.

Without dichroic reflectors, halogen lamps have a nearly identical infrared (IR) emission curve to incandescent bulbs. Dichroic reflectors eliminate the IR emission from the projected beam to provide a peak emission at 0.75 microns and 50% emission at 0.57 and 0.85 microns. To extend the life of the lamp reduced operating voltages are used. However, the spectral output will be different at lower currents.

Halogen bulbs require special care. The operation of the halogen cycle depends on a minimum bulb temperature. This limits the amount of derating of the input voltage to a level that will still maintain the critical bulb temperature. Since the gas operates at greater than atmospheric temperature, care must be taken with the bulb. It must not be scratched or handled. Any foreign substance on the glass, even finger oils, can cause local stress bulb breakage. It is recommended practice to clean the bulbs after installation but before use.

6.2.4.3 Discharge Tube. Light is generated by the electrical discharge in neon, xenon, krypton, mercury, or sodium gas or vapor. Depending on the pressure of the gas, the device yields more or less narrow-wavelength bands, some lying in the visible or the near-visible IR or ultraviolet (UV) region of the spectrum, matching the response curve of the human eye and/or of the used sensors (Figure 6.6).

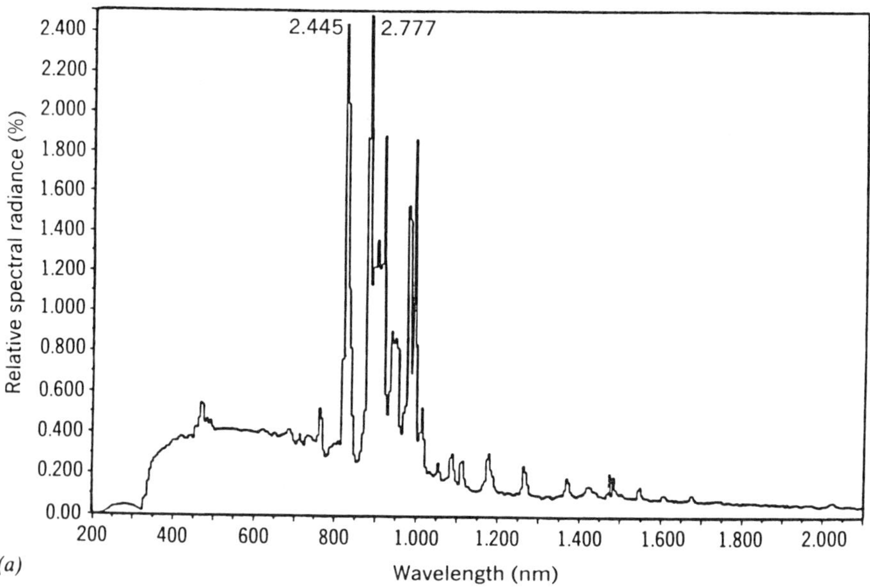

Figure 6.6 - Spectral distributions of some common discharge tube lamps. Top: Xenon lamp. Center: 400-W high-pressure sodium HID lamp, Bottom: mercury compact arc lamp.

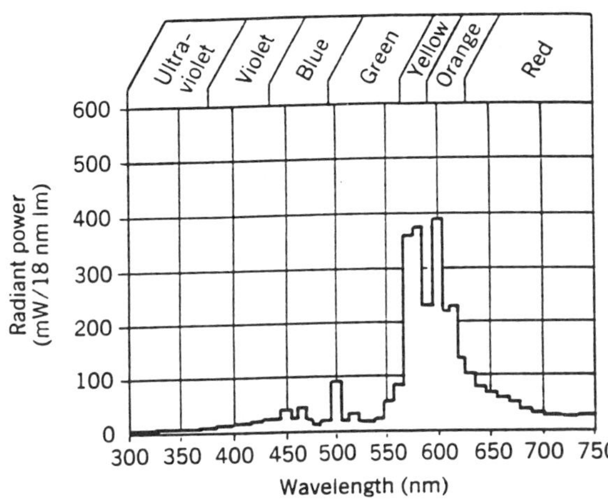
Ultra-violet
Violet
Blue
Green
Yellow
Orange
Red
Radiant power (mW/18 nm lm)
600
500
400
300
200
100
0
300 350 400 450 500 550 600 650 700 750
Wavelength (nm)

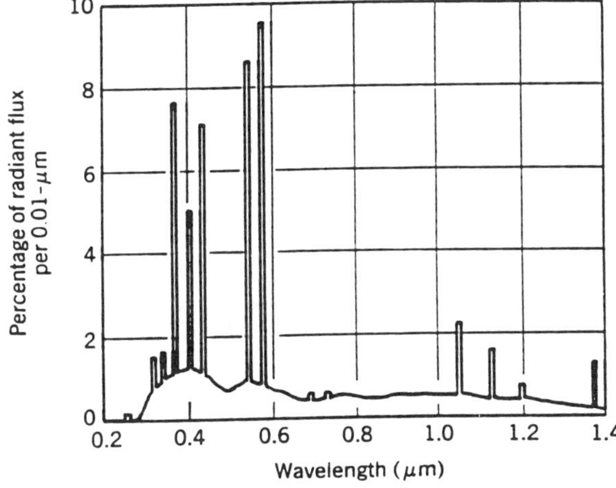
Percentage of radiant flux per 0.01-μm
10
8
6
4
2
0
0.2 0.4 0.6 0.8 1.0 1.2 1.4
Wavelength (μm)

(b)

Of particular interest in vision applications is the mercury vapor discharge tube, which generates, among others, two intense sharp bands, one at 360 nm and the other at 250 nm, both in the UV. These two lines often excite visible fluorescence in glasses, plastics, and similar materials. The envelope of the tube, made of fused quartz, transmits the UV. It is often doped with coloring agents, which absorb the cogenerated bands of the visible part of the mercury spectrum. Then, only UV is incident on the object, and the sensor sees exclusively the visible fluorescent pattern, which is the signature of the object. This source is often called "black light."

6.2.4.4 Fluorescent Tube. This device is a mercury discharge tube, where the 360-nm UV line excites the visible fluorescence of a special phosphor coating on the inside wall of the tube. In the widely used general-purpose fluorescent tube, very white and efficient light is generated. However, different colors can be obtained with different phosphor-coating materials. The tubes are manufactured with different geometries: long and straight, used as line sources, and circled or coiled, used as circularly or cylindrically symmetric sources.

Specialty shaped lamps can be fabricated where needed. Small-diameter tubes are notoriously unstable and, for critical vision application, require current stabilization. The light output from the lamp pulses at twice the power line frequency (120 Hz in the United States). The rate is indistinguishable to the human eye but can cause a problem if the camera is not operating at the line frequency. It is possible to operate the fluorescent lamps from high-frequency sources, 400 Hz and above. At these high frequencies the time of the phosphor is adequately long to give a relatively constant output.

Fluorescent lighting has some advantages in machine vision applications. Not only is fluorescent lighting inexpensive and easy to diffuse and mount, it also creates little heat and closely matches the spectral response of vision cameras. Its only drawbacks are a tendency toward flickering and of overall intensity. Because it provides diffuse light, the light level will be insufficient if one is using a solid-state camera and the part being inspected is very small or the lens aperture is small. Fluorescent lamps are useful for constructing large, even areas of illumination.

Aperture fluorescent tubes - lamps with a small aperture along the length of the tube - are well suited for line scan camera-based applications. Because of internal reflections within the tube, the output is significantly magnified. This can be critical given the short effective exposure times typically associated with line scan camera applications.

6.2.4.5 Strobe Tube. The strobe tube is a discharge tube driven by the short current pulse of a storage charge capacitor. It generates intense pulses as short as a few microseconds and, as such, can see a moving part as if it were stationary. Strobes are used in recording sporting events, ballistics work, and of course, in machine vision technology when looking at objects moving continuously on a production line. Accurate time synchronization is essential when using this technique. The arrival of an object in the view of the camera sensed by a photoelectric cell issues a trigger, which in turn generates the lighting pulse discharge. Immediately after this, the target of the sensor is scanned or read, yielding the video signal representing the passing object.

Peak emission is in the ultraviolet at 0.275 micron with a second significant peak at 0.475 micron, which has 50% emission at 0.325 and 0.575 micron. More than 25% of the relative emission is extended to 1.4 micron. For most applications a blue filter at the source or an infrared filter at the camera is required.

To improve stability, the discharge path is generally made relatively short. This approximates the geometry of a point source, and a reflecting and/or refracting diffuser is a must to reduce reflections and shadows. A serious drawback, however, is the instability of the electrical discharge in the tube. The exact path of the discharge varies from shot to shot, resulting in a nonreproducible lighting pattern. Some gray scale video detection requires sophisticated and expensive strobe stabilization techniques.

Strobe lamp systems, however, are expensive when compared to incandescent and fluorescent lamps. Unless the system has been carefully designed for use in a production environment, the reliability may be unacceptable; the modification of photographic strobe lamp systems is not adequate. Human factors must be taken into account. The repetitive flashing can be annoying to nearby workers and can even induce an epileptic seizure.

An alternative to strobes is shutter arrangements in the cameras, and today cameras are available with built-in shutters. A shutter arrangement, however, does not simply replace a strobe. Compared to the enormous peak power of the strobe, the shuttered light source has a much lower integrated light during the exposure time allowed by the motion of the object. In either case, image blur due to motion is not eliminated, only reduced to that associated with the flash time. Otherwise, it would be that due to the time associated with a full frame of a camera, or one-thirtieth of a second. Another alternative to a strobe, although an expensive one, is to use an intensifier stage in front of the camera that can be appropriately gated on or effectively "strobed."

6.2.4.6 Arc Lamp. Arc lamps provide an intense source of light confined to narrow spectral bands. Because of cost, short lamp life, and the use of high-voltage power supplies, these lamps are used only in special circumstances where their particular properties warrant. One such example is the backlighting of hot (2200° F) nuclear fuel pellets. The intense light from the arc lamp is sufficient to create the silhouette of the hot fuel pellet.

6.2.4.7 Light-emitting Diode. Semiconductor LEDs emit light generally between 0.4 and 1 micron as a result of recombination of injected carriers. The emitted light is typically distributed in a rather narrow band of wavelength in the IR, red, yellow, and green. White light arrangements are also available. The total energy is low. This is not a consideration in backlighting arrangements.

In front-lighting arrangements, banks of diodes can be used to increase the amount of light required. Light-emitting diodes are of interest because they can be pulsed at gigahertz rates (1 GHz = 1000 MHz), providing very short, high peak powers. Application specific arrangements of LEDs have been developed where specific LEDs can be turned on and off in accordance with a sequence that optimizes the angle of light as well as the intensity for a given image capture.

LEDs supply long-lasting, highly efficient and maintenance-free lighting which makes them very attractive for machine vision applications. LEDs can have lifetimes of 100,000 hours.

6.2.4.8 Laser. Lasers are monochromatic and "coherent" sources, which means that elementary radiators have the same frequency and same phase and the wavefront is perpendicular to the direction of propagation. From a practical point of view, it means that the beam can be focused to a very small spot with enormous energy density and that it can be perfectly collimated. It also means that the beam can easily be angularly deflected, and amplitude modulated either by electromechanical, by electro-optical, or by electroacoustical devices.

Lasers scanned across an object are frequently used in structured lighting arrangements. By moving the object under a line scan of light, all of the object will be uniformly illuminated, one line scan at a time. Infrared lasers have spectral outputs matched to the sensitivities of solid-state cameras. Filters may be required to focus the attention of the machine vision system on the same phenomenon being visually observed.

Several types of lasers have been developed: gas, solid-state, injection, and liquid lasers. The most popular ones and those most likely to be used in machine vision are as follows:

Type	Wavelength	Power Range
He-Ne gas	632.8 nm	0.5-40 mw
Argon gas	Several in blue-green	0.5-10 w
He-Cd vapor	Blue and UV	0.5-10 mW
Gas injection	Approximately 750 nm & IR	Up to 100 MW

Helium-neon (He-Ne) lasers are general-purpose, continuous-output lasers with output in the few-milliwatt range. This output is practical for providing very bright points or lines of illumination that are visible to the eye.

Diode lasers are small and much more rugged than He-Ne lasers (He-Ne lasers have a glass tube inside) and in addition can be pulsed at very short pulse lengths to freeze the motion of an object. The light from a diode laser does not have a very narrow angle like most He-Ne lasers but rather spreads quickly from a small point (a few thousandths of an inch typically), often requiring collection optics, depending on the application. The profile of the beam is not circular but elliptical.

The power ranges in diode lasers are up to 100 mW operating continuously and peak powers (which provide very high brightness) of 1 W or more. Because of the very localized nature of laser light, lasers present an eye hazard that should be considered. (A small 1-mW He-Ne laser directed into the eye will produce a very small point of light many times brighter than would result from staring directly at the sun.) There are Center for Devices and Radiological Health (CDRH) and OSHA standards relating to such applications, but they are not difficult to meet.

Infrared lasers include those used in laser machining such as carbon dioxide lasers and neodymium-YAG as well as a wide variety of less common lasers. Most of these lasers are capable of emitting many watts of power, making it possible to flood a large area with a single wavelength (color) of light. When used in conjunction with a colored filter to eliminate any other light, such a system can be immune to background light while being illuminated only as desired. Special IR cameras are needed to see IR light, and the resolution of these systems is less than visible camera systems. Infrared lighting can often be useful to change what is seen. For example, many colored paints or materials will reflect alike in the IR or may be completely transparent in the IR.

Ultraviolet lasers emit light wavelengths that are shorter than the blue wavelengths. The theoretical limit of resolution reduces to smaller features as the wavelength of light decreases. For this reason, high-resolution microlithography such as IC printing and high-resolution microscopy generally use UV light. Lasers such as excimer and argon can provide a very bright light for working at these short wavelength colors of light.

With a cylindrical lens or a fast deflector on the front of the laser, a line of light may be projected. As with all laser light, the beam has the advantage of being immune to the ambient light and to the effects of reflections or stray light. Also, the wavelength (typically red) is very stable to the camera under varying surface or color differences of the object being imaged. For example, a brown or black color will appear the same to the camera when illuminated by a laser beam. Laser beams, however, have a peculiar effect known as speckle. Speckle is almost impossible to control, and because of speckle, laser illumination is not normally recommended for high-magnification applications. Laser light can be recommended under following conditions:

1. When ambient or room lighting is difficult to control.
2. When the changing reflection of the part makes conventional light sources difficult to use.
3. When selective high-intensity illumination is required, that is, illumination of only a portion of the part, where flooding of light over the entire scene is determined to have disadvantages.
4. When beam splitters and prisms are used.

Such devices have a tendency to reduce intensity, but the laser beam is an already intense source:

1. As a substitute for shorter life light sources, such as fiber-optic quartz halogen illuminators. A good sealed laser will last 25,000 hr in the factory.
2. When a successive illumination of different points (or areas) is needed by angularly deflecting the laser spot from point (area) to point (area).
3. As a stable source of light that does not deteriorate with use. A fluorescent lamp, in contrast, loses 20% of its output over a period of time and even more with the accumulation of dust and contaminants.

6.2.5 Illumination Optics

6.2.5.1 Fiber-Optic Illuminators. Simply pointing a light source in the appropriate direction is an inefficient system and may require a large source to illuminate an area while leaving stray light in areas that should be dark. The use of fiber optics is one way of overcoming this problem by effectively moving the source closer.

When light propagating in an optically dense medium (refractive index greater than 1.0) reaches the boundary of an optically light medium such as air (refractive index 1.0) at an angle larger than about 50°, it is totally reflected (Figure 6.7). If the medium has the geometric shape of a thin rod, the reflected ray, as it further propagates, will be totally reflected at the next boundary, and so on. Hence, light entering the rod at one end is translated to the other end, much as water runs from one end of a pipe to the other.

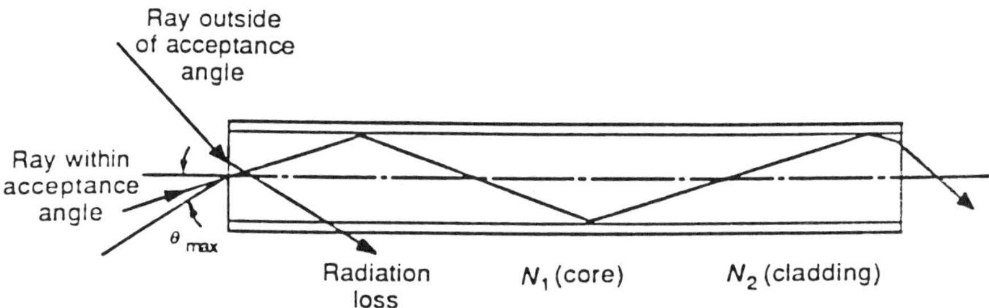

Figure 6.7 - Step index fiber.

A bundle of such thin fibers made of glass or plastic provides a channel for convenient translation of light to small constricted areas and hard-to-get-at places. The source of light is typically a small quartz halogen bulb. It should be coupled efficiently to the entrance end of the bundle and the bundle exit end efficiently coupled to the object to be illuminated. Efficiencies on the order of 10% are generally obtained with fiber-optic pipes 3-6 ft long.

The individual fibers at either end can be distributed in different geometries to produce dual or multiple beam splitting or different shapes of light sources such as quasi-point, line, or circle.

6.2.5.2 Condenser Lens. Fiber optics may not always give the desired results. The use of condenser and field optics (Figure 6.8) to transfer the light with maximum efficiency is the standard way to transfer the maximum amount of light to the desired area. The actual illuminance at the subject is affected by both the integrated energy put out by the particular light source (the luminance) and the cone angle of light convergent at the subject. For a constant cone angle of light at the subject, the actual size of the lens does not affect the illuminance. That is, a small lens close to the source will produce the same illuminance as a large lens further away that produces the same cone angle of light at the subject.

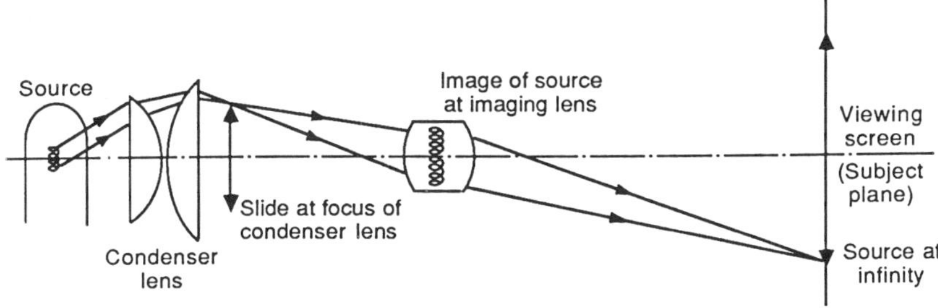

Figure 6.8 - Typical slide projector optical system.

A related variable is the magnification of the source. The maximum energy transfer is realized by imaging (with no particular accuracy) the source to the subject. The reflectors behind many sources do this to some slide projectors and similar systems use this principle by imaging the source to the aperture of the imaging lens to transfer the maximum amount of light through the system.

If the source is demagnified to an area, the cone angle of the light is higher on the subject side than the source side of the lens doing the demagnifying and, therefore, the light is made more concentrated. Conversely, if the source is magnified, the cone of light is decreased at the image, and the light is diminished (it may seem like more is being collected from the source, but the energy is spread over a larger area).

A standard pair of condenser lenses arranged with their most convex surfaces facing will do a reasonable job of relaying the image of the source some distance. If a long distance is required, a field lens can be placed at the location of this image to re-collect the light by imaging the aperture of the first condenser optic to the aperture of the second condenser optic without losing light (otherwise, the light would diverge from the image and overfill a second condenser pair the same size as the first).

The result at the second image is illuminance as high as if one very large lens had been used at the midposition where the field lens was located. In this manner, the light can be transferred long distances with small, inexpensive lenses (this is actually the type of relay system used in periscopes). When transferring light with such a system, it is important to collect the full cone angle of light at each stage; otherwise, the system will effectively have a smaller initial collection lens.

6.2.5.3 Diffusers. A diffuser (Figure 6.9) is useful when the nonuniform distribution of light at the source makes it undesirable to image the source onto the subject. In such a situation, a diffuser, such as a piece of ground or opal glass (ground glass maintains the direction of the light better), can first be illuminated at the source image plane, and then the diffuser is imaged out to the area of the subject. A diffuser (or as is sometimes used, a pair of diffusers separated by a small distance) will have losses since it scatters light over a wide angle.

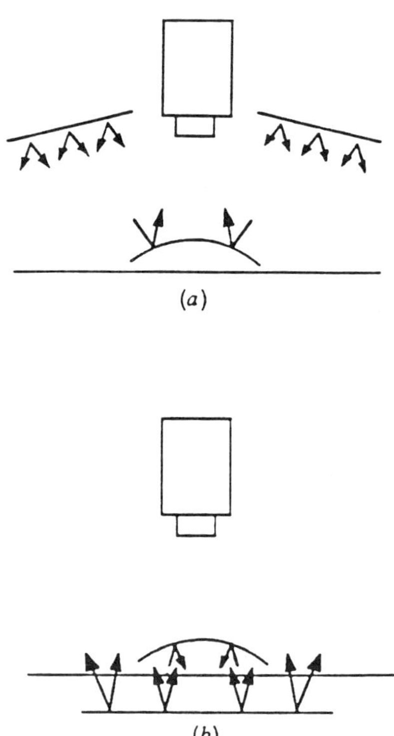

Figure 6.9 - (a) Diffuse front lighting for even illumination. (b) Diffuse backlighting to "silhouette" object outline.

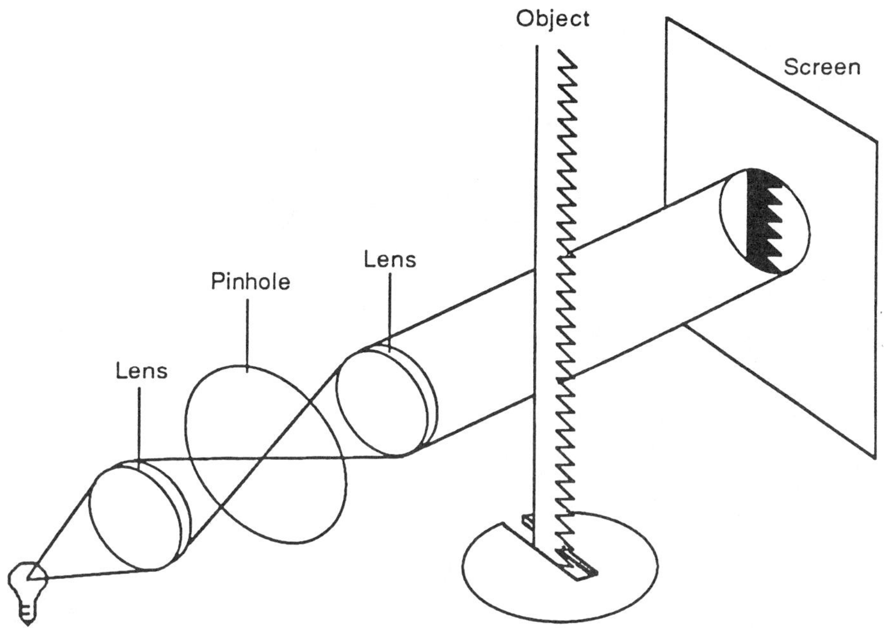

Figure 6.10 - Collimated light arrangement.

Many spotlights have a ribbed lens that actually serves to produce multiple-source images, each at a different size and distance to produce an averaging effect in the subject area. An alternative is to use a short length of a multimode, incoherent fiber-optic bundle. Such a fiber-optic bundle does not have the same fiber position distribution on one end as the other, so it will serve to scramble the uneven energy distribution at the image of the source to produce a new, more uniform source. Because of the random nature of such fiber bundles, this redistribution may not always be as uniform as desired.

6.2.5.4 Collimators. A third option to effectively deliver light is to move the image of the source far from the subject by collimating the light (Figure 6.10). There will still be light at various angles due to the physical extent of the source, but the image of the source will be at infinity. This last method is actually the most light-efficient method, but because of the physical extension of larger sources (it cannot all be at the focal point of the lens), this method is most appropriate for near-point sources such as arc lamps and lasers.

6.2.6 Interaction of Objects with Light

A visual sensor does not see an object but rather sees light as emitted or reflected by the object (Figure 6.11). How does an object affect incident light? Incident light can be reflected, front-scattered, absorbed, transmitted, and/or back-scattered. This distribution varies considerably with the composition, surface qualities, and geometry of the object as well as with the wavelength of the incident light.

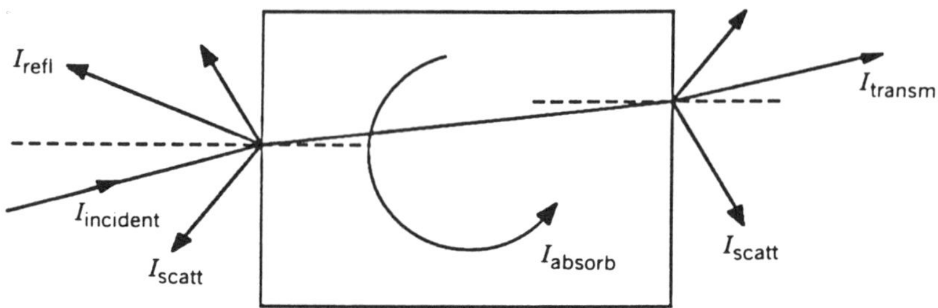

$$I_{inc} = I_{refl} + I_{front\ sc} + I_{absorb} + I_{back\ sc} + I_{transm}$$

Figure 6.11 -Interaction of objects with light.

6.2.6.1 Reflection. If the surface of an object is well-polished, incident light will bounce back (much as a ping-pong ball on a hard surface) at an angle with the normal equal to the angle of incidence. The reflected light may or may not have the same wavelength distribution as the incident light. In addition, reflection on some materials and at some angles of incidence may cause a change in the polarization of the light.

6.2.6.2 Scattering. If the surface of an object is rough, the light may bounce back, but over a wide angular range, both in front and on the back of the surface. In this case and when the area of incidence is large, the scatter may reradiate as a diffuse source (backlight box).

6.2.6.3 Absorption. Light energy is being used inside the body of the object to activate other processes (heating, chemical reaction, etc.). These processes are generally wavelength-dependent.

6.2.6.4 Transmission. After undergoing refraction at the interface (a slight change in the angular direction of the beam), light passes through and exits out of the object. Some light may also back-scatter at the exit interface.

6.2.6.5 Change of Spectral Distribution. Most of these processes are wavelength dependent and cause a change in the spectral distribution of the remaining beam. Consequently, the visual sensor can see an object differently colored and sometimes differently shaped depending on the component of reactive light at which it is looking.

6.2.7 Lighting Approaches

Lighting is dictated by the application, specifically, the properties of the object itself and the task, robot control, counting, character recognition, or inspection. If the application is inspection, the specific inspection task determines the best lighting: gaging, flaw detection, or verification. Similarly, the lighting may be optimized for the techniques used in the machine vision system itself - pattern recognition based on statistical parameters versus pattern recognition based on geometric parameters, for example. The latter will be more reliable with lighting that exaggerates boundaries.

The specific lighting technique used for a given application depends on the object's geometric properties (specularity, texture, etc), the object's color, the background, and the data to extract from the object (based on the application requirement).

It is the task of the lighting application engineer to evaluate the processes described in the preceding and how they apply in a particular application and to design a combination of lighting system and sensor that will enhance the particular feature of the object of inspection.

All this is rather complex. Practical analysis, however, is often possible by isolating and classifying the different attributes of different objects and by relating them to the five types of interaction processes.

6.2.7.1 Geometric Parameters: Shape or Profile of Object

Transmission and Backlighting. If the object is opaque to some portion of the visible spectrum or if it has some "optical density," transmission is an obvious method. When diffusely backlighted, the profile of a thin object is sharply delineated. For the case of a thick object, however, collimated light or a telecentric lighting system (described in the next section on optics) may be needed.

Structured Lighting. When the object is not easily accessible to backlighting or to collimated lighting, the object is very transparent (as in clear glass), or other constraints render the transmission method impractical, a special light system can sometimes be designed that will structure or trace the desired profile. The image sensor (Figure 6.12) then looks at the angular projection of the structure profile. The distortions in the light pattern can be translated into height variations. A very powerful method is to have a focused laser spot scan a sharp reflected or scattered profile of the object and have the camera look at this profile (Figure 6.13).

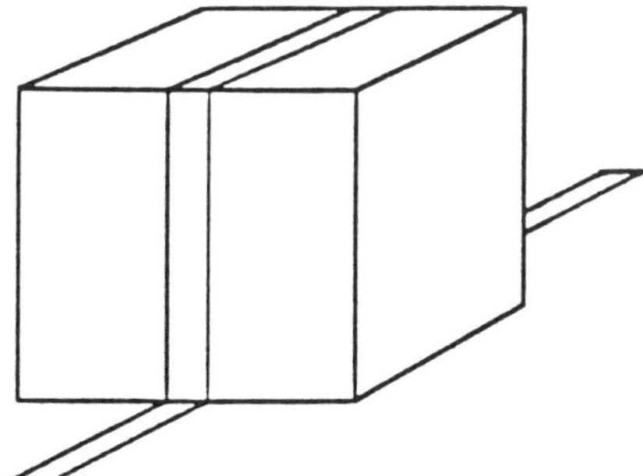

Figure 6.12 - Example of structured light.

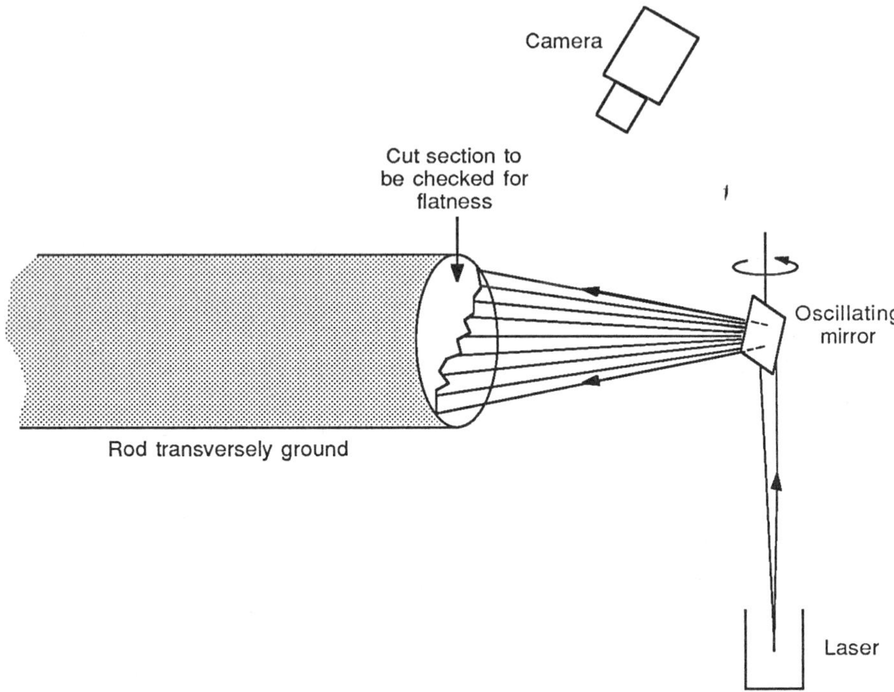

Figure 6.13 - Laser scanning "structured-light" example.

Cylindrical optics can be used to expand a laser beam in one direction to create a line of light. This generates a Gaussian line that is bright in the center and fades out gradually as you near both ends. An alternative approach is to use a diffraction grating arrangement. This approach yields better uniformity in intensity along the line axis and generally sharper-focused lines.

6.2.7.2 Front Lighting. Diffuse front lighting is typically the approach desired with a high-contrast object, such as black features on a white background. The field seen by the camera can be made very evenly illuminated. In applications where straight thresholding is used, a large extended source generally provides the easiest results. One of the most popular methods is to use a fluorescent ring light attached around the camera lens. If the object has low-contrast features of interest, such as blind holes or tappers on a machined part, diffuse lighting can make these features disappear.

Light Field Illumination: Metal Surfaces. Ground metal surfaces generally have a high reflectivity, and this reflectivity completely collapses at surface defects such as scratches, cracks, pits, and rust spots (Figure 6.14). This method can be used to detect cracks, pits, and minute bevels on the seat of valves, for example.

The method is excellent provided the surface is reasonably flat since any curvature will cause hot spots.

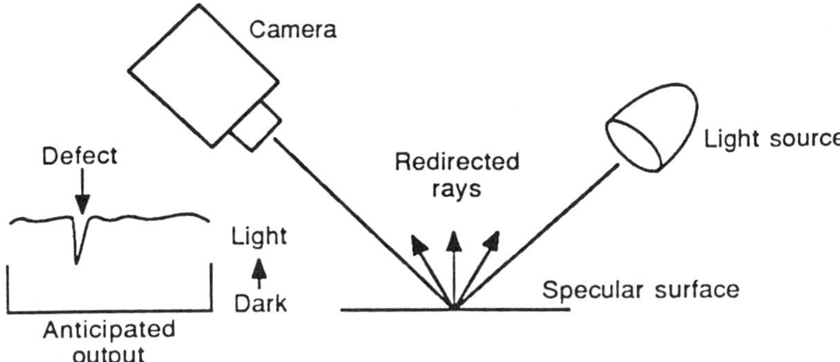

Figure 6.14 - Specular illumination (light field).

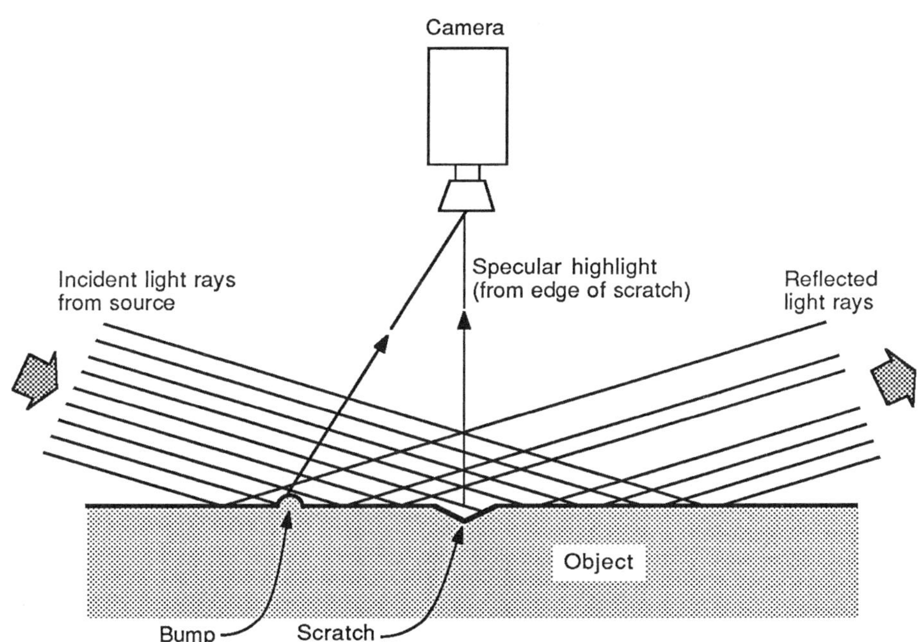

Figure 6.15 - Dark-field illumination.

Dark-Field Illumination: Surface Finish and Texture. This technique, widely used in microscopy, consists in illuminating the surface with quasicollimated light at a low grazing angle (Figure 6.15). The camera looks from the top at an angle that completely eliminates the reflection component. Hence, the field of view is completely dark, except for possible wide-angle scattering. Any departure from flatness, such as a bump or a depression, will yield a reflection component reaching straight into the sensor. A textured surface will be imaged with high contrast on the target of the sensor (Figure 6.16).

Figure 6.16 - Contrast enhancement using dark-field illumination.

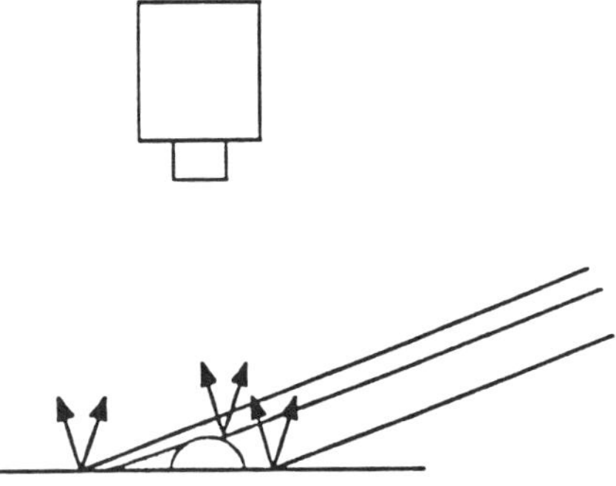

Figure 6.17 - Directional front lighting creating shadow.

Directional Lighting. Directional lighting is useful in cases where shadows are desirable (Figure 6.17). A shadow can be used to find a small burr on a flat surface or locate the edge of a hole. The best way to produce these effects is to make the source as effectively small as possible so that individual rays of light do not overlap. In the extreme case (such as a laser or restricted arc lamp source), the light can be collimated, that is, directed so that all of the rays of light are parallel to each other. A standard telescope objective lens will collimate light coming from a source at its designed focal length (within the limits of the aberrations already discussed).

Polarized Light. Polarized lighting is useful in reducing specular or shiny areas on dielectric materials, which reflect light just as a mirror and cause a very bright "glare" off a particular curved area. The areas immediately surrounding the curved area reflect their light away from the camera and appear very dark. It is easiest to obtain even illumination from an object when the object is diffuse such that it reflects light independently of the direction of illumination (a so-called Lambertian surface) and over a wide area. If the surface is truly specular, the specular light must be used. If the problem is just one of glints, the specular glints can often be removed by polarizing illumination light and then viewing the subject through a second, crossed polarizer. This method removes the bright, specularly reflected light because that light remains polarized. Truly diffuse surfaces, however, depolarize the light so that the second polarizer in the viewing system only blocks half of the diffuse light and passes the rest. Even with a uniformly specular object, this effect can be used if a diffuse background is used. In such a case, the polarizer removes any light reflecting from the subject and leaves a bright background. This object subtraction method is equivalent to a back-illumination system in its effect, but it can be used where backlighting may not be practical (perhaps a heavy machined part on a conveyor belt).

In dielectrics such as glass, about 5% of light is reflected at the interface air to glass (or 10% for a two-sided sheet). To that extent, glass shares the reflection properties of metal. Since glass is a dielectric material, it causes polarization changes on light.

When one looks at a picture frame from an angle of about 57°, one sees a glare that often kills the contrast of the framed image. This glare is the transverse component of polarization of the incident light, whose component is substantially reflected while the longitudinal component is not.

Figure 6.18 shows the path of a beam of light originating from point S and incident on the interface at an angle Φ. The T component is the refracted component at an angle Φ', further transmitted through the glass. The R component is the reflected component bouncing back at an angle equal to the angle of incidence. The two components of polarization parallel to the plane of incidence are depicted by bars in the plane of the paper; the transverse polarization is indicated by small dots. From Figures 6.18(a) and 6.18(b), it is seen that at 57° the reflected component contains exclusively transverse polarized light.

This means that if we were to use light polarized in the plane of incidence and at the critical angle of 57° for glass (called the Brewster angle), there would be no reflection at all, except where the angle of incidence departs from the value of 57°, as it does on a scratch or on most other surface defects (Figure 6.19). The polarizing effect at a dielectric interface, while present also for metals, is easily

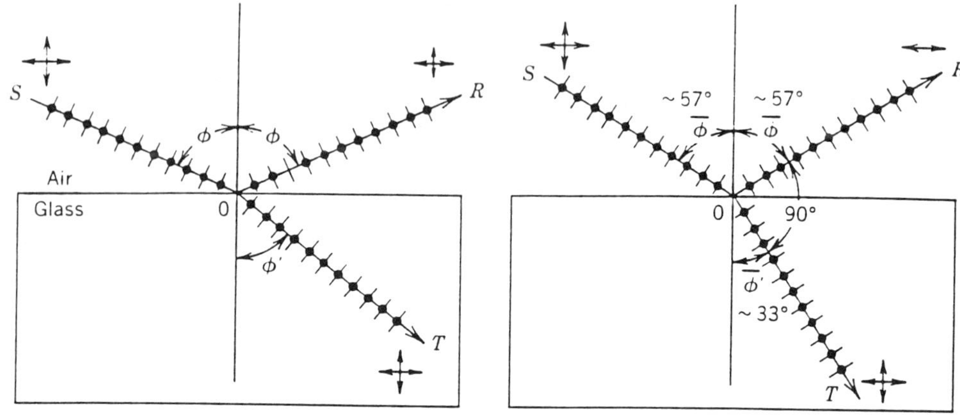

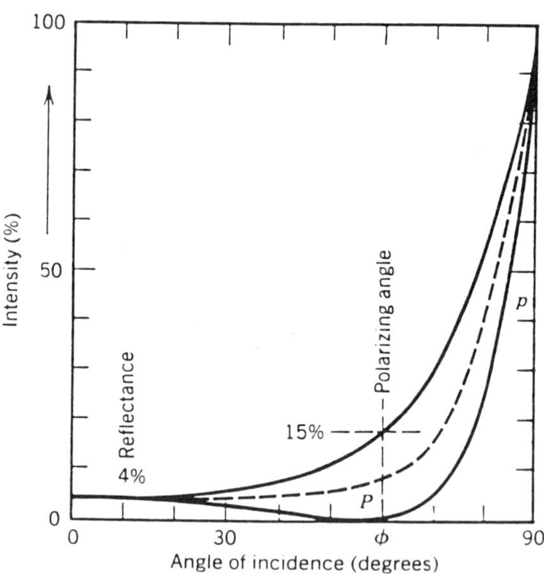

Figure 6.18 - Polarization effects.

lost in the intense reflectivity. In practice, it has very little or no application for metals.

In some cases, polarized light can be used to illuminate a subject from the exact same direction as the camera with minimal losses by the use of a polarizing beam splitter (a beam splitter that reflects one direction of polarization but passes the other). A similar system without polarization will lose 75% of the light just from going twice through a 50:50 beam splitter. A polarizing system will only lose the amount of light "depolarized" by the subject. The optics to accomplish this are not inexpensive (a few hundred dollars) but can be useful if light level is a particular limitation.

Not all light sources are easy to polarize. Fiber-optic sources, small light sources, and small fluorescent lights (less than 12 in.) are relatively easy to polarize. Uniformly polarizing flood lamps, for example, is more difficult. Another challenge is that for many applications an arrangement of lights is required. Establishing the optimal polarizer direction for each light source requires systematic

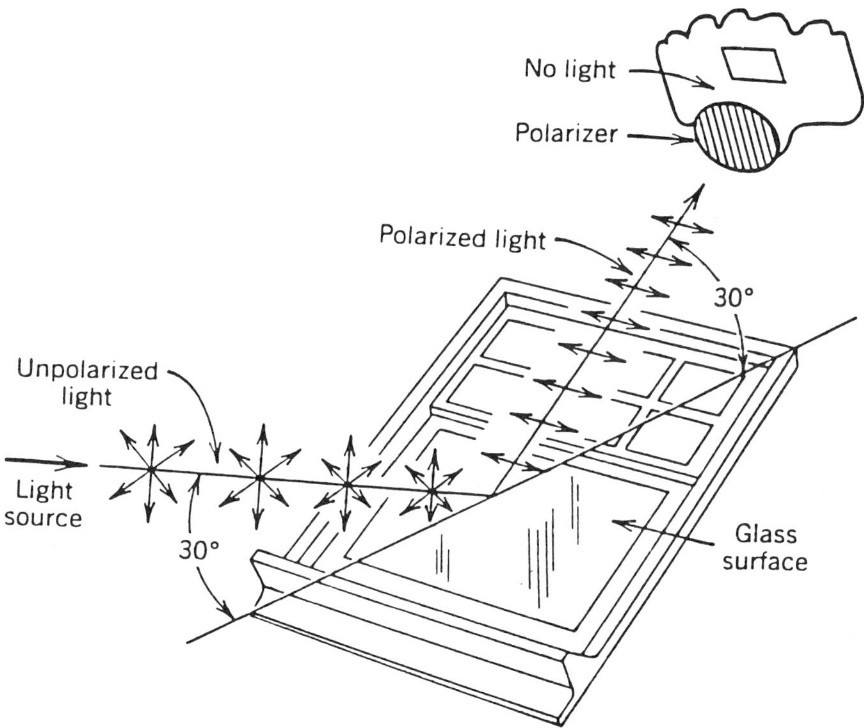

Figure 6.19 - Setup to eliminate glare from glass.

experimentation. Significantly, for some applications, glare is the very feature that results in high contrast and is desirable.

Chromaticity and Color Discrimination by Filters. The vision sensor can sometimes be rendered insensitive to some colors and therefore enhance other colors by using a light source deprived of that color or wavelength or by filtering out the unwanted colors out of the camera field. In the first case, the object does not reflect or scatter that component because it is not present. In the second case, the filter absorbs the reflected or scattered unwanted component before reaching the target of the sensor. In both cases, the sensor is rendered insensitive to the unwanted color.

Another reason to use a single-color region for a particular inspection function is that different colors can be used to perform multiple inspection functions at the same time. This is especially effective if a full-image inspection is needed and some form of structured light is also to be used. The structured light color can be removed from the full-image inspection (particularly if a laser is used), and the overall illumination can be filtered out by the structured light camera (e.g., by using a blue flood of the image scene light and a He-Ne laser structured light) to produce a higher signal-to-noise ratio. This type of approach opens up many possibilities for complicated inspection tasks.

Commercial filters exist that have reproducible specified color absorption bands and can be used either at the source or at the sensor.

6.2.8 Practical Lighting Tips

Lighting selection requires individual judgment based on experience and experimentation. An optimum lighting system provides a clear image, is not too bright or dark, and enables a good vision system to distinguish the features and characteristics it is to inspect.

It is noted that each of the techniques used to control light intensity (filters, polarizers, shutters) impacts on the structure of light wavelength, angle of incidence, or degree of detection as influenced by the object's geometric and chromatic properties. Throughout illumination analysis it must be understood that lighting must be adequate to obtain good signal-to-noise properties out of the sensor and that the light level should not be excessive so as to cause blooming, burn-in, or saturation of the sensor.

The objective in general is to obtain illumination as uniform as possible over the object. It is noted that "lighting" outside the visible electromagnetic spectrum may be appropriate in some applications - X-ray, UV, and IR - and that machine vision techniques can apply equally as well to such applications.

Lighting is a critical task in implementing a machine vision system. Appropriate lighting can accentuate the key features on an object and result in sharp, high-contrast detail. Lighting has a great impact on system repeatability, reliability, and accuracy.

Many suppliers of lighting for machine vision applications now offer what might be called "application specific lighting arrangements." These are lighting arrangements that have been refined and optimized for a single specific application; e.g., blister packaging inspection in the pharmaceutical industry, ball grid array packages in the semiconductor industry, etc. Some are optimized for imaging uneven specular objects.

6.3 IMAGE FORMATION BY LENSING

6.3.1 Optics

The optics creates an image such that there is a correspondence between object points and image points where the image is to be sensed by the sensor, as well as contribute to object enhancement. Except for the scaling or magnification factor, in an ideal optical system, the image should be as close to a faithful reproduction of the 2D projection of the object. Consequently, attention must be paid to distortions and aberrations that could be introduced by the optics.

Many separate devices fall under the term "optics." All of them take incoming light and bend or alter it. A partial list would include lens, mirrors, beam splatters, prisms, polarizers, color filters, gratings, etc. Optics have three functions in a machine vision system:

1. Produce a two-dimensional image of the scene at the sensor. The optics must place this entire image area (called the field-of-view or FOV) in focus on the sensor's light sensitive area.
2. Eliminate some of the undesired information from the scene image before it arrives at the sensor. Optics can perform some image processing by the addition of various filters. Examples include: using a neutral density filter to eliminate 80% of the light in an arc welding application to prevent sensor burnout, using a filter in front of the sensor which only allows light of a specific color to pass, and using polarizer filters to eliminate image glare (direct rejections from the lights).
3. Optics can be used in lighting to transfer or modify the light before it arrives at the scene in the same manner as optics is used between the scene and sensor; items 1 and 2 above.

6.3.2 Conventional Imaging

6.3.2.1 Image Formation in General. An imaging system translates object points (located in the object plane) into image points (located in the image plane), where the image is to be sensed by the sensor. Except for the scaling or magnification factor, it is assumed that, in an ideal conventional imaging system, the image should be a faithful reproduction of the object. This is being done unconsciously in the human eye, and it is being done according to a programmed design in a vision camera.

What is the mechanism of imaging by a thin lens (Figure 6.20)? All rays emerging from object point A_o, after being refracted at the surface of the lens, converge at point A_i. Some of these rays have been traced on the figure. Hence, A_i images in A_i. And that is true for all points of the object plane. Hence, the object plane is sharply translated, or "imaged," in the image plane; the only difference is the scaling or magnification.

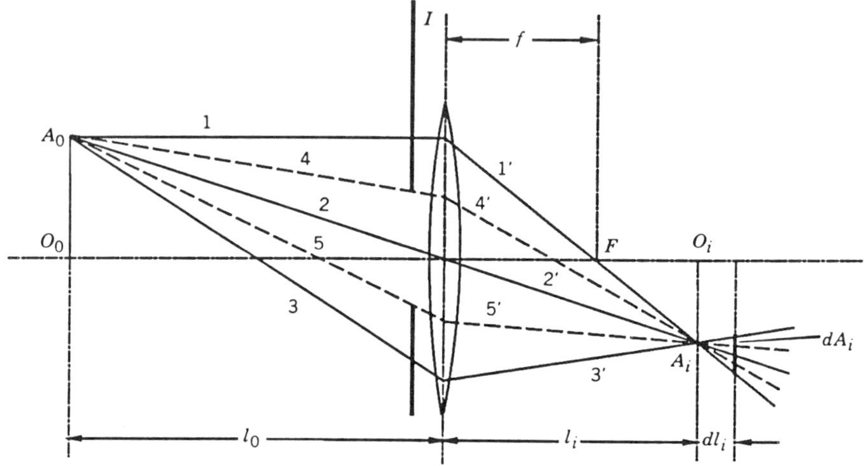

Figure 6.20 - Imaging properties of thin lens.

6.3.2.2 Focusing. Viewing Figure 6.20 again, if we depart slightly from l_i, say we look at the image at distance $l_i + d_i$, we still have an image. Point A_o, however, is no longer imaged into a single point A_i but into a small area dA_i, the so-called circle of confusion. The image is no longer sharp, it is no longer in focus, and it has lost contrast because light is now spread over an area. The smaller we take the iris aperture of the lens, the narrower is the solid angle of the light bundle, and the longer we can take d_i and still keep an image. Hence, closing the stop aperture of a lens increases the "focal depth," or the depth over which the image remains in acceptable focus. Of course, it also decreases the amount of light incident on the sensor.

It is interesting to note that sometimes good use can be made of defocusing. If the wanted detail of an object is very bright, the aperture stop of the lens can be brought up to the point where the bright feature slightly saturates the camera. At that point a slight defocusing of the lens will keep the desired detail equally bright, while decreasing all other details of the scene.

6.3.3.3 Focal Length. Focal length can make an image smaller or larger. A short focal length reduces the image size while a longer length enlarges the image. The focal length specifies the distance from the final principal plane to the focal point (Figure 6.20). The final principal plane is not necessarily located at the back of the lens (nor is it necessarily behind the first principal plane). The focal point of a lens is the point at which the lens will cause a set of parallel rays (rays coming from infinity) to converge to a point. Since the focal length is generally given, a collimated beam can be focused to a point, which will be the focal point of the lens. Alternatively, often a back focal length of a lens is specified, which is the distance from the rear element of the lens to the focal point.

The plane at which the parallel rays appear to bend from being parallel toward the focal point is called the principal plane. Most lenses typically have multiple principal planes. The separation of the principal planes and their positioning can be used to determine the optimum use of a lens (the best object to image distance ratio).

Lenses with a focal length of more than 25-mm are telephoto lenses. These lenses make the object appear closer (larger) than it really is and show less of the normal scene; that is, they have a smaller field of view than the standard 20-mm lens (the approximate focal length of the eye).

Lenses with a focal length shorter than 15 mm are wide-angle lenses. They make the object appear further from the camera (smaller) and have a larger field of view than the standard 20-mm lens. While some might consider zoom lenses as the answer to avoid determining the most appropriate focal length lens, a zoom lens suffers from a slight loss of definition, less efficient light transmission, and higher cost.

6.3.3.4 f-number. The f-number (f-stop) of a lens is useful in determining the amount of light reaching the target. Standard f-stops (often mechanical click stops) are for f-numbers of 1.4, 2, 2.8, 4, 5.6, 8, 11, 16, etc. Each f-stop changes the area of the lens aperture (which changes as the diameter divided by 2 squared) by a factor of 2. Changing the lens setting by one f-stop changes the amount of light available to the camera by a factor of 2. As the f-number increases, the amount of light is reduced. If more light is needed on the sensor, a lower f-number lens should be used. The f-number is determined by dividing the focal length by the diameter of the aperture (adjustable opening) of the lens. For example, a lens with a 50-mm focal length and 10-mm aperture diameter will be a f/5 lens.

Some cameras are equipped with lenses that automatically adjust for illumination variances using electronic lens drive circuits. These may or may not be valuable for a machine vision application. While they may compensate for lighting changes, it may be that those lighting changes are the very variable the application is designed to detect - a change in color saturation, for example, or the reflectance property stemming from a texture change.

6.3.3.5 Other Parameters in Specifying Imaging System

1. Working Distance. The distance between the object and the camera (Figure 6.20). In some applications, the working distance is severely restricted by lack of accessibility to the object, lighting needs, or other constraints.

2. Field Angle. The angle subtended by an object set in focus. The field angle is approximately the angle whose tangent is the height of the object divided by the working distance. The maximum field angle is obtained when camera is focused to infinity. The minimum field angle depends on the distance to the image of the principal plane of the objective. This is usually a function of the amount of extension rings inserted between objective and camera. Extender rings fit between the camera and the lens and magnify the image by increasing the distance between the lens and the image plane on the sensor, or the image distance. This allows the lens to focus in closer to the object while reducing the field of view. Extender rings can cause a loss of uniformity of illumination, reduction in the depth of field and resolution, and increased distortion. Significantly, not all lenses are designed to be used with tubes, and in such cases the back focal plane distance is critical for m performance, especially in gauging applications.

The maximum field angle of standard closed-circuit television (CCTV) camera objectives, as listed by manufacturers, is listed in Table 6.1. Its minimum is listed in the table, as calculated for the case of the maximum reasonable amount of extension rings set at 4 times the focal length.

3. Field of View. The area viewed by a camera at the focused distance, which is of course a significant parameter as it affects the resolution. The object should fill as much of the field of view as possible to capitalize on the resolution properties of the sensor and on the computing speed of the associated processor.

The field of view can be adjusted by adjusting the camera's distance from the object (also known as the working distance). The greater the distance from the part, the larger the field of view; the smaller the distance, the smaller the field of view. Alternatively, one can change the focal length of the lens. The longer the lens focal length, the smaller the field of view; the shorter the focal length, the larger the field of view.

The best field of view for any application is determined from the size of the smallest detail one wants to detect. This in turn is factored into the number of resolvable elements in the sensor. For example, if a sensor can resolve a scene into 512 X 512 pixels and the smallest detail to be detected is 0.01 X 0.01 in., the maximum field of view is 5 X 5 in. Significantly, sampling theory suggests that reliable detection requires more than one sample of the phenomenon, that is, a minimum of two samples. Consequently, the ideal *maximum field* of view would be 2.5 X 2.5 in. or less.

TABLE 6.1 Field Angles (degrees) for COSMICAR Objectives

Focal Length (mm)	Maximum Field Angle	Minimum Field Angle
8.5	72	20
12.5	53	14
25	27	7
35	20	5
50	15	3.5
75	9.5	2.4

What this says conversely is that the minimum-sized defect should cover an area of 2 X 2 pixels. For a 512 X 512-array sensor/processor, the smallest detail the system can detect is 4/262,144, or 0.0015%, of the field of view. For a 256 X 256 arrangement, the smallest detail would be 4/65,536, or 0.006%. This, of course, reflects a detail with sufficient contrast. Significantly, detection of an edge in space for purposes of gauging (measuring between two edges) can be done to a subpixel using a variety of signal processing techniques.

4. Magnification. The ratio of a linear dimension l_i on the image (Figure 6.20) to the corresponding dimension l_o on the object:

$$M = l_i/l_o$$

In general, the shorter the focal length of the objective and/or the shorter the working distance, the higher the magnification. The magnification commonly used in machine vision covers a wide range, extending from 0.001 times for large objects such as vehicles to 10 times in the case of microscopic specimens. Significantly, typical sensors used in machine vision systems have an 8 X 6-mm photoactive area format; that is, the object image must be made to "fit" an 8 X 6-mm field. At this time it should become clear that working distance, field of view, and magnification are all interdependent and somehow integrated in the notion of field angle.

One consideration about magnification that must be kept in mind is that unless special telecentric optics are used, any change in the working distance (distance from object to lens) will result in a proportional change in the magnification:

Where D is the distance to the object. The working distance can change due to vibration or other factors related to staging the object. This can be important in gaging applications where such changes in working distance contribute to the error budget. Similarly, it should be understood that for objects with three-dimensional features in the working distance space, the features will be viewed with pixels of different sizes due to magnification effects. So, too, a pixel will have different sizes at the top and bottom of a bin when viewing the bin from above.

5. Resolution. A term often used for two different concepts depending on the application field. In machine vision technology, it is generally defined as the ratio to the linear field of view of the width of the sharpest edge that can be reliably gauged by the imaging system. While it should equal, ideally, the ratio of the pixel size to the linear size of the sensor, or approximately 1/400 in practice, it degenerates to about 1/100 because of diffraction losses, aberrations in the optics, lighting deficiencies, and limited camera bandwidth. It should be noted that by repeating a measurement and resorting to arithmetic averaging techniques or other signal-processing-enhancing techniques, the processed resolution accuracy can sometimes be improved by up to one order of magnitude, but at the cost of compute power and typically processing speed.

This concept of resolution should not be confused with the concept of detection resolution, as used in many other fields of application and sometimes also in machine vision. The human eye and even camera sensors can easily detect stars in the sky but could not possibly resolve the diameter of a star by resolving its two edges. Similarly, machine vision systems are sometimes required to simply detect the presence of an object (or of a defect in an object), irrespective of its shape or size.

Of special consequence in gaging applications is the theoretical limit of resolution (Rayleigh limit) for an optical system. The resolving power of the lens for viewing distant objects may be expressed as

$$r = 1.22N\lambda$$

Where r is linear resolution, λ is the wavelength, and N is the f-number of the lens. For example, for light of 0.5 micron and a lens f-number of f/2, the resolution limit is $r = 1.22(2)(0.5) = 1.4$ *microns,* or approximately 50 microinches.

6. Depth of Field. The area in front of and behind the object that appears to be in focus on the image target. The depth of field is related to the depth of focus by

Depth of field—depth of focus
Magnification

It varies with the aperture and is greater when

1. The lens is stopped down (a small aperture or a high f-number)
2. The camera is focused on a distant object (a greater working distance) or
3. There is a longer focal length lens.

When stopped down for sharp focus, more illumination may be required to guarantee that the sensor is operating at an optimal signal-to-noise ratio.

As the aperture is increased, eventually the area surrounding the object will go out of focus. Pixel sizes can vary as a result of depth of field and focus conditions. This is especially critical in gauging applications.

One note of caution with respect to depth of field is that to the eye, things sometimes appear sharper on a monitor because background images appear smaller. This may be interpreted as greater depth of field than the optical arrangement will convey to the machine vision system. In other words, because of our perceptions, when viewing a monitor the system may not actually be operating on the optimal images from the point of view of depth of field.

There is a good deal of confusion, even among professionals, when trying to further quantitatively define the amount of defocusing that can be tolerated while remaining within the "depth of focus." The reason is, of course, that the amount of acceptable defocusing depends on the field of application.

As discussed in the preceding, defocusing results in both a loss of resolution and a loss of contrast. We suggest that loss of contrast is paramount in most applications in the machine vision field and that the depth of field is exceeded when contrast is reduced by some 75% or when light originating from an object point falls on four pixels instead of on a single one. Since defocusing is usually evenly distributed on the two axes, this translates in a dilution of the light from one object point into two pixels on a linear axis of the sensor array. The concept of depth of field is extremely important in the general field of vision and in all of its applications. Further theoretical discussion, however, is beyond the scope of this book.

6.3.3.6 Distortion and Aberrations in Optics. It is easy to derive the focal length of the lens needed to image certain fields of view at a specified working distance l^o and at a specified magnification M. Solving two equations immediately leads to the desired specifications f and l_I:

$$1/l_i + 1/l_o = 1/f \text{ and } l_i/l_o = M$$

Unfortunately, these formulas are valid only within the very restricted conditions of ideally thin lenses, no spherical distortion, no field curvature, and using monochromatic light.

While these simple formulas are extremely useful in clarifying the concepts of imaging and in obtaining some feeling and a guide as to what will be needed, optimization of one parameter generally leads to gradual degeneration of one or several others and to the so-called system aberrations.

This makes it very difficult to design an imaging system that will optimize all parameters mentioned earlier in relation to a specific type of application while keeping distortions and aberrations under control.

Geometric Distortions. Geometric distortions reduce the geometric fidelity of the image and result in the magnification varying with distance from the optical axis of the lens. Two common distortions are pincushion and barrel (Figure 6.21). These distortions are observed as changes in magnification along the lens axes. With pincushion distortion magnification increases as the distance from the center increases. Barrel distortion has the effect of decreasing magnification away from the image center. Distortion of lenses must be understood, especially in gauging applications.

Aberrations. A number of aberrations commonly found in lenses and their effects are briefly discussed:

1. Chromaticity. The wavelength dependence of the refractive index of glass causes the focal length of any single lens to vary with the color of the refracted light (Figure 6.22).

2. Sphericity. The focal point of a spherical refracting or reflecting surface varies in proportion to the departure of the ray from the paraxial region (Figure 6.23).

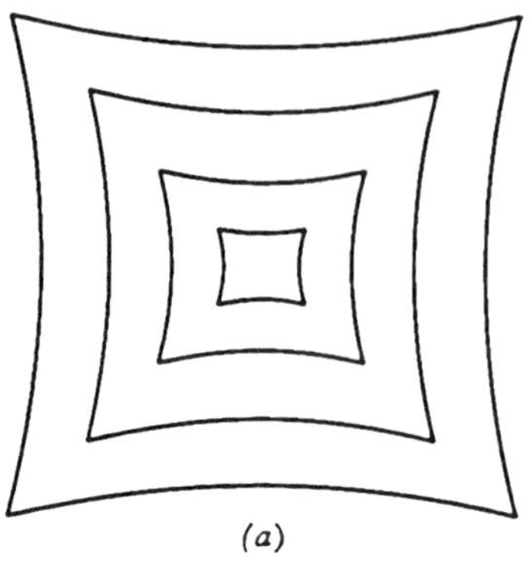

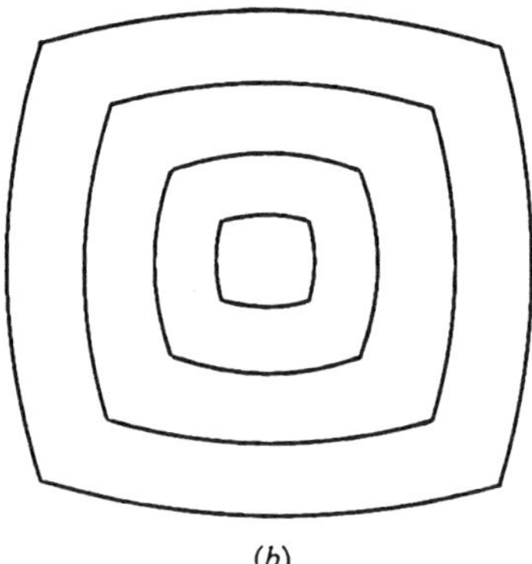

Figure 6.21 - Lens distortions: (a) pincushion distortion; (b) barrel distortion.

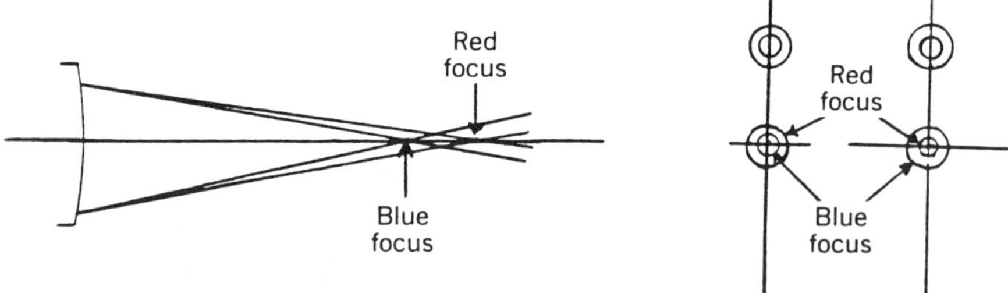

Figure 6.22 - Chromatic aberration.

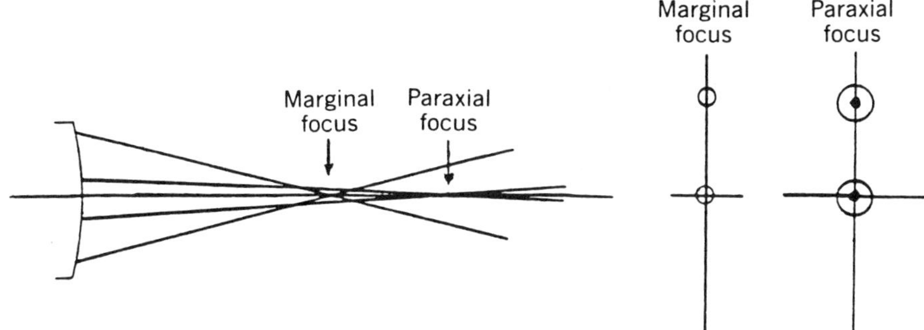

Figure 6.23 - Spherical aberration.

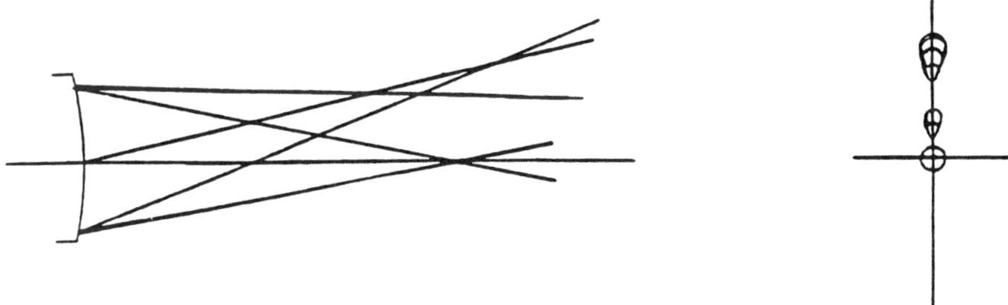

Figure 6.24 - Coma aberration.

3. Coma. It derives its name from the comet-like (Figure 6.24) appearance of the image of an object point located off-axis.

4. Astigmatism. A second-order aberration causing an off-axis object point to be imaged as two short lines at right angles to each other (Figure 6.25).

5. Field Curvature. A radially increasing blur caused by the image plane being tangential to the focused surface (Figure 6.26).

The preceding four types of aberrations affect both field resolution and field linearity. They increase when departing from the near-axial region, at larger numerical apertures. They also increase when departing from the paraxial condition, at shorter focal lengths.

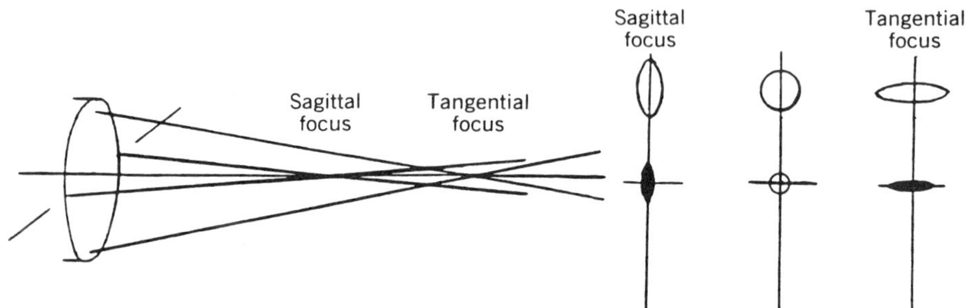

Figure 6.25 - Astigmatism aberration.

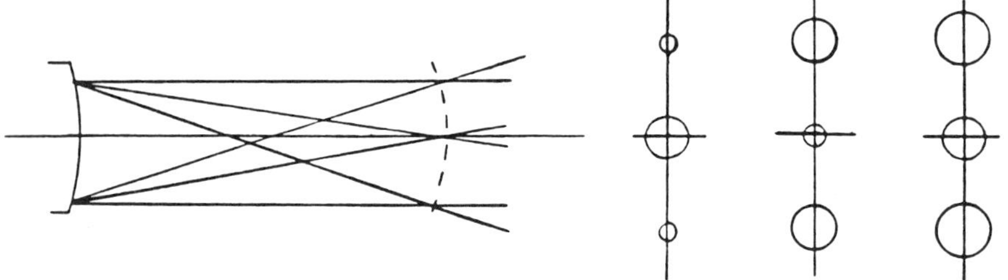

Figure 6.26 - Petzval field curvature aberration.

6. Vignetting. A uniformly illuminated field is imaged to a field that is uneven in illumination (Figure 6.27).

Different types of aberrations are corrected by substituting single lenses, systems with multiple lenses tending to compensate for each other. No single design offers a universal better objective. The design of the composite objective should be selected as the best compromise for a particular application. A properly corrected good objective may produce a substantially aberrated image when used at improper object or image distances.

6.3.3.7 Different Types of Objectives. Different types of objectives have been designed that are optimized for different types of applications. Their description and a few comments will serve as a guide in selecting the type of objective most suitable for a particular application.

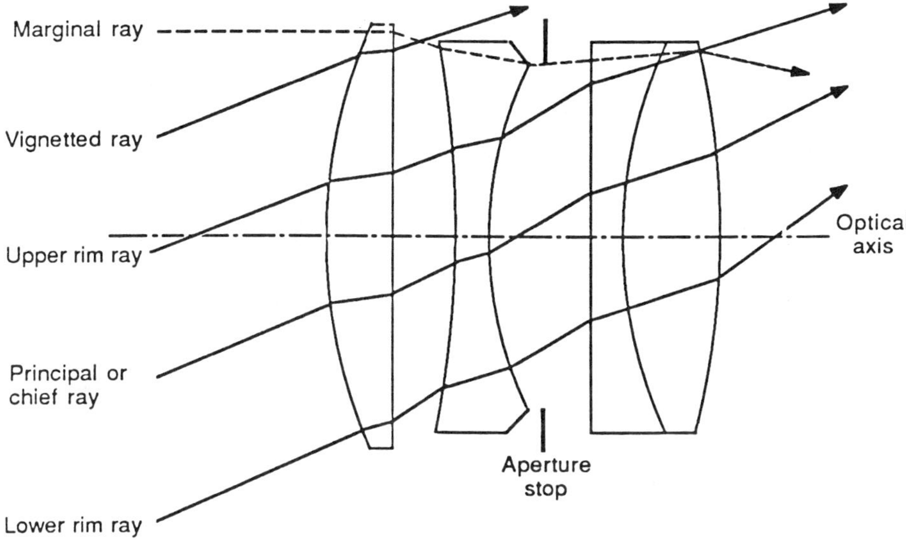

Figure 6.27 - Vignetting.

Video Camera Objectives. These are primarily designed for TV closed circuit surveillance applications. The primary requirements are to provide maximum light level (large numerical aperture or low f-stop number) and a recognizable image. Aberrations are of secondary importance, and they are really suitable only for low-resolution inspection work such as presence or absence. They are available in a wide range of focal lengths, from 8.5 mm (wide-angle lens), through the midrange, 25-50 mm, to 135 mm (telelens). In general, wide-angle objectives (8.5-12.5 mm) have high off-axis aberrations and should be used cautiously in applications involving gauging.

For most of these objectives, a 5-10% external adjustment of the image distance is provided, allowing one to focus anywhere within the range of working distances for which the objective has been designed. This range can be increased on the short end by inserting so-called extension rings of various thicknesses between the objective and the camera body. By further increasing the image length l_i, they automatically increase magnification, or the size of the imaged object. The use of extension rings offers great flexibility. It should be remembered, however, that commercial objectives have been corrected only for the range of working distances indicated on the barrel. The abusive use of extension rings can have a disastrous effect on image distortion. These remarks on the use of extension rings apply not only to video camera but also to any type of objective.

35-mm Camera Objectives. These are made specifically for 35-mm photographic film with a frame much larger than the standard video target size. Their larger format provides a better image, particularly near the edge of the field of view. Their design is usually optimized for large distances, and they should not be used at distances much closer than that indicated on the lens barrel, meaning that no extension ring other than the necessary C-mount adapter should be added between objective and camera body. They are widely available from photographic supply houses with various focal lengths and at reasonable cost.

Reprographic Objectives. We classify under this heading a number of specialized high-resolution, flat-field objectives designed for copying, microfilming, and IC fabrication. Correction for particular aberrations has been pushed to the extreme by using as many as 12 single lenses. This is generally done at the price of relatively low numerical aperture and the need of a high level of light.

Copying objectives are generally of short focal length and have high magnification. At the other end of the spectrum, reducing objectives used in exposing silicon wafers through IC masks have very small magnification, extremely high resolution, and linearity. Their high cost (up to $12,000) may be justified in applications where such parameters are of paramount necessity.

Microscope Objectives. These are designed to cover very small viewing fields (2-4 mm) at a short working distance. They cause severe distortion and field curvature when used at larger than the designated object distance. They are available at different quality grades at correspondingly different price levels. They mount to the standard CCTV camera through a special C-mount adapter.

Zoom Objectives. Allowing instantaneous change of magnification, zoom objectives can be useful in setting up a particular application. It is to be noted that the implied "zooming" function is only effective at the specified image distance, that is, at the specified distance between the target and the principal plane of the objective. In all other conditions, the focus needs to be corrected after a zooming step. The addition of any extension ring has a devastating effect on the zooming feature. Also, the overall performance of a zoom objective is less than a standard camera lens, and its cost is much higher.

6.3.3.8 Miscellaneous Comments

Cylindrical Lens. Cylindrical lens arrangements (Figure 6.28) can unfold cylindrical areas and effectively produce a field-flattened image that can be projected to the sensor image plane.

Mirrors may be used to provide the projection of 360° around an object into a single camera. One's imagination can run wild if one recalls their experiences with mirrors and optics in the fun houses of amusement parks. The distortion of images may, in a machine vision application, be a valuable way to capture more image data. On the other hand, sensor resolution may become the limiting factor dictating the actual detail sensitivity of the system.

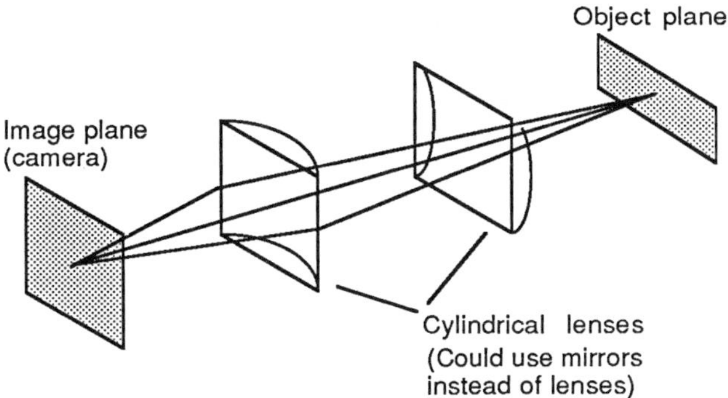

Figure 6.28 - Cylindrical lenses.

Mounts:

1. C Mount. The C mount lens has a flange focal distance of 17.526 mm, or 0.690 in. (The flange focal distance is the distance from the lens mounting flange to the convergence point of all parallel rays entering the lens when the lens is focused at infinity.) This lens has been the workhorse of the TV camera world, and its format is designed for performance over the diagonal of a standard TV camera.

Generally, this lens mount can be used with arrays that are 0.512 in. or less in linear size. However, due to geometric distortion and field angle characteristics, short focal length lenses should be evaluated as to suitability. For instance, an 8.5-mm focal length lens should not be used with an array greater than 0.128 in. in length if the application involves metrology. Similarly, a 12.6-mm lens should not be used with an array greater than 0.256 in. in length.

If the lens-to-array dimension has been established by using the flange focal distance dimension, lens extenders are required for object magnification above 0.05. The lens extender is used behind the lens to increase the lens to image distance because the focusing range of most lenses is approximately 5-10% of the focal length. The lens extension can be calculated from the following formula:

Lens extension = focal length/object magnification

2. U Mount. The U-mount lens is a focusable lens having a flange focal distance of 47.526 mm, or 1.7913 in. This lens mount was primarily designed for 35-mm photography applications and is usable with any array less than 1.25 in. in length. It is recommended that short focal lengths not be used for arrays exceeding 1 inch. Again, a lens extender is required for magnification factors greater than 0.05.

3. L Mount. The L-mount lens is a fixed-focus flat-field lens designed for committed industrial applications. This lens mount was originally designed photographic enlargers and has good characteristics for a field of up to 2 1/4 in. The flange focal distance is a function of the specific lens selected. The L-mount series lenses have shown no limitation using arrays up to 1.25 in. in length.

Special Lenses. There are standard microscope magnification systems available. These are to be used in applications where a magnification of greater than 1 is required. Two common standard systems for use with a U system are available, one with 10 magnification and one with 4 magnification.

Cleanliness. A final note about optics involves cleanliness (Figure 6.29). In most machine vision applications some dust is tolerable. However, dirt and other contaminants (oily films possibly left by fingerprints) on the surface of lenses, mirrors, filters, and other optical surfaces contribute to light scattering and ultimately combine to reduce the amount of light transmitted to image area.

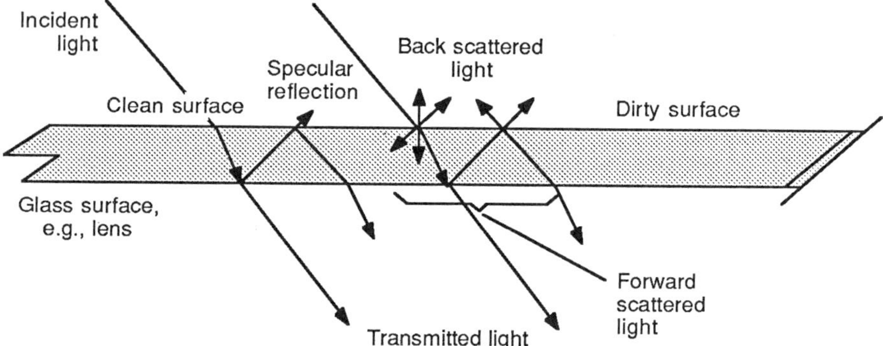

Figure 6.29 - Transmission, reflection and scattering of light at optical surface.

6.3.4 Practical Selection of Objective Parameters

6.3.4.1 Conventional Optics. The best guide in selecting an objective for a particular application is undoubtedly experience and feeling. An experienced engineer will generally borrow a rough estimate of the focal length needed in a particular geometry from previous experience either in CCTV techniques or in photography. Examination of the thus obtained image will immediately suggest a slight correction in the estimate, if necessary. The purpose of this tutorial would not fully be achieved, however, if the case of a novice engineer were not considered, having little or no such experience.

It is best to start the design from the concept of field angle (the angle whose tangent is approximately equal to the object height divided by the working distance), the only single parameter embodying the other quantities specifying an imaging geometry: field of view, working distance and magnification. The maximum field angle φ_{max}, or the angle of view when the camera is focused at infinity, is listed in the manufacturer's specifications. Table 6.1 lists φ_{max} values for COSMICAR CC objectives. The minimum field angle φ_{min} depends on the distance of the principal plane of the lens to the image plane. It has been calculated and listed in the table for the case of the maximum reasonable amount of extension rings set at 4 times the focal length.

The following steps should be followed:

1. From Table 6.1 determine the range of objectives that will cover the field at the specified working distance.
2. Adjust the image distance by adding extension rings to obtain the desired magnification.
3. If the total length of extension rings arrived at by performing step 2, dangerously approaches the value of 4 times the focal length, and/or particularly if some image aberration begins to show up, switch to the next shorter focal length objective and repeat step 2.

6.3.4.2 Aspherical Image Formation. The concept of an image as being a faithful reproduction of an object is, of course, immaterial in those machine vision applications where only presence, consistent shape, color, or relative position are involved. An elongated object will not fill the field of the 4/3 aspect ratio of a conventional CCTV. In order to make more efficient use of the full capability of vision hardware and vision algorithms; it is sometimes desirable to use a different magnification for the two-field axis (Figure 6.28). A simple such arrangement is to use a conventional spherical objective, conferring the same magnification in the two directions. A cylindrical beam expander is then added to change the magnification, as desired, in one image axis only.

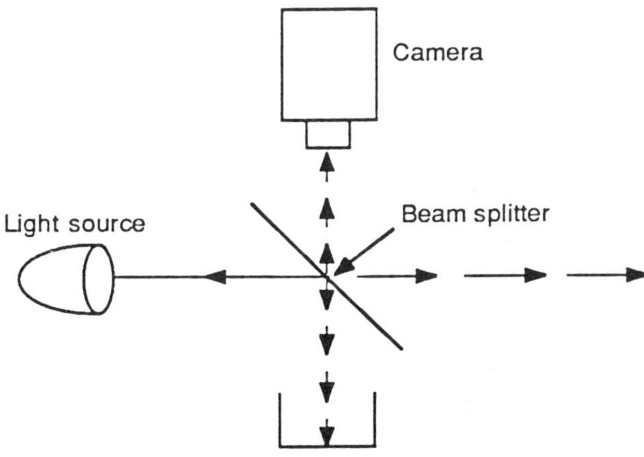

Figure 6.30 - Beam splitter.

6.3.4.3 Telecentric Imaging. A telecentric optical arrangement is one that has its entrance or exit pupil (or both) located at infinity. A more physical but less exact way of putting this is that a telecentric system has its aperture stop located at the front or back focal point of the system. The aperture stop limits the maximum angle at which an axial ray can pass through the entire optical system. The image of the aperture stop in object space is called the entrance pupil and the image in image space is called the exit pupil.

If the telecentric aperture stop is at the front focus (toward the object), the system is considered to be telecentric in image space; if the telecentric stop is at the back focus (toward the image), then the system is telecentric in object space. Doubly telecentric systems are also possible.

Since the telecentric stop is assumed to be small, all the rays passing through it will be nearly collimated on the other side of the lens. Therefore, the effect of such a stop is to limit the ray bundles passing through the system to those with their major axis parallel to the optical axis in the space in which the system is telecentric. Thus in the case of a system that is telecentric in image space, slight defocusing of the image will cause a blur, but the centroid of the blur spot remains at the correct location on the image plane. The magnification of the image is not a function (to first order, for small displacements) of either the front or back working distance, as it is in non-telecentric optical systems, and the effective depth of focus and field can be greatly extended.

Such systems are used for accurate quantitative measurement of physical dimensions. Most telecentric lenses used in measurement systems are telecentric in object space only. Its great advantages are that it has no z-dependence and its depth of field is very large. Hence, telecentric lenses provide: a constant system magnification; a constant imaging perspective and solutions to radiometric applications associated with delivering and collecting light evenly across a field of view.

Large telecentric beams cannot be formed because of the limiting numerical aperture (D/f), and the large loss of light at the aperture. In practice, telecentric lenses rarely have perfectly parallel rays. Descriptions such as "telecentric to within 2 degrees" mean that a ray at the edge is parallel to within 2 degrees of a ray in the center of the field.

The use of telecentric lenses is most appropriate when:

The field of view is smaller than 250 mm.

The system makes dimensional measurements and the object has 3D features or there are 3D variations in the working distance (object-to-lens distance).

The system measures reflected or transmitted light and the field of view (with a conventional lens) is greater than a few degrees.

6.3.4.4 Beam Splitter. A beam splitter arrangement (Figure 6.30) can be a useful way to project light into areas that are otherwise difficult because of their surroundings. In this arrangement a splitter only allows the reflected light to reach the camera.

6.3.4.5 Spilt Imaging. Sometimes we are called on to look at two or more detailed features of an object that are not accessible to a single camera. Examples are two opposite corners of a label for proper positioning or, front and back labels on a container. Two or more cameras can be used in a multiplexed arrangement, where the video data of each camera is processed in succession, one at a time. This, of course, can only be done with a corresponding increase in inspection time.

Two or more camera fields can also be synthesized into a single split field using a commercial TV field splitter. The compromises here are a loss of field resolution and more complex and expensive hardware. Both these methods have drawbacks and are not always practical. An alternative method consists in imaging the two (or more) parts of interest on the front end of two (or more) coherent (image quality) fiber-optic cables and, in turn, reimage two (or more) other ends of the cables into a single vision camera. The method, while providing only moderate quality imaging, provides extreme flexibility.

6.4 IMAGE SCANNING

Since, eventually, a correlation must be established between each individual point of an object and the reactive light emitted by that light, and since that correlation must be a one-to-one correlation, some sort of sequential programming is needed.

6.4.1 Scanned Sensors: Television Cameras

In one method an object is illuminated, and all of its points are simultaneously imaged on the imaging plane of the sensor. The imaging plane is then read out, that is, sensed, point by point, in a programmed sequence. It outputs a sequential electrical signal, corresponding to the intensity of the light at the individual image points. The most common scanned sensor is the CCTV camera. This is discussed in greater detail in the next chapter.

6.4.2 Flying Spot Scanner

The same correlation can be established if we were to reverse the functions of light source and sensor. In this method, the sensor is a point source detector, and it is made to look at the whole object, all object points together (Figure 6.31). These points are now illuminated one by one by a narrow pencil of light (a scanned CRT or, better, a focused laser) moving from point to point according to a programmed (scanned) sequence. Here, again, the sensor will output a sequential signal similar to that of the first method. In a flying spot scanner, extremely high resolutions of up to 1 in 5000 can be achieved.

6.4.3 Mixed Scanning

When dealing with two- or three-dimensional objects moving on a conveyor line, a method of mixed scanning is often adopted using the mechanical motion of the object as a flying spot scanner in the direction of motion and a scanned sensor

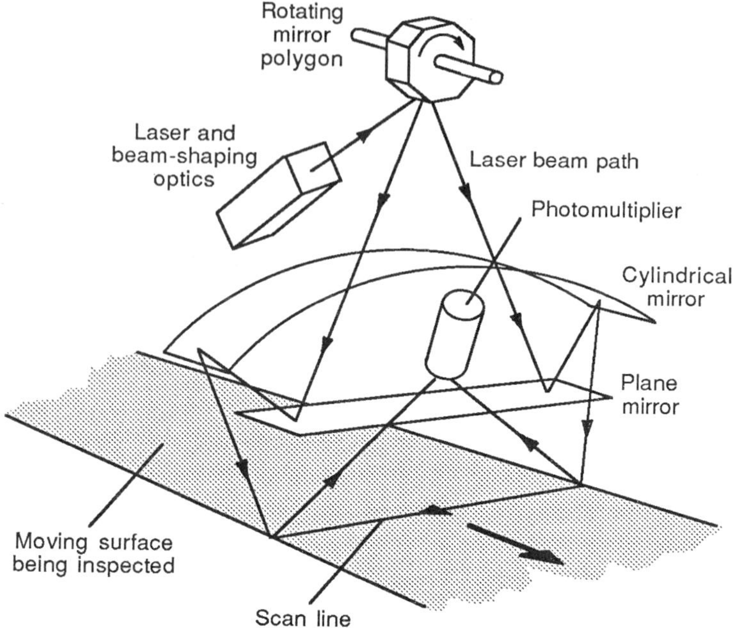

Figure 6.31 Flying spot scanner for web surface inspection.

for the transverse axis. A popular system, for example, capable of achieving high resolution uses a line scanner, a one-dimensional array of individual photosensors (256-12,000), in one axis and a numerical control translation table in the other.

REFERENCES

OPTICS/LIGHTING

Abramawitz, M. J., "Darkfield Illumination," American Laboratory, November, 1991.

Brown. Lawrence, "Machine Vision Systems Exploit Uniform Illumination," Vision Systems Design, July, 1997.

Forsyth, Keith, private correspondence dated July 9, 1991.

Gennert, Michael and Leatherman, Gary, "Uniform Frontal Illumination of Planar Surfaces: Where to Place Lamps," Optical Engineering, June, 1993.

Goedertier, P., private correspondence, January 1986.

Harding, K. G., "Advanced Optical Considerations for Machine Vision Applications," Vision, Third Quarter, 1993, Society of Manufacturing Engineers.

Higgins, T. V., "Wave Nature of Light Shapes Its Many Properties," Laser Focus World, March, 1994.

Hunter Labs, “The Science and Technology of Appearance Measurement,” manual from Hunter Labs. Reston, VA.

Kane, Jonathan S., "Optical Design is Key to Machine-Vision Systems," Laser Focus World, September, 1998.

Kaplan, Herbert, "Structured Light Finds a Home in Machine Vision," Photonics Spectra, January, 1994.

Kopp, G. and Pagana, L. A., "Polarization Put in Perspective," Photonics Spectra, February, 1995.

Lake, D., "How Lenses Go Wrong - and What To Do About It," Advanced Imaging, June, 1993.

Lake, D., "How Lenses Go Wrong - and What To Do About It - Part 2 of 2," Advanced Imaging, July, 1993.

Lapidus, S. N., “Illuminating Parts for Vision Inspection,” *Assembly Engineering* March 1985.

Larish, John and Ware, Michael, "Clearing Up Your Image Resolution Talk: Not So Simple," Advanced Imaging, April, 1992.

Mersch, S. H., “Polarized Lighting for Machine Vision Applications,” Conference Proceedings, the Third Annual Applied Machine Vision Conference, Society of Manufacturing Engineers, February 27-March 1, 1984.

Morey, Jennifer L., "Choosing Lighting for Industrial Imaging: A Refined Art," Photonics Spectra, February, 1998.

Novini, A., “Before You Buy a Vision System”. *Manufacturing Engineering,* March 1985.

Schroeder, H., “Practical Illumination Concept and Techniques for Machine Vision Applications,” Machine Vision Association/Society of Manufacturing Engineers Vision 85, Conference Proceedings.

Smith, David, "Telecentric Lenses: Gauge the Difference," Photonics Spectra, July, 1997.

Smith, Joseph, "Shine a Light," Image Processing, April, 1997.

Stafford, R. G., “Induced Metrology Distortions Using Machine Vision Systems,” Machine Vision Association/Society of Manufacturing Engineers Vision 85, Conference Proceedings.

Visual Information Institute “Structure of the Television Roster,” Publication Number 012-0384, Visual Information Institute, Xenia, OH.

Wilson, Andrew, "Selecting the Right Lighting Method for Machine Vision Applications," Vision Systems Design, March, 1998.

Wilson, Dave, "How to Put Machine Vision in the Best Light," Vision Systems Design, January, 1997.

7

Image Conversion

7.1 TELEVISION CAMERAS

7.1.1 Frame and Field

Generally, machine vision systems employ conventional CCTV cameras. These cameras are based on a set of rules governed by Electronic Industries Association (EIA) RS-170 standards. Essentially, an electronic image is created by scanning across the scene with a dot in a back-and-forth motion until eventually a picture, or "frame," is completed. This is called raster scanning.

The frame is created by scanning from top to bottom twice (Figure 7.1).

This might be analogous to integrating two scans of a typewritten page. With one scan all the lines of printed characters are captured as organized; with a second scan all the spaces between the lines are captured. Each of these scans corresponds to a field scan. When the two are interleaved, a double-spaced letter page results, corresponding to a frame.

A field is one-half of a frame. The page is scanned twice to create the frame, and each scan from top to bottom is called a field. Alternating the line of each of the two fields to make a frame is called interlace. In a traditional RS-170 camera, the advantage of the interlaced mode is that the vertical resolution is twice that of the noninterlaced mode. In the noninterlaced mode, there are typically 262 horizontal lines within one frame, and the frame repeats at a rate of 60 Hz. Today

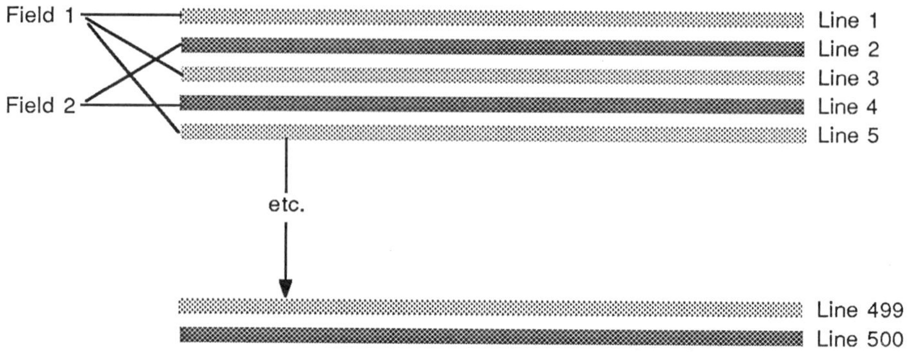

Figure 7.1 - Interlaced-scanning structure (courtesy of Visual Information Institute).

non-RS-170-based cameras are available that operate in a progressive scan or non-interlaced mode providing full "frame" resolution at field rates - 60 Hz. Depending on the camera and the imaging sensor used, the vertical resolution could be that of a full field or a full frame.

In all cases the camera sweeps from left to right as it captures the electronic signal and then retraces when it reaches the end of the sweep. A portion of the video signal called sync triggers the retrace. Horizontal sync is for the retrace of the horizontal scan line to the left of beginning of the next line. Vertical sync is for the retrace of the scan from bottom to top.

In the United States where the EIA RS-170 standard rules, a new frame is scanned every 30th of a second; a field is scanned every 60th of a second (Table 7.1). In conventional broadcast TV there are 525 lines in each frame. In machine vision systems, however, the sensor often used may have more or less resolution capability although the cameras may still operate at rates of 30 frames per second. In some parts of the world the phase-alternating line (PAL) system is used, and each frame has 625 lines and is scanned 25 times each second.

In machine vision applications where objects are in motion, the effect of motion during the 30th-of-a-second "exposure" must be understood. Smear will be experienced that is proportional to the speed. Strobe lighting or shutters may be used to minimize the effect of motion when necessary. However, in RS-170 format synchronization with the beginning of a frame sweep is a challenge with conventional RS 170 cameras. Since the camera is continually sweeping at 30 Hz, the shutter/strobe may fire so as to capture the image partway through a field. This can be handled in several ways. The most convenient is to ignore the image signal on the remainder of the field sweep and just analyze the image data on the next full field. However, one sacrifices vertical resolution since a field represents half of a frame.

TABLE 7.1 Television Frames

	Scan lines in Two Fields	Scan lines per Field	Field Rate (Hz)	Frame Rate (Hz)
Interlaced system				
526/60/2:1	525	262.5	60	30
Noninterlaced system				
263/60	526	263	59.88	59.88

In some applications where the background is low and the frame storage element is part of the system, the "broken" field can be pieced together in the frame store and an analysis conducted on the full frame. However, there is still uncertainty in the position of the object within the field stemming from the lack of synchronization between the trigger activated by the object, the strobe/shutter, and the frame. For example, for machine vision applications this means the field of view must be large enough to handle positional variations and the machine vision system itself must be able to cope with positional translation errors as a function of the leading edge of object detection.

In response to this challenge, camera suppliers have developed products with features more consistent with the requirements of high-speed image data acquisition found in machine vision applications. Cameras are now available with asynchronous reset, which allows full frame shuttering by a random trigger input. The readout inhibit feature in these cameras assures that the image data is held until the frame grabber/vision engine is ready to accept the data. This eliminates lost images due to timing problems and makes multiplexing cameras into one frame grabber/vision engine possible.

In some camera implementations, in asynchronous reset operating mode, upon receipt of the trigger pulse, the vertical interval is initiated, and the previously accumulated charge on the active array transferred to the storage register typically within 200 microseconds. This field of information, therefore, contains an image that was integrated on the sensor immediately before receiving the trigger. The duration of integration on this "past" field can be random since the trigger pulse can occur anytime during the vertical period. This would result in an unpredictable output. Hence, this field is not used.

After the 200 microsecond period, the active array is now ready to integrate the desired image, typically by strobe lighting the scene. Integration on the array continues until the next vertical pulse (16.6 msec) or the next reset pulse, whichever occurs first. Then the camera's internal transfer pulse will move the information to the storage register and begin readout. With this arrangement only half of the vertical resolution is applied to the scene.

7.1.2 Video Signal

The video signal of a camera can be either of the following:

Composite Video. Format designed to provide picture and synchronization information on one cable. The signal is composed of video (including video information and blanking, sometimes called pedestal and sync).

Noncomposite Video. Contains only picture information (video and blanking) and no synchronization information. Sync is a complex combination of horizontal and vertical pulse information intended to control the decoding of a TV signal by a display device.

Sync circuits are designed so a camera and monitor used for display will operate in synchronization, the monitor simultaneously displaying what the camera is scanning. Cameras are available that are either sync locked or genlocked. A camera with sync lock requires separate driving signals for vertical drive, horizontal drive, subcarrier, blanking, and so on. Genlock cameras can lock their scanning in step with an incoming video signal; a feature more compatible with the multicamera arrangements often required in machine vision applications. Genlock optimizes camera switching speeds when multiplexing cameras but is not required to multiplex cameras.

7.1.3 Cameras and Computers

When interfacing a camera to a computer, the camera can be either a slave or a master of the computer. When the camera is the master, the camera clock dictates the clocking of the computer, and the computer separates the sync pulses from the video composite signal generated in the camera. When the camera acts as the slave, the computer clock dictates the clocking of the camera via a genlock unit.

When the camera is the master, cameras of different designs can be interfaced to the same system as long as they all conform to EIA RS-170 standards. However, only one camera at a time can be linked to the computer. When the camera is the slave, using genlocking, multiple cameras can be interfaced to the same computer.

7.1.4 Timing Considerations

Timing considerations should be understood. Typically, a horizontal line scan in a conventional 525-line, 30-Hz system is approximately 63.5 microseconds (Figure 7.2). However, 17.5%, or 11 microseconds, of this time represents horizontal blanking. The pixel time is therefore (63.5 - 11)/512, or approximately 100 nsec. In other words, a machine vision system with these properties generates a pixel of data (2, 4, 6, or 8 bits) every 100 nsec.

The actual pixel-generating rate dictates the requirement of the master clock of the camera. In this instance the master clock has to be 1/100 nsec, or 10 MHz. The selection of the number of pixels per line is limited by the ability of the TV system to produce data at the rate demanded. While 10 MHz is typical and special systems can generate data at a rate of 100 MHz, the practical limit is 40 MHz.

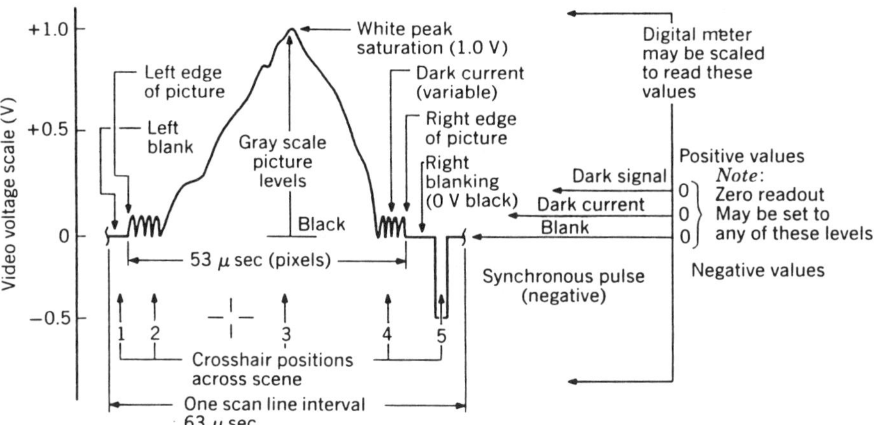

Figure 7.2 - Camera timing considerations (courtesy of Visual Information Institute).

The vertical rate is a function of the number of lines in the image. If 260 lines are developed at 60 Hz, each line must be scanned in 1/15,600 sec, or 64 microseconds. Conventional TV operates at horizontal rates of 15.75 KHz, which is determined by

Frame rate X number of lines per frame = 30 x 525 = 15.75 KHz

Vertical blanking is typically 7.5% of the vertical field period.

What all this means is that in an RS-170 camera, the two halves of the picture which would be stitched together by the computer are not really taken at the same moment in time. There are a few milliseconds between the shifting of all the "odd" lines and all the "even" lines. Any movement in the field of view of the camera will cause the scene to change slightly between reading out these fields.

Progressive scanning cameras are available. These do not operate to the RS-170 standard of two interlaced fields combined to form a single frame. Instead the lines of video on the imager are read sequentially, one line at a time from top to bottom. Hence, the full vertical resolution of the imager can be obtained, typically at field reading rates - 60 Hz.

7.1.5 Camera Features

RS-170 cameras come with a variety of features that may or may not have a positive impact on their use in machine vision applications. An automatic black level circuit that maintains picture contrast throughout the light range can be important in machine vision applications. Many include circuits (automatic light range, automatic light control, automatic gain control, etc.) to assure the maximum sensitivity at the lowest possible light level. These provide automatic adjustment as a function of scene brightness.

In applications where the video output is a measure of light intensity, the automatic gain control should be disabled. Under certain light conditions the camera may present images with better contrast.

Gamma correction, or correction to compensate for the nonlinearity in the response of the phosphor of a cathode ray tube (CRT-based display), is designed to give a linear output signal versus input brightness. Both automatic light control and gamma correction circuits are designed to optimize the property of the display monitor for human viewing. In machine vision applications these circuits may distort the linearity of the sensor, defeating a linear gray scale relationship that may be the basis of a decision. Disabling gamma correction can increase the contrast between the dark image and the bright image.

The resolution performance of a camera is influenced by the operating mode. Of fundamental importance is adequate illumination for a high signal-to-noise ratio. The closer to 100% contrast, the closer a machine vision system performance is maximized.

Most camera specifications designate minimum illumination. This is the minimum amount of light that will produce a field voltage video signal. For solid-state cameras the sensitivities range from 0.2- to 10-ft candles. To operate the camera under the widest range of scene illumination, it should be kept as close as possible to the bottom range specified by the camera manufacturer.

Other features commonly found in cameras include the following:

Automatic Light Range (ALR) or Automatic Light Control (ALC). Circuit that ensures maximum camera sensitivity at the lowest possible light level and makes automatic adjustment as a function of scene brightness.

Automatic Black Level. Circuit that maintains picture contrast constant throughout light range.

Automatic Gain Control. Amplifier gain is automatically adjusted as a function of input.

Automatic Lens Drive Circuit. A variable gain amplifier that automatically changes gain in lieu of AGC. Requires compatible lens furnished with camera.

Where low light level applications are involved, solid-state imagers can be combined with image intensifiers, devices that act as amplifiers of the light image. Using front-end image intensifiers, images can be gated to record snapshots of rapidly occurring events. Image intensifiers can be delivered with special spectral sensitivity properties, thereby providing a degree of filtering. Cooled solid state imagers enhance S/N and dynamic range performance permitting longer-term exposures to handle low light level applications. While such cameras are used in scientific imaging applications, they have been used sparingly in machine vision applications.

7.1.6 Alternative Image-Capturing Techniques

Machine vision systems also employ other means of capturing image data. Linear pickup devices include linear solid-state arrays; either charge coupled device (CCD) or photodiode, and mechanically scanned laser devices. These provide single lines of video in high spatial detail in relatively short periods of time. Linear array cameras are frequently used in a fixed arrangement or to capture information as an object passes the array, such as inspecting products made in continuous sheets or webs. Two-dimensional information can be detected in a manner analogous to viewing a train passing a barn as if viewing it through slats in a barn. As the train passes, only that part of the train equivalent to the viewing area through the slats can be observed in one instance.

Similarly, a linear array can capture the two-dimensional information of a moving object (Figure 7.3). Significantly, the object motion must be repeatable, or travel speed well-regulated, or motion/camera synchronization available, to obtain repeatable two-dimensional images from the buildup of independent linear images. An alternative to the object moving would be to move the linear array camera across the object in a repeatable fashion.

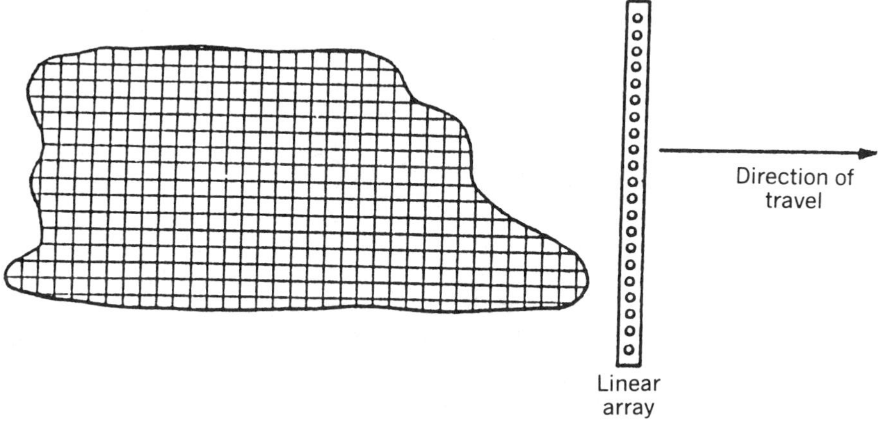

Figure 7.3 - Using linear array to capture "image" data. Each row in typical detection grid represents successive scans by same sensing element. Each column represents information gathered by entire array on one scan.

In either case, by positioning the linear array of sensors perpendicular to the axis of travel, image data can be captured. Information is gathered from the array at fixed intervals and stored in memory. Each successive scan of the linear array produces information equivalent to one row of a matrix camera.

With linear arrays (Figure 7.4), pixels are scanned from one end of the array to the other, producing a voltage-time curve with voltage proportional to light: Scanning rates can be as high as 60-80 MHz, though typical rates are on the order of 20 MHz. At these rates, depending on the velocity of the object, since the data is being collected serially, there will be a positional variation as the object is mapped to the array.

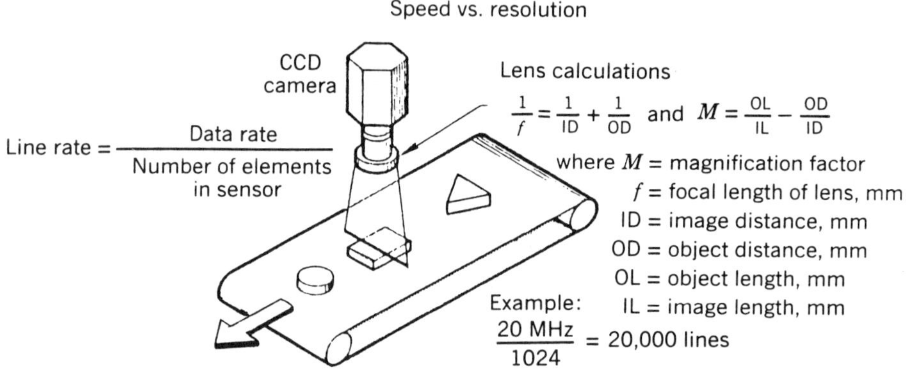

Figure 7.4 - Factors in applying linear arrays.

Linear sensor arrays are available with up to 12,000 pixels. Given a 20-MHz scan rate, it will take 0.1 msec to scan 2048 pixels. The pixel dwell time is therefore a function of the number of pixels in the array and the scanning rate.

Linear array cameras do not produce a picture display directly. This can present a challenge when trying to determine if an object is in the field of view and in focus. With memory as inexpensive as it has become, many systems based on line scan cameras store selected contiguous horizontal pixels and a number of contiguous lines in the direction of travel, for example. This stored image data can then be displayed for visual diagnostics, etc.

Linear array cameras are available that can capture a two-dimensional scene. These employ a mirror-galvanometer arrangement where the mirror scans the scene one line at a time across the linear array. These cameras found widespread use in document scanners. However, they are too slow for machine vision applications. The 1728 X 2592 versions take 0.5-2.0 sec to capture a scene at full resolution.

Alternative scanner arrangements involve using flying spot scanner arrangements such as lasers with a single photodetector sensor. Laser scanners exist that provide both linear and area scans. As a consequence of time versus position in space, an image can be created that can be operated on as with any image.

7.2 SENSORS

Vidicons were used in the early machine vision systems. As much as anything, their instability contributed to the failure of the early machine vision installations.

The development of solid state sensors, as much as anything, is what made reliable machine vision possible. Several types of solid-state matrix sensors are available: charge coupled device (CCD), charge injection device (CID), charge prime device (CPD), metal-oxide semiconductor (MOS), and photodiode matrix. Charge injection devices and CPDs have a somewhat better blue-green sensitivity than CCDs and MOS units. Since solid state cameras have extended red infrared sensitivity, they generally incorporate IR absorbing filters to tailor sensitivity to the visible spectrum. Solid-state cameras are very small, lightweight, rugged, have fixed spatial resolution elements, and are insensitive to electromagnetic fields.

Solid-state sensors are composed of an array of several hundred discrete photosensitive sites. A charge proportional to incoming photon levels is stored electronically and depleted by photon impingement, but the site selection is performed electronically and the sites are physically well defined, leading to superior geometric performance.

All solid-state sensors provide spatial stability because of the fixed arrangement of photosites. This can mean repeatable measurements in machine vision applications as well as a reduction in the need for periodic adjustments and calibration. Typically, the photosites are arranged to be consistent with the 4: 3 aspect ratio adopted by the TV industry. While in the early solid state cameras the actual pixels themselves had a rectangular shape 4:3 ratio, in response to the requirements of the machine vision industry, imaging chips are now available with square pixels. The size of the array is typically on the order of 8 X 6 mm. As noted, imagers are available with a 1:1 aspect ratio or square pixels, which may be of value in gaging applications where both vertical and horizontal pixel dimensions are the same.

All visible solid-state sensors experience a phenomenon called highlight smear. It occurs as an unwanted vertical bright line coincident with every pixel in the highlight portions of a scene. It is experienced to different degrees in different types of sensors.

7.2.1 Charge-Coupled Devices

CCD imaging is a three-step process:

1. Exposure which converts light into an electronic charge at discrete sites called pixels.
2. Charge transfer, which moves the packets of charge within the silicon substrate.

3. Charge-to-voltage conversion and output amplification.

The CCD falls into two basic categories: interline transfer and frame transfer (Figures 7.5 and 7.6). Frame transfer devices use MOS photocapacitors as detectors. Interline transfer devices use photodiodes and photocapacitors as detectors.

Area image sensor block diagram

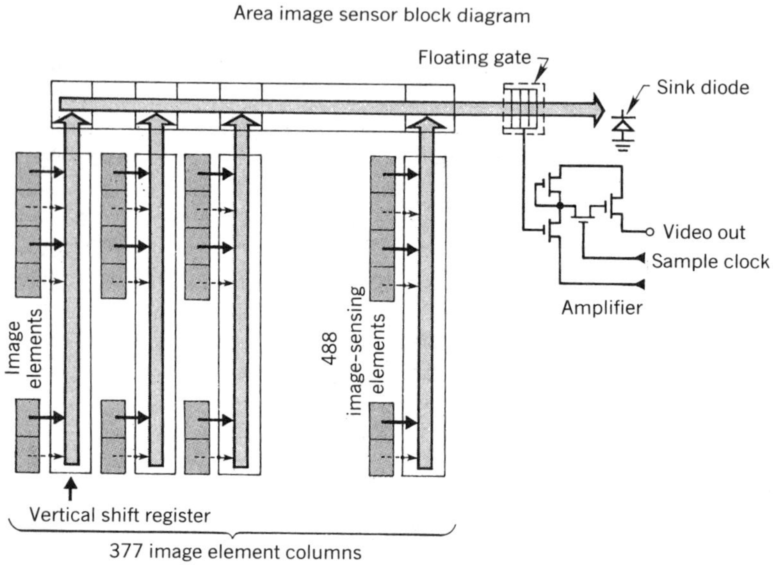

Figure 7.5 - Interline transfer CCD.

Frame transfer CCD

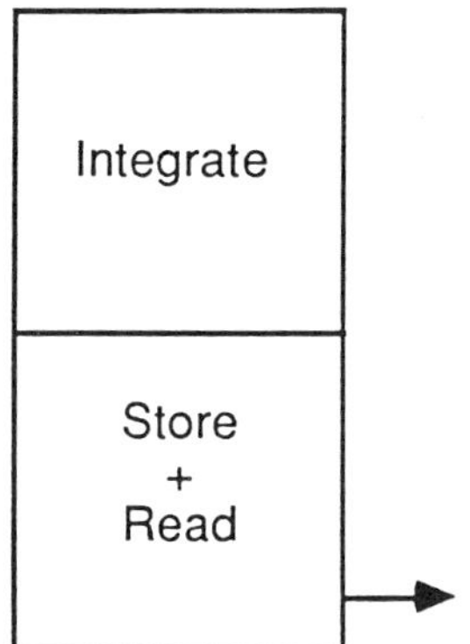

Figure 7.6 - Frame transfer CCD (Courtesy of Cidtec).

In interline transfer the charge packets are interleaved with the storage and transfer registers. The charge packets are transferred line by line to the output amplifier in two separate fields. In this design, each imaging column has an optically encoded shift register adjacent to it. In the imaging column, separate sites are defined under each photogate. Charge transfer takes place, and a horizontal shift register serially reads out the data.

This organization results in an image-sensitive area interspersed with an insensitive storage area. Consequently, the sensitivity is reduced for a given area. While the capability exists to defeat interlace and present contiguous and adjacent data vertically, horizontally one still has interspersed insensitive areas. This presents increased challenges where subpixel processing is required. It also results in a lower overall sensitivity or quantum efficiency.

In the frame transfer (Figure 7.6) organization, the device has a section for photon integration (the image register) and another similar section for storage of a full frame of video data. By appropriate clocking, the whole frame of information is moved in parallel into the storage section, transferred one line at a time into the horizontal register, and then finally transferred horizontally to the output stage. A second frame of image data is collected as the first frame is read out.

Because the parallel register is used for both scene detection and readout, a mechanical shutter or synchronized strobe illumination must be used to preserve the scene integrity. Some implementations of frame transfer imagers have a separate storage array, which is not light sensitive to avoid this problem. This does yield higher frame rate capability. Performance is compromised, however, because integration is still occurring during the image dump to the storage array, which results in image smear. Because twice the silicon area is required to implement this architecture, they have lower resolutions.

This organization makes the whole image area photosensitive and allows separation of the integration time from the readout time. Since most of these devices have been developed with TV transmission in mind, they generally operate in an interlaced mode. This requires the use of a full frame of memory to restore spatial adjacency for further processing. The simplicity of the frame transfer design yields CCD imagers with the highest resolution.

Both frame transfer and interline transfer CCDs are well suited to operation with strobes and avoid smear because integration time and readout times are independent and signal transfer from the image area to the readout area is much faster than field sweep times. The two-stage readout mechanisms permit synchronization without requiring strobe synchronization. Strobe exposure times as low as 1 microsecond are possible with some CCD cameras. A shutter in the camera eliminates stray light concerns that could otherwise affect pixel signals during readout.

7.2.2 Matrix-Addressed Solid-State Sensors

Matrix-addressed devices such as the CID, MOS, CMOS and photodiode can be read sequentially, which facilitates high-speed signal processing. As in the frame transfer CCD, they are devices with contiguous photosites, with intersite spacing a function of each manufacturing approach.

Charge Injection Device. The CIDs consist of a matrix of photosensitive pixels whose readout is controlled by shift registers. A CID operates by injecting a signal charge in each row, row by row, until all rows in a frame are read out. Given a "take picture" command, all CID pixels may be immediately accessed via the row and column load devices and emptied of signal charges. The inject signals (normally applied periodically) are inhibited (with charge-inhibit operation), placing the CID in a light-integrating mode. The imager can be exposed by flash or shutter for from 1 microsecond to 1 second duration. All photon-generated signal charges are collected and stored until readout occurs at the pixel and/or frame rate. Longer integration time may require cooling to avoid the influence of dark current.

Unlike other sensing technologies, CID sensors can be nondestructively sampled. Sensor locations can be randomly accessed, a technique useful in avoiding bleeding due to optical overloads on given site locations. An injection-inhibit mode allows interruption of camera readout upon command, synchronous stop motion of high-speed events, and integration of static low-light imagery. When injection inhibit is removed, a single field of information may be stored externally for further processing or monitor display.

Metal-Oxide Semiconductor. The MOS readout mechanisms are read directly from the photosite. Some offer zigzag pixel architecture with alternate rows of pixels offset by one-half pixel. With a special clocking arrangement involving the simultaneous readout of two horizontal rows, higher resolutions are achieved.

Standard MOS sensors suffer from low sensitivity and random and fixed pattern noise. They also have a tendency to experience lag due to incomplete charge transfer.

Charge Prime Device. Hybrid MOS/CCD sensors using charge priming techniques of the column electrodes overcome the noise limitations of MOS sensors and improve dynamic range. These use a CCD register for horizontal scanning. The charge primed coupler allows a priming charge to be injected into the photoelectric conversion area, forming a bias level so a transferable signal is maximized. When the pixel data is transferred, only a fraction of the signal charge is removed, leaving the priming charge in the coupler for the next cycle.

Complementary Metal Oxide Sensor: CMOS. There are two basic types of CMOS imagers, either passive or active pixels. In passive pixel versions, photo-generated carriers are typically collected on a p-n junction and passively transferred to a sensing capacitor through multiplexing circuitry and an analog integrator. For an active CMOS pixel, in addition to multiplexing circuitry, MOS transistors are included in each pixel to form a buffer amplifier with a sensitive floating input capacitance. The advantage of the active CMOS pixel is in exhibiting gain as the analog charge packet is transferred from the large photosite into a potentially much smaller sensing capacitance. The analog signal is further buffered to minimize susceptibility to noise sources. Such sensors also lend themselves to random pixel accessing or accessing only certain specific blocks of pixels. A bonus of CMOS is that the manufacturing procedure lends itself to the addition of circuitry to perform other functions such as analog-to-digital conversion, image compression, automatic exposure control, mosaic color processing, adaptive frame rate functions, etc.

Photodiode Matrix Sensors. Photodiode matrix sensors consist of individual photodiodes arranged in a square pattern. These sensors can be operated in interlaced or noninterlaced modes. The photodiode detector is more sensitive and has more uniform and better spectral response as well as a higher quantum efficiency.

7.2.3 Line Scan Sensors

Line scan sensors, or linear arrays, are devices composed of a single line array of photosites, which may be of CCD or photodiode construction. As with matrix array photodiode arrangements, photodiode detectors have somewhat more sensitivity, more uniformity, and better spectral response with higher quantum efficiency.

Linear array sensors are useful wherever objects are in motion and where high resolution is important. Linear arrays with up to 12,000 photosites are available. One drawback of the linear array is the absence of a display output. This makes it necessary to employ an oscilloscope for setup purposes such as to see if the scene is in focus. There are cameras that include dynamic memory so "picture" data can be displayed when combined with digital/analog converters. The display used may limit the resolution of the image displayed. Where motion is involved, repeatability is important, and conveyor speeds must be well regulated.

7.2.4 TDI Cameras

TDI or time-delay and integration cameras consist of an interconnected set of CCD rows, referred to as stages. Essentially they are ganged line scan arrays, typically 2048 X 96 pixels. They are used in applications that might otherwise use line scan arrays and provide a signal-to-noise improvement that is theoretically the square root of the number of lines in the array.

After an initial exposure is taken at the first stage, and while the object is moving toward the second stage, all accumulated charges are transferred to corresponding cells in the second stage. Another exposure is now taken, but this time using the second row of CCDs to accumulate more charges. This process is repeated as many times as there are stages, which may be as many as 256.

The output signal is in a format equivalent to that produced by a line scan CCD. Synchronization between object motion and charge transfer is essential for proper TDI operation and to achieve the desired improvement in S/N.

7.2.5 *Special Solid-State Cameras*

Certain CCD cameras are available with special designs to capture image data at extraordinarily high rates, up to 12,000 pictures per second. Multiple outputs are used to partition the imager into blocks so that data can be read in parallel. If two outputs are used, the effective data rate increases by a factor of two. The more parallelism used, the less bandwidth required for each output.

7.2.6 *Performance Parameters*

Resolution. Regrettably, the term *resolution* reflects concepts that have evolved from different industries for different types of detectors and by researchers from different disciplines. The TV industry adopted the concept of TV lines; the solid-state imaging community adopted pixels as equivalent to photosites; and the photographic industry established the concept of line pairs, or cycles per millimeter:

1 cycle = 1 line pair = 2 TV lines = 2 pixels

One cycle is equivalent to one black and white (high-contrast) transition, or two pixels, and it represents the minimum sampling information required to resolve elemental areas of the scene image.

Another way to view resolution is as a measure of the sensor's ability to separate two closely spaced (high-contrast) objects. This should be distinguished from detectivity, the ability to detect a single object or an edge in space.

In cameras, resolution is usually specified as "limiting," or as the MTF (modulation transfer function) at a specific spatial frequency. Specification sheets that refer to resolution or response without specifying the MTF present limiting resolution. The MTF is a measure of the output amplitude as a function of a series of sine-wave-modulated inputs, a sine wave spatial frequency response. Spatial frequency refers to the number of cycles per unit length. The main usefulness of the MTF is that it permits the cascading of the effects of all components (optics as well as camera) of an imaging system in determining a measure of the overall system resolution. The system MTF is the product of all the component MTFs.

Among other things, this means that it is desirable to sample at least twice the resolution of the optics to avoid or minimize the aliasing or blurring effects.

Contrast transfer function (CTF) is another term frequently encountered when dealing with resolution. It represents the square-wave spatial frequency amplitude response. This is often used because it is easier to measure.

Limiting resolution represents the point on the MTF curve where the spatial frequency has fallen to 3% as measured with a bar pattern having 100% contrast, expressed as the number of TV lines (black and white) that can be resolved in a picture height. It represents the minimum spaced discernible black and white transition boundaries. Measurement is made at high light levels so noise is not a limitation. Another way to view limiting resolution is as the spatial frequency at which the MTF just about equals the sensor noise level. Limiting resolution is not very useful except for purely visual display considerations.

A pixel is not strictly a unit of resolution unless the sensor is a solid-state device or the output of a frame grabber; that is, a pixel is a discrete unit. Even in a solid-state array, however, resolution is not exactly equal to the photosite array because one experiences photosite "bleeding" and crosstalk of photons from one photosite to a neighboring photosite. This is a function of photosite spacing and the fabrication process, among other things. Pixel size, resolution, and the minimum detectable feature are not equal. The minimum detectable feature can be as small as two times the pixel size. However, besides the array properties, this is a function of process variables: optics, lighting, circuitry, and image-processing algorithms.

In the TV industry, the units of resolution are expressed as TV lines per picture height. This is a measure of the total number of black and white lines occurring in a dimension equal to the vertical height of the full field of view. The horizontal resolution generally refers to the number of TV lines equivalent to the length of that segment of a horizontal scanning line equal to the height of the vertical axis.

Aspect Ratio and Geometry. Generally, 3(V) X 4(H) aspects are standard, with 1 X 1 available. The standard format for 2/3 in. sensors is 0.26 (Vertical) X 0.35 (Horizontal) in., and for 1-in. sensors, it is 3/8 (V) X 1/2 (H) in. (input field of view).

Geometric Distortion and Linearity. This is not usually a problem with solid-state sensors. To reduce distortion, sensors with formats greater than 2/3 in. one should utilize 35-mm format lenses to minimize optical distortion (standard CCTV C-mount lenses are 16 mm format).

Detectivity. The ability to sense a single object, for example, an edge. In solid-state sensors an edge can be detected within a single pixel because edges typically fall on a number of contiguous pixels.

Quantum Efficiency (Figure 7.7). The number of photoelectrons per second per number of incident quanta at a specific wavelength per second. The quantum efficiency of an interline transfer imager is on the order of 35% given the fill factor, while that of a frame transfer imager is in the 70-80% range with its nearly 100% fill factor.

Responsivity. Output current for a given light level.

Dynamic Range. This parameter is a measure of the range of discrete signal levels (gray scale steps) that can be detected in a given scene with fixed illumination levels and no automatic light level or gain controls active in the sensor. Dynamic range is determined by the relationship between signal saturation and the minimum acceptable signal level or the level just above the system noise.

There are several means of defining and measuring dynamic range. For example,

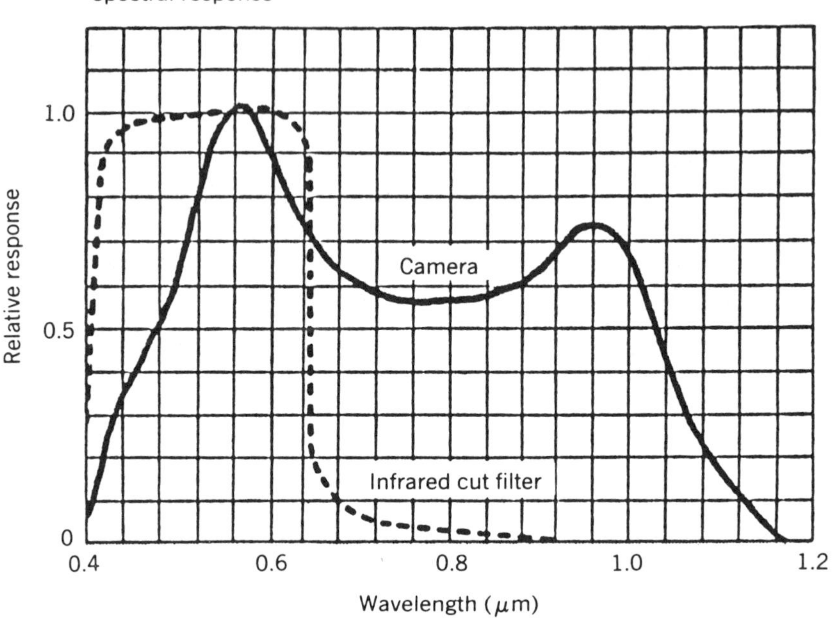

Figure 7.7 - Typical solid-state sensor spectral response.

$$R = [6(L - D)]/e,$$

where L is the peak light signal in the scan (above blanking), D is the black level (above blanking), and e is the peak-to-peak signal noise level.

Because of the nature of computers, 8 bits (256 gray shades) is typically used. Cameras typically used in machine vision do not contain a full 8 bits of true information because of noise introduced by system circuitry.

Sensitivity. This is a measure of the amount of light required to detect a change of the pixel value - one gray level increase. Sensitivity is typically determined by three factors: sensor quantum efficiency, circuit noise and light integration.

Fixed Pattern Noise. The sensitivity/noise variation from pixel-to-pixel that stems from photosite manufacturing variations and variations in the amplifiers that pick up column or row data. This "noise" shows up in the same place for every picture.

Gamma. Slope of signal output characteristic as a function of uniform faceplate illumination (plotted on a log-log plot). In general, the signal output may be expressed as S = KE, where E is the input illumination level. Most vidicons have a gamma of about 0.6-0.7, while silicon-based sensors usually are linear, and gamma is 1.

Saturation. Point at which a "knee" is observed in the log-log plot of the signal output versus faceplate illumination.

Signal-to-Noise Ratio. Ideally a function of photoconduction current resulting from the image on the photosensor:

(a) Limitation results from amplifier noise.
(b) Signal current is not only dependent on faceplate illuminance but also on the area of the active raster.

Shading or Signal Uniformity. Signal uniformity (constant level over the entire field of view) can be important for unsophisticated processing techniques (i.e., level slice or binary video). Solid-state sensors have pixel-to-pixel signal uniformity within about 10%. However, in some solid-state sensors one can experience "dead" pixels, or pixels with virtually no sensitivity. Cameras often substitute the average value of neighboring pixels for the dead pixel. Optics and lighting can also affect signal uniformity across the sensor and must be considered in any application.

Color Response. CCDs have no natural ability to distinguish or record color. Color cameras are available that are based on three chips or an integral color filter array. In the three-chip cameras, optics are used to split the scene onto three separate image planes. A CCD sensor and corresponding color filter is placed in each of the imaging planes. Color images can then be detected by synchronizing the outputs of the three CCDs, reducing the frame rate back to that of a single sensor camera.

In the cameras based on integral color filter arrays, filters are placed on the chip itself using dyed photoresist. While this approach reduces camera complexity, each pixel can be patterned only as one primary color. This reduces overall resolution and increases quantization artifacts. These filters also decrease the amount of light typically passing less than 50%.

Testing for spectral response usually involves equipment such as monochromators. However, when given color samples are available, relative response measurement can be made. In general, vidicons are blue-green sensitive, while solid-state sensors are red-green sensitive (with manufacturer-dependent exceptions).

Time Constants ("Lag, Stickiness, Smear"). In integrating sensors, scene motion can cause two basic types of distortion. [An *integrating sensor* is basically a parallel-input (i.e., photons, area) storing medium, with serial readout (the scanning beam or *XY* address in solid-state devices).] *Smear* is usually the result of image motion between readouts of a given location (i.e., motion taking place in 1/30 sec). The presence of lag or stickiness is evidenced by "tails" on bright moving objects and an impression of "multiple exposures" in moving scenes. Solid-state sensors generally require no more than one or two frames to stabilize a new image.

Dark Current. The current flow present when the sensor is receiving no light, it is a function of time and temperature (Figure 7.8). Dark current is not a factor if cameras are swept at 30 Hz, but it is a factor in operating cameras at a slow scan rate or at high temperatures. In the latter case, thermoelectric coolers may be required. Dark current is a strong function of temperature (doubling approximately every 8 degrees centigrade). The amount of dark current is directly proportional to the integration time and the storage time.

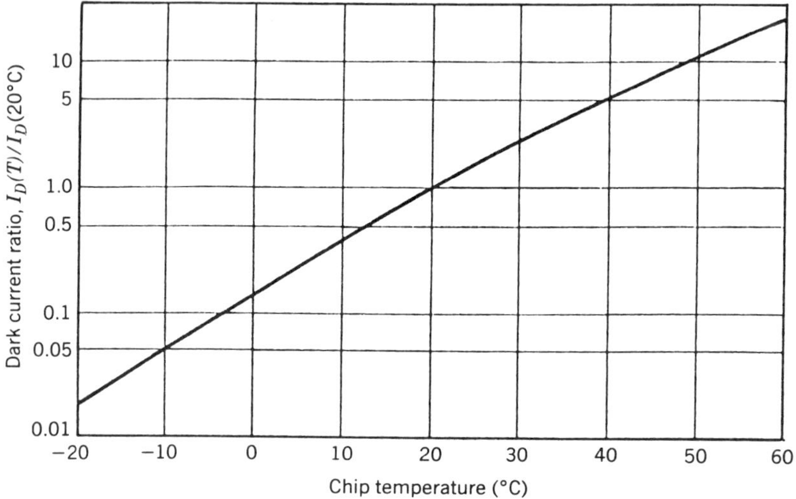

Figure 7.8 - Temperature versus dark current.

Blooming. This phenomenon is experienced when saturated pixels influence contiguous pixels, ultimately causing them to saturate and resulting in the defocusing of regions of the picture. Virtually all cameras today come with anti-blooming or overflow drain structures built into the imager itself. A side benefit of incorporating overflow drains is the ability to use that feature to implement electronic exposure or shutter control.

Aliasing. In solid-state cameras, aliasing is experienced due to the fact that the image is formed by an array of picture elements rather than a continuous surface. Consequently, there are discontinuities between picture elements where light is not detected. This becomes more noticeable when viewing scenes with lots of edges.

Crosstalk. This is a phenomenon experienced in solid-state matrix array cameras, especially those operating in an interline transfer arrangement where half of the pixel integrates with the remaining half used as a storage site. It is due to signal electrons generated by long-wavelength (>0.7 micron) photons. The electrons tend to migrate to undesired storage sites, causing degraded resolution.

7.3 CAMERA AND SYSTEM INTERFACE

Having obtained an electronic analog signal (Figure 7.9) corresponding to the image input, the next step is for the image processor to take the video signal (possibly massage it at this point with various analog preprocessing circuitry) and convert it to a stream of digital values. This process is called digitizing, digitalizing, or sometimes sampling/quantizing (Figure 7.10).

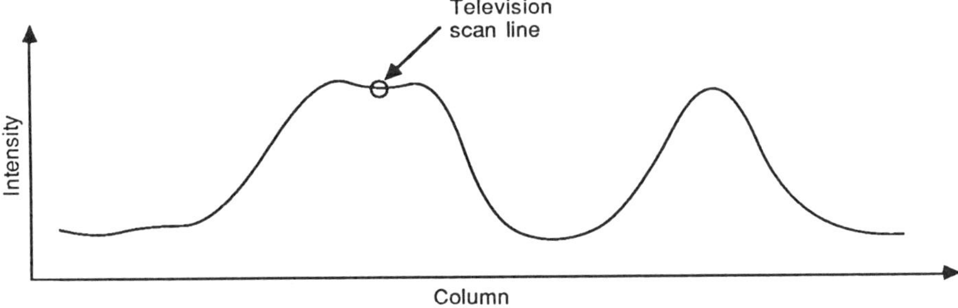

Figure 7.9 - Typical analog TV scan line.

7.3.1 A/D Converter

The output signal from the sensor consists of individual voltage samples from each photosite element and the special signals to tell what voltage sample corresponds to which element. This information is placed end-to-end to create the analog electrical signal (Fig. 7.9).

A digital computer does not work with analog electrical signals. It needs a separate number (electrically coded) for each intensity value of each element, along with a method of knowing which intensity value corresponds to what sensor element.

This transformation from an analog signal to an ordered set of discrete intensity values is the job of the A/D converter. Although an analog signal can have any value, the digital value a machine vision system uses can only have an integer value from 1 to a fixed number, N. Typical values of N are 2, 8, 64, 256. In computer terminology, this corresponds to storage areas of 1 bit, 3 bits, 6 bits, and 8 bits.

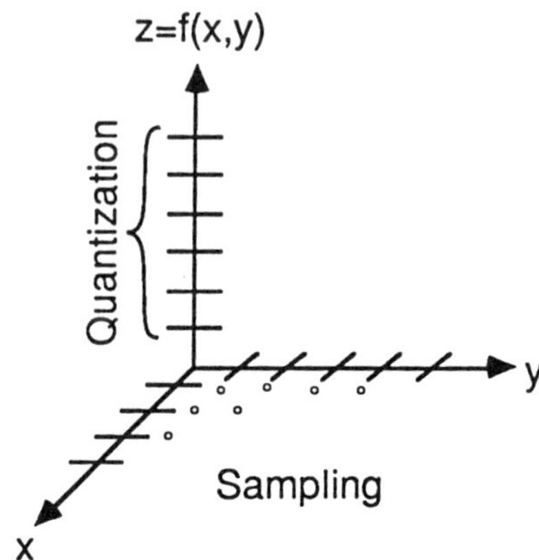

Figure 7.10 - Digitized picture representation in "3D".

Figure 7.11 shows an analog signal being digitized into a 2-bit storage area. Each output digital value is the closest allowable value (0, 1, 2, 3) to the analog signal. Some changes in the analog signal are NOT present in the digitized signal because the changes are smaller than half the voltage difference between the allowed digital values.

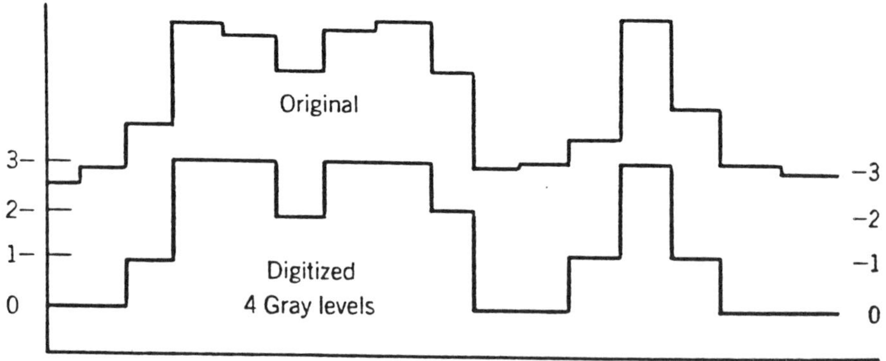

Figure 7.11 - Digitizing analog signal into 2 bits.

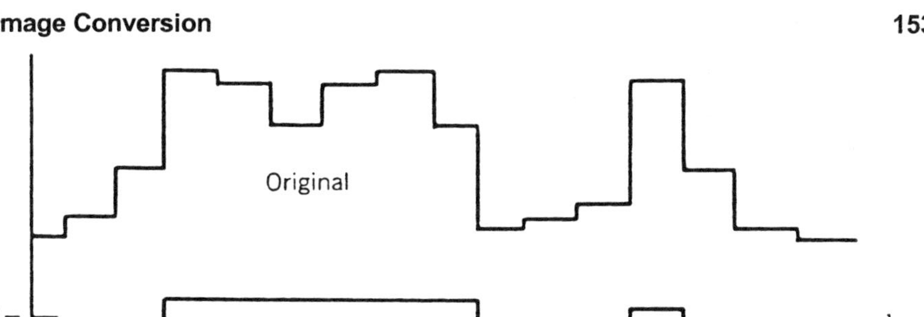

Figure 7.12 - Digitizing analog signal into 1 bit.

Figure 7.12 shows the same analog signal being digitized into a 1-bit storage area. The output digital values can only have values of 0 or 1. Even less of the analog signal features are present than in Figure 4. From these two figures, several important concepts exist.

The entire process of converting analog signals to digital values is called *digitization.* The number of possible digital values is important. More values in the digital signal having more information about the analog signal and, therefore, the original image. As these intensity values range from black, the lowest value, to white, the highest value, they are called gray level values.

The number of possible gray level values a system has is stated in one of two ways:

1 "This system has N gray values or levels,"

2· "This system has 8 bits of gray," where N = 2

The conversion of N values to B bits is given below. A system with 2 gray levels (1 bit of gray) is called a binary system.

NUMBER OF BITS:

1 2 3 4 5 6 7 8

NUMBER OF GRAY LEVELS:

2 4 8 16 32 64 128 256

The actual gray value is a function of the integration of four variables: illumination, viewpoint, surface reflectance and surface orientation. The surface reflectance of an object is determined by such surface characteristics as texture and color material. The resulting distribution of light intensities forms an image.

7.3.2 Digitization

What follows is a description of the digitization process taken from a paper entitled "Understanding How Images Are Digitized" given by Stanley Lapidus at the Vision 85 Conference.

The sampling process is illustrated in Figure 7.13. The top graph represents the cross section of a cut through a three-dimensional illumination surface, which could be taken as a representation of an optical image. In such a three-dimensional illumination, surfaces x and y are the coordinate axes of the plane in which an object is viewed, and z is the intensity of the light falling on the object. Here z = 0 represents black, or total darkness, and z very large represents a strong light intensity. Since the top graph represents a cross section of such an *x, y,* z surface, the graph's vertical axis is z, the light intensity, and the horizontal axis is the x-wise extent of the three-dimensional illumination surface. The middle graph represents the sampling points in the x direction of the image plane. The bottom graph represents sampled, discrete, gray level values that correspond to the light intensity values of the top graph.

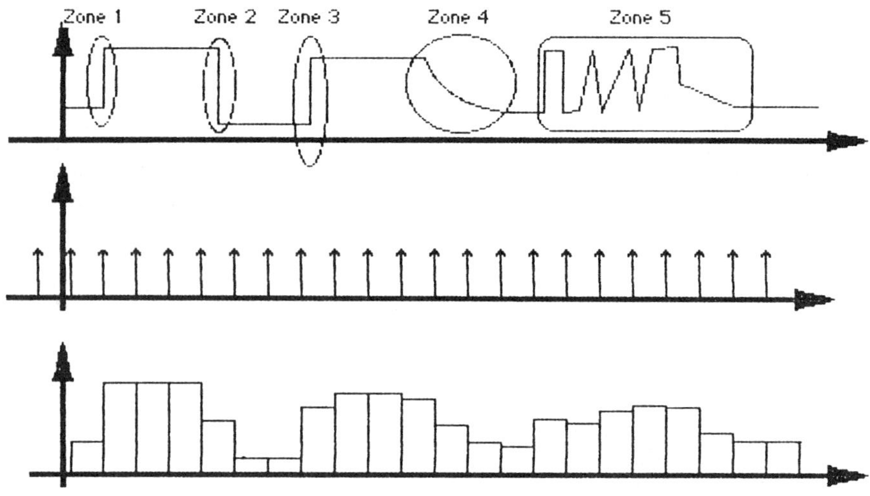

Figure 7.13 - Sampling process.

Looking at the top graph, some areas are white, some are black, and some are of gradually decreasing gray shades (remember, this is an illustration of an intensity profile; the result of a part interacting with a lighting environment).

Looking at the figure:

The first zone is exactly along a pixel boundary.

The second zone is halfway between two pixels.

The third zone is offset.

The fourth zone is gradually sloping.

The fifth zone shows aliasing.

Each of the five zones illustrates a different phenomenon encountered in the sampling process.

Zone 1. In the leftmost transition, the pixel boundary occurs exactly on the transition. The pixel immediately to the left of the transition is dark; the abrupt pixel to the right of the transition is light. The change from dark to light occurs at a single pixel boundary.

Zone 2. The transition occurs exactly in the middle of a pixel. This pixel has a gray value that is halfway between the value of its neighbor to the left and the value of its neighbor to the right. This is because half of the pixel area is black and half of the pixel area is white, averaging to middle gray.

Zone 3. The transition occurs about one-quarter of the way over. As a result, the area bounded by the pixel is mostly dark, and a dark gray value results for this pixel.

Zone 4. In real-world machine vision systems, edges are rarely as abrupt as in the first three cases. Real edges go from one gray value to another over an area spanned by a number of pixels. This is due to physical limitations of the camera, lens, and digitizing element in the front end. This case is illustrated in zone 4. The transition is transformed to a staircase where each step is a measure of the average intensity in the pixel. Note that in the original image, the step *is not* a smooth function. Real-world edges have glare, shadow, and other anomalies that keep edges from being sloping straight lines.

Zone 5. This zone illustrates the problem of aliasing. Aliasing occurs when the grid of pixels is too coarse for the intricacy with which gray scale transitions occur in the image. This causes some dark-to-light transitions to be swallowed up. For reliable imaging, care must be taken to prevent aliasing. A similar problem was first encountered in radar, satellite, and digital telephone technologies a few decades ago. Scientists have developed some powerful techniques and tools to address this problem with sampled data. A particularly famous and useful tool is the so-called sampling theorem.

In gray scale systems we see how the positioning of a transition or edge strictly inside of a pixel causes the pixel to take on an intermediate gray value that falls between the values of the pixels on either side of the pixel in which the transition occurs.

For binary systems, a pixel that contains a strictly internal transition will be turned into a black or white pixel depending on the average intensity within the pixel. If the average intensity is greater than some threshold, the pixel will go white; if the intensity is less, it will go black. This means that changes in the threshold or the light intensity will cause a black zone to get wider or narrower (change its size) if edges are not abrupt and do not occur over a number of pixels.

Establishing the correct threshold is shown in Figure 7.14. Figure 7.14a shows the effect of a setting that is too low so that more pixels are assigned to white than should be. Figure 7.14b shows the effect of a threshold set too high so too many pixels are assigned to black. In both cases information is lost. Figure 7.14c reflects the properties one obtains with an appropriate setting.

One must be careful not to allow thresholds to vary in real applications, but preventing variations in light or other variables that can affect the results of a fixed threshold for real applications is difficult. Consequently, in many applications the ideal is an adaptive threshold, one that bases the setting on the results of the immediate scene itself or the next most immediate scene.

One such technique involves computing a binary threshold appropriate for a scene by analyzing a histogram or plot of the frequency of each gray shade is experienced in the scene. Another approach, computing a threshold on a pixel-by-pixel basis using the gray level data in a small region surrounding the pixel, is a local adaptive thresholding tactic.

There is a relationship between the signal-to-noise ratio *(SIN)* of an analog signal and the required number of gray levels:

S/N (peak to rms) = 6.02R + 10.79 dB

where R is the resolution of the A/D converter. That is, 6 bits = 47 dB and 8 bits = 59 dB. Increased quantizing resolution generally improves performance because the digital form more accurately represents the analog signal. Gamma or

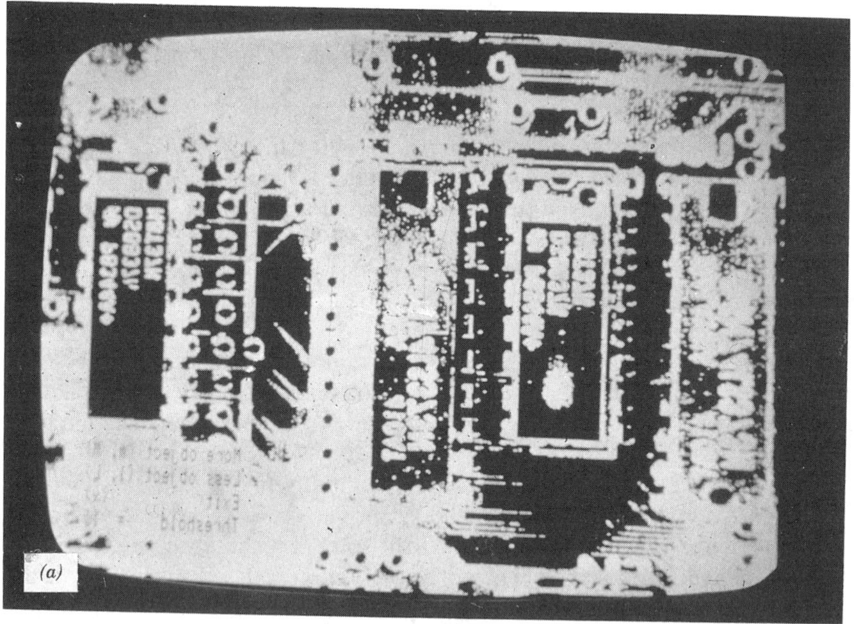

Figure 7.14 - Setting a threshold (courtesy of Cognex).

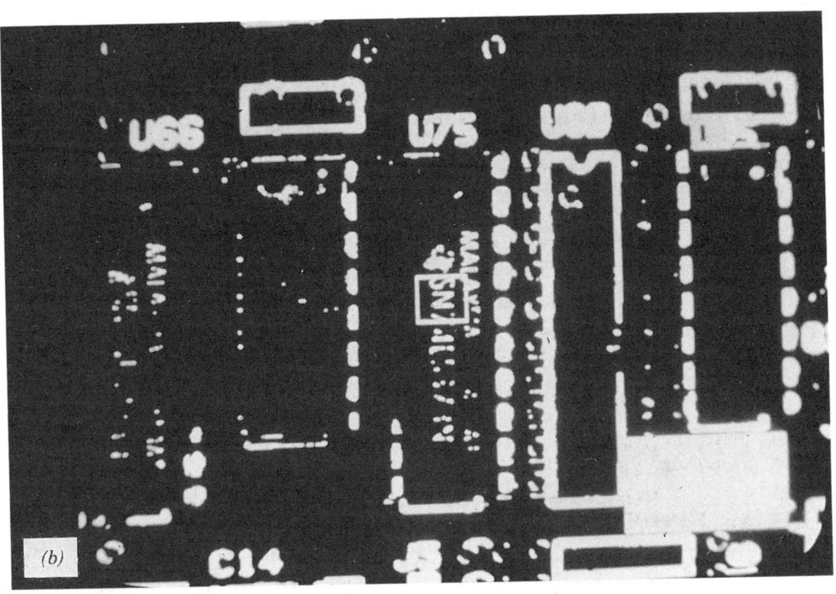
U66
U75
C14

(b)

U66
U75
MALAYSIA
SN74LS373N
410AT
P8344A+
DS8837N
N8T37N
C14

(c)

linearity properties may introduce quantizing error that may reduce the effective number of bits. Noise is also a function of spatial resolution inasmuch as greater bandwidth produces greater noise. Hence, achievable gray level resolution decreases as spatial resolution requirements increase.

As far as machine vision applications are concerned, as speeds of pixel processing increase, filterage of noise from incoming signal becomes more important.

It is not clear in machine vision applications whether 8 bits is better than 6 bits. Solid-state cameras today only offer *S/N* on the order of 45-50 dB. On the other hand, cameras are improving, and since, for the most part, 8 bits versus 6 bits does not impact significantly on the speed, the 8-bit and even 10-bit capacity may be an advantage.

7.3.3 What Is a Pixel?

The word, pixel, is an acronym for picture element. Any one of the discrete values coming out of the A/D is a pixel. This is the smallest distinguishable area in an image. An abbreviation for pixel is the pel, a second order acronym.

As the output of A/D is the input to the computer, the computer is initially working with pixels. These pixels can be thought of as existing in the computer in the same geometry as the array of the elements in the sensor.

7.4 FRAME BUFFERS AND FRAME GRABBERS

Having digitized the image, some systems will then perform image-processing operations in real time with a series of hardwired modules. Others store the digitized image in random-access memory (RAM). Boards exist that combine these functions (digitizing and storage) and are called frame grabbers (Figure 7.15).

In those systems that do not use frame buffers, many capitalize on data compression techniques to avoid the requirement for the large amount of RAM. One approach, "image following," saves the coordinates and values of nonwhite pixels. Run length encoding techniques may be used. Vector generation is another binary storage technique that records the first and last coordinates of straight-line segments.

In frame grabber systems, the image frame can be considered as a bit-mapped image storage with each pixel in the image frame represented by an N-bit word indicating the light intensity level. The size of the image frame is set by the $n \times m$ pixel matrix. As resolution of the image increases by a factor of 2, the size of the buffer memory increases by 4.

Some frame grabbers have some capabilities to perform simple image processing routines using an arithmetic logic unit and lookup table in the data path between the A/D converter and two or more frame buffer memories. Logical operations could include frame addition to average several frames, frame subtraction

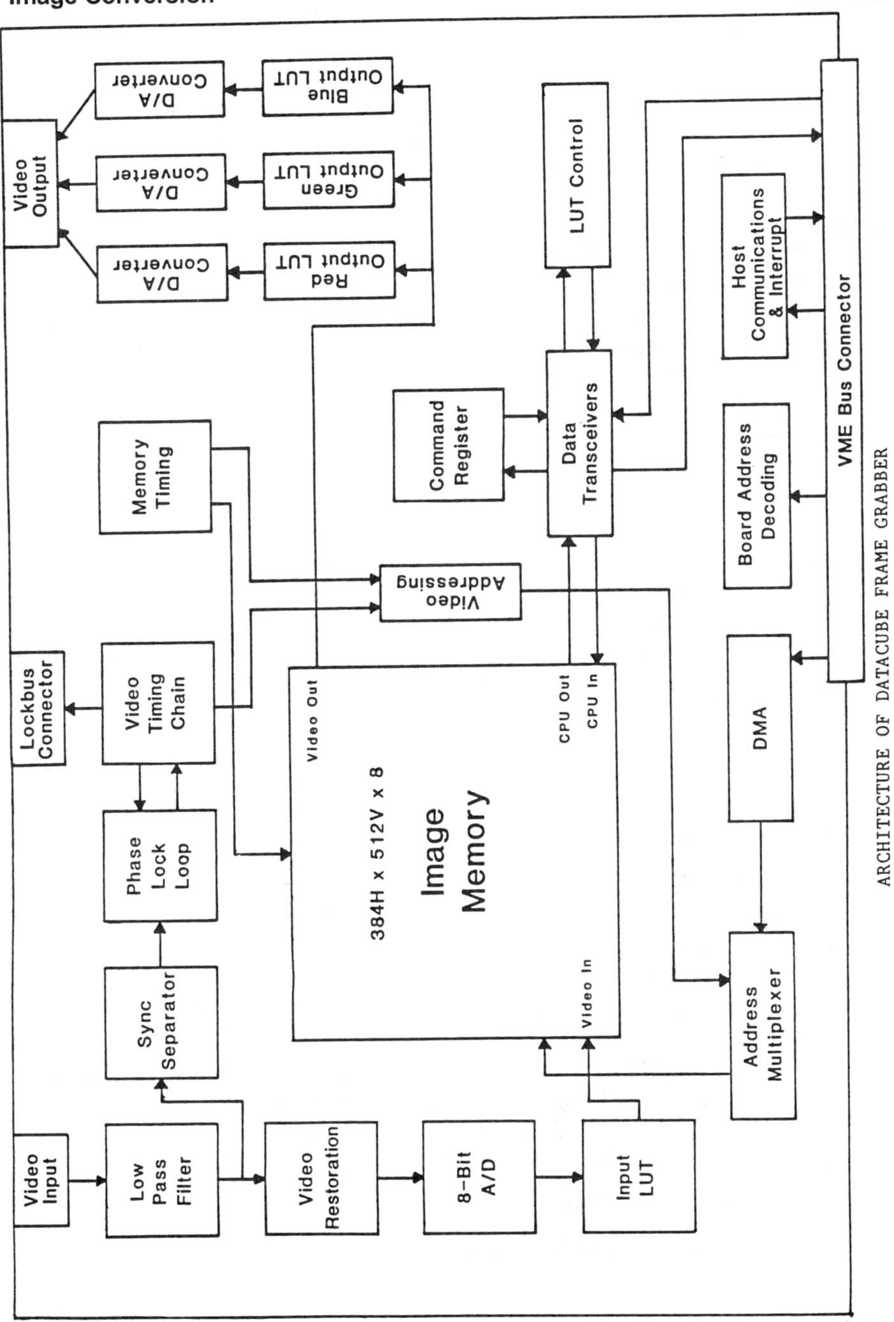

Figure 7.15 - Functions on typical frame grabber board (courtesy of Datacube).

and remapping of pixels, or transforming the gray level of every pixel into a new gray level regardless of the gray values of the other pixels in the image.

Frame addition can correct for low light levels and improve the signal-to-noise ratio of an image. Subtraction can eliminate background data to reduce the amount of data required for processing. Remapping can be used for contrast enhancement or segmentation into a binary picture.

The frame grabber's "back end" usually has the capacity to accept pixel data from the image memory and convert the digital signal to an analog signal. Synchronizing information is also incorporated, resulting in an RS-170 video signal that can be fed to a standard video monitor for display of the process image.

Many frame grabbers are designed to plug the data right into a personal computer. In these cases the output interface constraint is addressed by adding buffer memory on the frame grabber to store data during PC bus interruptions.

A major consideration is the camera interface. Accurate synchronization to the camera's fast pixel clock and low pixel jitter are important parameters in most machine vision applications. Some frame grabbers have the capability to handle asynchronously acquired image data. This feature is appropriate for applications that involve high speed or the acquisition of inputs from multiple cameras. Some frame grabbers are also designed to handle analog video data from non-standard sources such as line scan cameras or TDI cameras or area cameras with progressive, variable scan rate and multiple tap outputs.

In other words, not all frame grabbers are equal. The specific application will dictate which frame grabber design is most likely to lead to a successful deployment.

7.5 DIGITAL CAMERAS

Cameras are available that operate in non-standard format and deliver 8-bit (or higher) digital data. This data can be input directly into a computer thereby eliminating a frame grabber or may still feed a frame grabber with the capability of handling a digital input. The digital cameras typically have higher resolution than analog cameras and improved accuracy of image information.

A digital camera skips the preprocessing steps inherent in an analog camera, and it adds no timing information to the video signal. Instead it uses an internal analog-to-digital converter (ADC) to digitize the raw analog signal from each pixel on the imager. The camera then outputs the digitized value for each pixel in a parallel digital form. Accuracy improves since a digital camera is much less sensitive to surrounding electrical noise.

A digital camera operates in a progressive scan mode scanning a complete image. Exposure and scanning are typically under computer control.

7.6 SMART CAMERAS

These are cameras with embedded intelligence. In effect they are camera-based, self-contained general-purpose machine vision systems. A smart camera consists of a lens mount, CCD or CMOS imager, integrated image and program memory, an embedded processor, a serial interface and digital I/O (input/output). As microprocessors improve, smart cameras will only get smarter. Typically an integrated Windows^{TM}-based software environment is available for designing specific application solutions. Generally these cameras can be connected to a local area network.

7.7. Sensor Alternatives

Besides capturing images based on sensors that handle the human visual part of the electromagnetic spectrum, it is possible to use sensors that can capture image data in the ultraviolet, infrared or X-ray region of the spectrum. Such sensors would be substitutes for conventional imagers. Often the sensors embody the same principles but have been "tampered with" to make them sensitive to the other spectral regions. Figure 7.16 depicts an X-ray based machine vision system to automatically inspect loose metal chips, granules or powder materials for foreign objects.

Figure 7.16 - Guardian system from Yxlon uses X-ray based machine vision techniques to sort foreign material.

REFERENCES

"Inspection Vision Systems Getting A Lot Smarter," *MAN*, August, 1998.

Beane, Mike, "Selecting a Frame Grabber to Match System Requirements," *Evaluation Engineering*, May, 1998.

Bloom, L., "Interfacing High Resolution Solid State Cameras to Digital Imaging Systems," *Digital Design,* March 25, 1986.

Boriero, Pierantonio and Rochon, Robert, "Match Camera Triggering to Your Application," *Test & Measurement World*, August, 1998.

Chocheles, E. H., "Increased A/D Resolution Improves Image Processing," *Electronic Products, October* 15, 1984.

Fossum, Eric, "Active-pixel sensors challenge CCDs," *Laser Focus World*, June, 1993.

Harold, P., "Solid State Area-Scan Image Sensors vie for Machine Vision Applications," *EDN*, May 15, 1986.

Hershberg, I., "Advances in High Resolution Imagers," *Electronic Imaging,* April 1985.

Higgins, Thomas V., "The technology of image capture," Laser Focus World, December, 1994.

Hori, T., "Integrating Linear and Area Arrays with Vision Systems," *Digital Design, March* 25, 1986.

Jacob, Gerald, "A Look at Video Cameras for Inspection," *Evaluation Engineering*, May, 1996.

Lake, Don, "Beyond Camera Specmanship: Real Sensitivity & Dynamic Range," *Advanced Imaging,* May, 1996.

Lake, Don, "Solid State Color Cameras: Tradeoffs and Costs Now," *Advanced Imaging*, April, 1996.

Lapidus, S. N., "Gray-Scale and Jumping Spiders," SME/MVA Vision 85 Conference.

Lapidus, S. N., "Understanding How Images are Digitized," SME/MVA Vision 85 Conference.

MacDonald, J. A., "Solid State Imagers Challenge TV Camera Tubes," *Information Display,* May 1985.

Meisenzahl, Eric, "Charge-Coupled Device Image Sensors," *Sensors*, January, 1998.

Pinson, L. J., "Robot Vision: An Evaluation of Imaging Sensors," The International Society for Optical Engineering (SPIE) Robotics and Robot Sensing Conference, August 1983.

Poon, Steven S. and Hunter, David B., "Electronic Cameras to Meet the Needs of Microscopy Specialists," *Advanced Imaging*, July, 1994.

Rutledge, G. J., "An Introduction to Gray Scale Vision Machine Vision," SME/ MVA Vision 85 Conference.

Sach, F., “Sensors and Cameras for Machine Vision,” *Laser Focus/Electro-Optics,* July 1985.

Silver, W. M., “True Gray Level Processing Provides Superior Performance in Practical Machine Vision Systems,” Electronic Imaging Conference, 1984, Morgan Grampian.

Stern, J., “CCD Imaging,” *Photo Methods,* May 1986.

Titus, Jon, "Digital Cameras Expand Resolution and Accuracy," *Test & Measurement World*, June, 1998.

Visual Information Institute, “Structure of the Television Roster,” Publication No. 012-0384, Visual Information Institute, Xenia, OH.

Wheeler, Michael D., "Machine Vision The Next Frontier: Network Cameras, *Photonics Spectra*, February, 1998.

Wilson, A., “Solid State Camera Design and Application,” *Electronic Imaging,* April 1984.

Wright, Maury, "Digital Camera Interfaces Lead to Ubiquitous Deployment, *EDN*, January 15, 1998.

Yamagata, K., et al, “Miniature CCD Cameras: A New Technology for Machine Vision,” *Robotics Age,* March 1985.

8

Image Processing and Decision-Making

8.1 IMAGE PROCESSING

Image processing may occur in either the hardware or software. Image processing hardware makes sense when large numbers of images are to be processed repetitively by the same set of algorithms. Hardware implementation is faster than software execution but with less flexibility. Most systems perform some image-processing operations in hardware and some in software.

Most of today's machine vision systems manipulate images in the spatial domain. An alternative, mentioned only in passing, is to operate in the temporal domain, specifically the Fourier transform of the image. When applied to an image, this transform extracts the amplitude and phase of each of the frequency components of the image. Significantly the phase spectra contains data about edge positions in an image. The reason Fourier transforms are not generally used in machine vision is because of the large computational requirements. Advances in array processors and system architectures may change this in the near future.

Image processing is typically considered to consist of four parts (Figure 8.1).

Enhancement/Preprocessing

Operations using the original image to create other images, finally resulting in an image(s) that contains only the desired information.

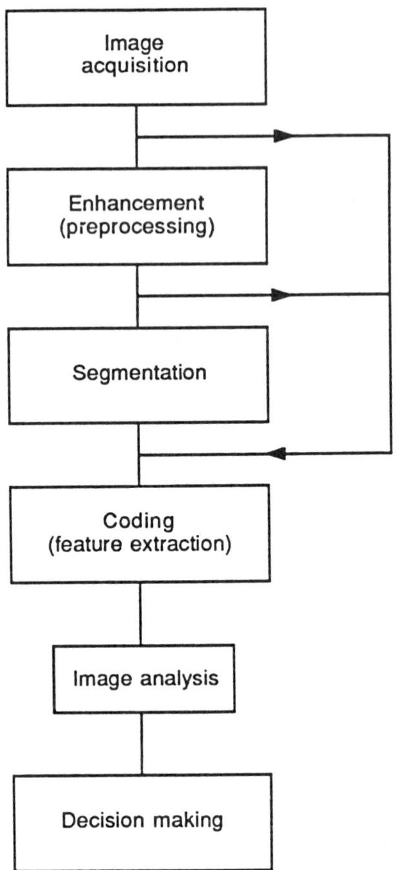

Figure 8.1 - Block diagram depicting process steps in machine vision.

Segmentation

Process of separating objects of interest (each with uniform attributes) from the rest of the scene or background, partitioning an image into various clusters.

Coding/Feature Extraction

Operations that extract feature information from the enhanced and/or segmented image(s). At this point, the images are no longer used and may be deleted.

Image Analysis/Classification/Interpretation

These are operations that use the extracted feature information and compare the results with known standards. This step answers the question what the system was purchased for and outputs the results to the appropriate device(s).

8.2 IMAGE ENHANCEMENT/PREPROCESSING

Enhancement techniques transform an image into a "better" image, or one more suitable for subsequent processing to assure repeatable and reliable decisions. There are three fundamental enhancement procedures: pixel or point transformations, image or global transformations, neighborhood transformations.

8.2.1 Pixel Transformations

There are a number of image enhancement routines that can be applied to improve the content of the image data before coding and analysis. Contrast and brightness enhancements alter an image's gray scale (Figure 8.2). Single pixel operators transform an image, pixel by pixel, based on one-to-one transformations of each pixel's gray level value. Some such operations include:

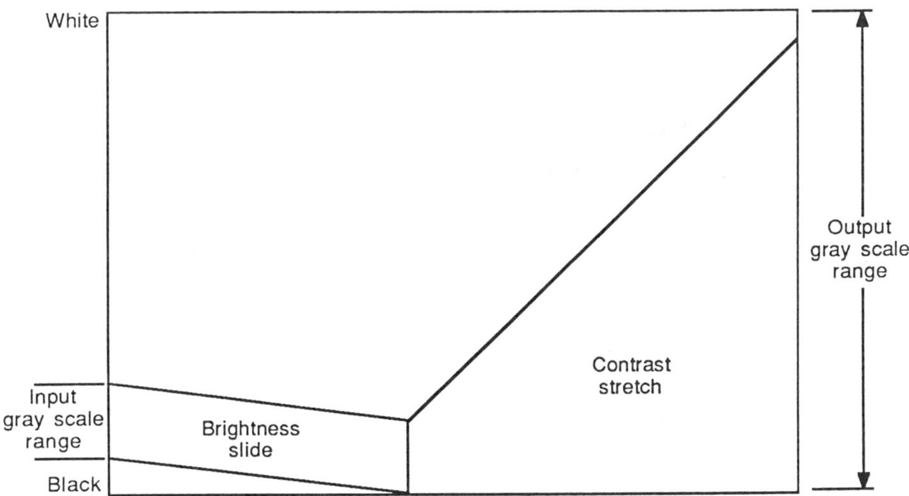

Figure 8.2 - Brightness sliding/contrast stretching.

Scaling. Multiplying each pixel by a constant. Scaling is used to stretch the contrast, normalize several images with different intensities, and equalize an image's gray scale occupancy to span the entire range of the available gray scale.

Addition or subtraction of a constant to each pixel. Brightness sliding involves the addition or subtraction of a constant brightness to all pixels in the image. This shifts the entire frequency distribution of gray level values. This technique can be used to establish a black level by finding the minimum gray scale value in the image, and then subtracting it.

Inverting. Replacing each pixel by its gray value complement. White becomes black and vice versa, and grays invert light to dark.

8.2.2 Global Transformations

In this case, an operation is performed on an entire image. Geometric scaling, translation, rotation, and perspective manipulations of images are typically used prior to combination or enhancement operations. Often these operations are used in order to bring two images of similar content into a register. Patching of several images into one composite may be facilitated by these operations, analogous to cutting and pasting several prints together into one.

Geometric operations based upon image arrays are usually handled by software routines similar to those used by computer graphics packages. Of major importance to images, though, is the concept of spatial interpolation. For instance, when an image is rotated, its square input pixel grid locations will generally not fall onto square grid locations in the output image. The actual brightness in the output pixel locations must be interpolated from the spatial locations in which the rotated pixels were calculated to land. Differing interpolation schemes can produce rotated images of varying quality.

Other global transformations include:

Smoothing in time. Performed by averaging many frames of image data. This tactic can be used for the suppression of noise within an image.

Subtraction. Performed by subtracting one image from another. Can be used to detect changes that have occurred during the time period between the capture of the two frames. This technique is used to remove the effects of a busy background, or to detect detail associated with a moving part or to remove scene objects contained in one image and not in another.

Multiplication. The multiplication of two image matrixes can be used to correct for constant geometric distortions as might be due to the optical design, for example. In a similar manner, one could normalize the nonuniformity in sensor sensitivity. Images with different X and Y scale factors can be corrected or normalized using these tactics.

8.2.3 Neighborhood Transformations

These take two forms, binary and gray scale processing. In both cases, operators transform an image by replacing each pixel with a value generated by looking at pixels in that pixel's neighborhood.

Binary Neighborhood Processing. A binary image is transformed, pixel by pixel, into another binary image. At each pixel, the new value is generated by the old value of each pixel in the neighborhood. This can be thought of as a square matrix passing over the old image. At each pixel, all values inside the matrix are combined, giving one new value. The matrix moves to the next pixel, and the process repeats until the new image is generated. The following examples are demonstrated for the 3 X 3 case:

Dilation (growing). The pixel in the output image is white if any of its eight closest neighbors is white in the input image. This has the effect of growing the white regions out in all directions, and closing interior holes.

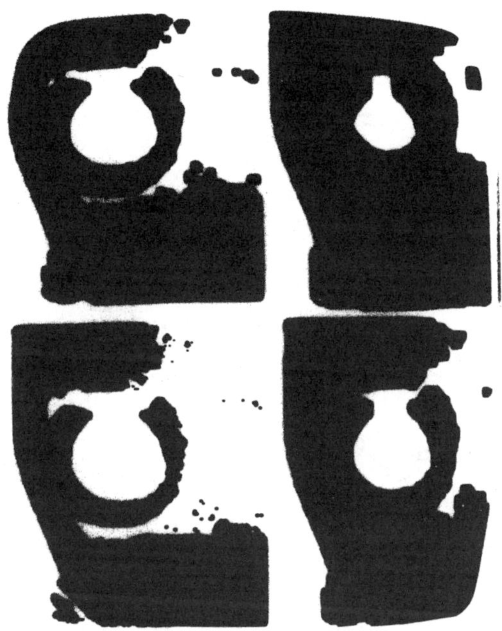

Figure 8.3 - A few properties of dilation.

A few properties of dilation can be seen in Figure 8.3. It takes away single, dark pixels within the image, but it also destroys narrow lines and fine detail. In this respect, it resembles an averaging or defocusing process.

A single dilation by itself is not a very significant step. Several steps of this type, however, can achieve quite a bit. A simple example is an application for checking that two holes are a minimum distance apart. A good approach would be to create binary image of the two holes, dilate each hole by ½ minimum distance, and check to see if the two holes are connected. This will work for any orientation or location of the holes in the image. An interesting note is that dilating a binary image in all directions, then exclusive OR'ing it with the original image, gives a nice binary silhouette of the white areas of the image.

Erosion (shrinking). The pixel in the output image is white if ALL of its eight closest neighbors are white in the input age. Erosion can be used to check things like minimum width of a given feature. To do this, erode the feature the proper number of times, and if it goes away, it was too narrow. Also, erosion is good for removing stray white pixels in a black area (see Fig 8.4a-b).

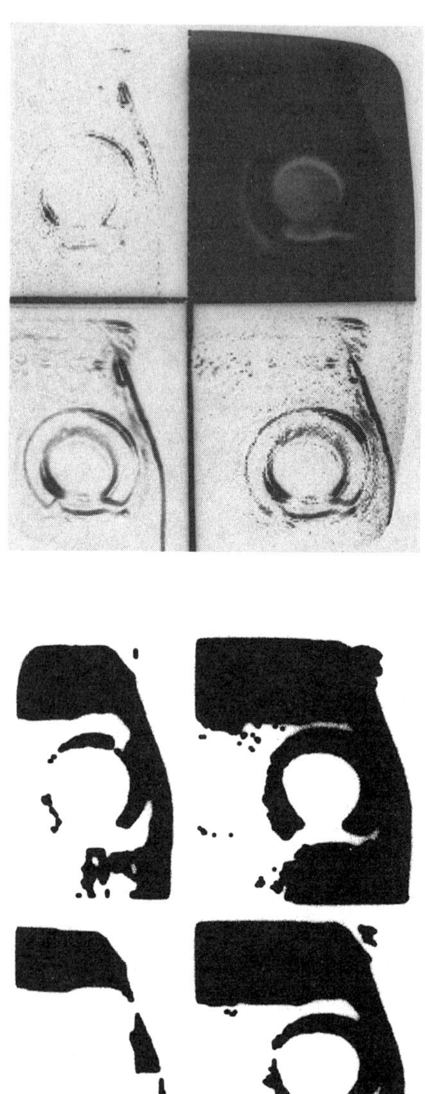

Figure 8.4 a-b - Examples of results of erosion.

Binary neighborhood operators are not commutative; dilation, then erosion, will not equal erosion, then dilation. Once erosion is used to remove a single black pixel surrounded by white, dilation will not replace it. There is no "seed' pixel for dilation to grow from.

Single Point Removal. If all eight neighboring pixels are of one color, make the middle pixel that color. If a * b * c * d * e* f* g * h * i = 1, then e (new) = e (old), to remove black pixels; if a + b + c + d + e + f + g + h + i = 0, then e (new) = 0, else e (new) = e (old), to remove white pixels. This has the effect of eliminating single white pixels in a field of black, and vice versa. The ability to do this is especially important in neighborhood operators, where one single pixel may grow into a large area if dilations are used.

Skeletonization. A combination of several binary neighborhood operators that reduce all white areas in the scene to single-pixel wide "skeletons." This is useful for dealing with edges and silhouettes, and for finding centroids and axes of shapes. Fine detail along edges is lost.

Gray Scale Neighboring Processing. A neighborhood is passed over the input image, except that the image is gray scale, the coefficients assigned to the neighborhood are real numbers, and the function is generally arithmetic.

8.2.4 Spatial Filters

Spatial filtering operations create an output image based on the spatial frequency content of the input image. In view of the computational intensity of temporal domain processing such as Fourier transforms, spatial filtering is typically performed as follows: on a pixel-by-pixel basis, an output image is generated based on a pixel's brightness relative to its immediate neighboring pixels. Where a neighborhood's pixel brightness makes rapid transitions from light to dark or vice versa, the image may be said to contain high frequency components. A neighborhood of slowly varying pixel brightness represents low-frequency components.

Spatial filtering is carried out using spatial convolutions. For each input pixel within an image, an output pixel based on a weighted average of it and its surrounding neighbors is calculated. Typically, a three pixel by three pixel neighborhood is used for this calculation, although larger neighborhoods may be used for added flexibility.

In one nearest neighbor operation involving nine contiguous pixels carrying out a spatial convolution, nine weighting coefficients are defined and labeled A-I (Figure 8.5). The brightness of the input pixel being evaluated and its eight immediate neighbors are each multiplied by their respective weighting coefficients. These products are summed, producing a new, spatially filtered output pixel brightness. This operation is applied systematically to each pixel in the input image, resulting in a spatially filtered output image.

Edge Detection

For pixel E, the Sobel Value is computed as:

$$GRAD\ X = (A+2B+C) - (G+2H+1)$$

$$GRAD\ Y = (C+2F+1) - (A+2D+G)$$

And the Sobel Value is:

$$Sobel = \sqrt{GRAD\ X^2 + GRAD\ Y^2}$$

A	B	C
D	E	F
G	H	I

Figure 8.5 - Spatial convolution/neighborhood operation.

Figure 8.6 - Result of edge enhancing filtering operations; boundary image (courtesy of General Scanning/SVS)

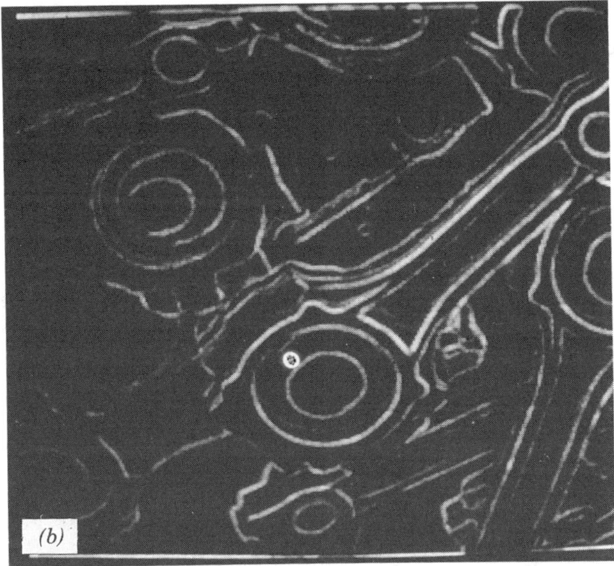

Figure 8.6 – Continued.

A spatial filter attenuates or accentuates the two-dimensional frequency content of an image. These operations may be used to bring out an image's high-frequency details, yielding a sharper image. High pass filters have little effect on low frequency information - areas in the image of fairly uniform brightness - but accentuate the high frequency data, such as edges.

Alternatively, the high-frequency details may be attenuated, yielding a low-pass image of little detail. This has a smoothing or blurring effect. Of great importance to machine vision systems is the ability to bring out the edge detail in an image. Edge enhancers [Figures 8.6(a) and 8.6(b)] not only accentuate high-frequency data, but unlike high-pass filters that leave the low-frequency image data unchanged, they also eliminate it. Edges and other high-frequency data including noise are highlighted. Edge enhancement can allow the image-processing system to make edge-to-edge boundary distance measurements.

A widely used image enhancement operator is the median filter. The median filter operator smoothes the noise in an image and tends not to cause blur because it replaces each pixel in the image with the middle value of the pixel's local neighborhood. A mode filter uses the most frequently occurring pixel value rather than the median. One variant of this filter is the rank value filter. This operator counts each pixel value within a window using the corresponding weight specified in a mask kernel. It then sorts the values from least to greatest and selects the value at the specified rank or index.

As noted above, in some systems image enhancements are Boolean logical combination of images to insert portions of one image into another. Image combination of two images is done pixel by pixel over the entire image space; corresponding spatially located pixels from each input image are combined to form a composite output pixel at the same spatial location in the output image. These techniques are associated with image algebra and mathematical morphology.

8.3 SEGMENTATION

A scene can be segmented by windows, regions or boundaries.

8.3.1 Windows

Windows are established to isolate only those areas in a scene with the attributes of interest, a hole for example (Figure 8.7). Only those pixels in the windows are processed, reducing the total number of pixels processed in a frame to a more manageable number, making it possible to handle more vision/decisions per unit time. The pixels in the windows can be processed in the same way the entire scene might have been processed: representations established or features extracted.

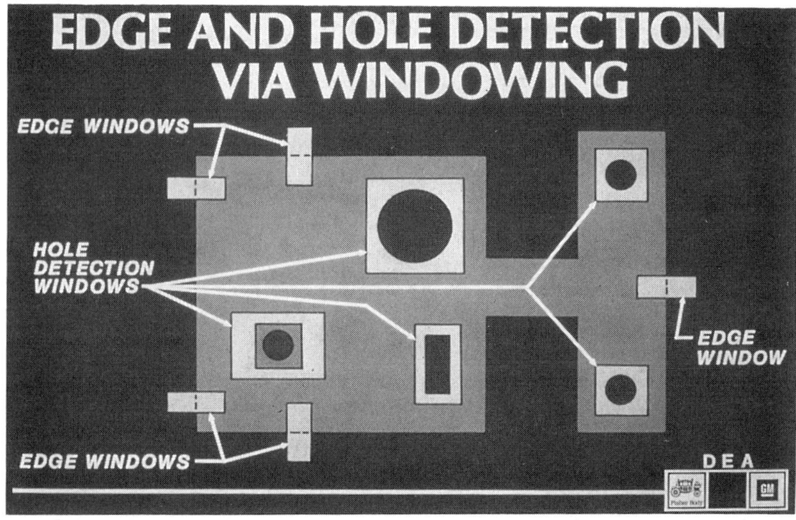

Figure 8.7 - Sample of the application of "windows" over specific features of interest.

Versions of these systems come with fixed or adaptive windows. In the case of fixed windows, the parts have to be repeatedly positioned, as the windows will always be set up in the exact same locations in two-dimensional space as when they were set up during training. Systems with adaptive windowing capabilities can compensate for translation, and in some cases, rotational errors. By training the system to recognize a reference attribute on the part, such systems first search for that attribute, and then establish the windows in accordance with the fixed relationship established during training. In some systems, the object itself becomes its own window. This is referred to as a "perfect hull" and sometimes is called hardwired connectivity.

8.3.2 Region Segmentation

This is the process of partitioning an image into elementary regions (adjacent pixels) with a common property (such as specific gray level or gray level range), and then successively merging adjacent regions having sufficiently small differences in the selected property until only regions with large differences between them remain. A popular execution of this segmentation is based on using thresholding techniques to establish a binary image.

Thresholding is the process of assigning "white" (maximum intensity) to each pixel in the image with gray scale above a particular value, while all pixels below this value become "black". That particular value is the threshold and is a gray scale value. Areas that are lighter than the threshold become white; areas darker than the threshold become black. The resulting image, consisting of only black and white, is called a binary image. Thresholding was the first segmentation technique used, and almost all systems use it to some extent. It has a simplifying effect on the image. The number of pixels in the image does not change, but each pixel can now have one of only two values, usually written 1 or 0.

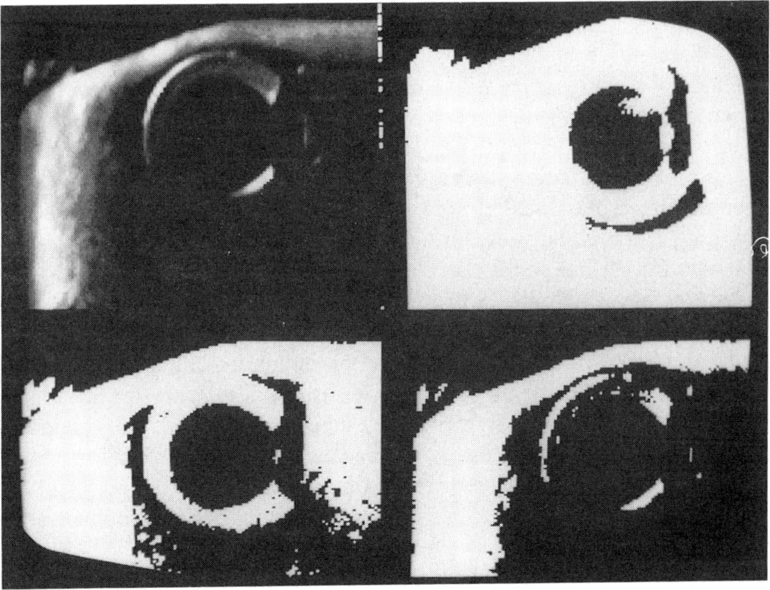

Figure 8.8 - Gray scale image.

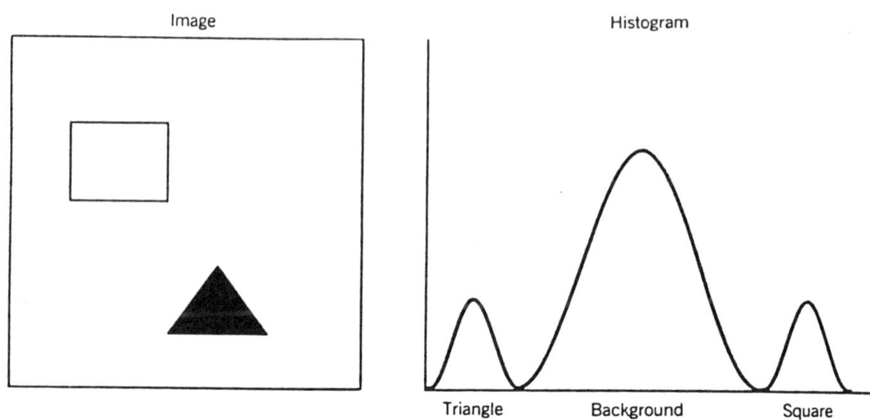

Figure 8.9 - An eight-bit system.

Figure 8.8 shows a gray scale image (on an 8-bit, 256 level system), and the same image at three different thresholds: 72, 100 and 130. Notice that as the threshold increases, the white area shrinks until it just contains the brightest pixels of the input image. In this image, thresholding does a fairly good job of enhancing the outer edges of the metal part but does not give an accurate description of the bushing. Because of the variation in part brightness, the bushing blends into the part. Since we are using one threshold for the entire image, this is an example of global thresholding. For global thresholding to work well, brightness must not vary too much over the surface of the part.

Many systems have two thresholds: upper and lower. White is assigned to pixels with values between the two. Note Figure 8.9 for an 8-bit system. It can be seen that setting the upper threshold to 255, and manipulating the lower, will duplicate the single threshold case; in practice, this is usually done for simplicity unless dual thresholds are required. Dual thresholding is useful for situations where a medium gray feature is of interest; for instance, a gray part on a black and white background. Selecting a lower threshold between black and gray, and an upper between white and gray, will produce a threshold image of a white part on a black background.

It is apparent that thresholding is a powerful simplifying process, but it is very sensitive to one parameter: the threshold. Identical images (in gray level) may have very different results when different threshold values are used. Therefore, unwanted variation in the intensity of the image will cause spurious results in the threshold image.

This leads to a key question: How to select the threshold? The simplest approach is to choose a threshold ahead of time. For minimum sensitivity, the threshold should be halfway between light and dark nearby in the image. - note: light and dark nearby in the image, not 255 and 0. Depending on the image capture setup, light may be only 25% of the white level, or dark may be light gray, not black. The optimum threshold setting depends on light and dark levels in the particular image. For this reason, a strategy of adaptive thresholding is often used.

Adaptive Thresholding

Adaptive thresholding is a technique of choosing the threshold based on the image's gray level values in some way. It is ideal for coping with variations in light and part appearance. There is no general method for choosing a threshold for an arbitrary image; something must be known about the image to define the process. So, automatic thresholding techniques tend to be specialized for a certain image or type of images. Although the techniques for choosing a threshold actually feature extraction methods, they are discussed here.

Pixel Counting. If there is a region that is known to be at least XX% white, one can place a window there and count the number of white pixels above a certain threshold. If there are too few white pixels, increase the threshold and try again; if too many white, decrease the threshold. This process repeats until a suitable pixel count is made. Counting the number of white pixels within a window is called "pixel counting." This technique is very simple and fast, unless many iterations need to be made. The disadvantages are part location cannot be too inaccurate, execution time is unpredictable since it depends on the number of iterations, and most importantly, it assumes that the brightness within the window is representative of the entire image. Another consideration is how to pick the initial threshold.

Max/Min/Average Gray Scale. A technique that does not require iterations is to set the threshold according to some combination of max/min/average gray scale value within a window. The most common is to place a window that contains both light and dark, and find the maximum and minimum gray scale values; the threshold is then set to the average of the two. This method is good if large image-to-image variations are expected, but is even more sensitive to local disturbances in intensity than pixel counting. This method works well for smaller windows.

Repeated Thresholding. Similar to pixel counting, this is also an iterative process. Assume that some feature is known that should be present in the image in a certain place. Threshold the picture and look for that feature. If it is not found, use a new threshold, either higher or lower, depending on whether one sees white or black where the feature is expected. Part mislocation must be well known for this method to work.

Histogramming. A powerful method is based on histogramming. Histogramming is building a bar of frequency of occurrence versus gray scale value for all pixels within a window. For an 8-bit system, there will be 256 bars with the height of each equal to the number of pixels having that gray level value (Fig. 8.10).

Dark is to the left and white is to the right. Knowing this, one can identify "bumps" in the histogram as belonging to certain areas or objects as in the image of Figure 8.9.

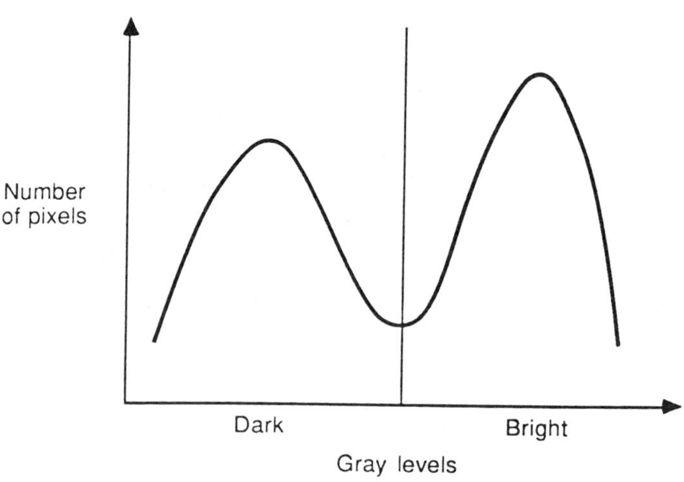

Figure 8.10 - Histogram of frequency plot of gray values.

The advantage for threshold selection is obvious. If one wanted to examine the square, one would choose a threshold (gray level value) between the upper two "bumps" (T_a. - Fig 8.11). If one is interested in the triangle, choose a threshold between the two (T_b - Fig. 8.11). Choosing a threshold within one of the bumps will lead to a 'breakup" of the object (Fig. 8.11).

The classic use of a simple light on dark image (or dark on light) leads to what is called a "bi-modal" histogram. To choose the threshold, find the midpoint between the two lobes (Fig. 8.12).

Unfortunately, histograms in the real world are rarely this simple. Because objects vary in brightness, and contrast in the image is always less than we want, the lobes of the histograms tend to overlap.

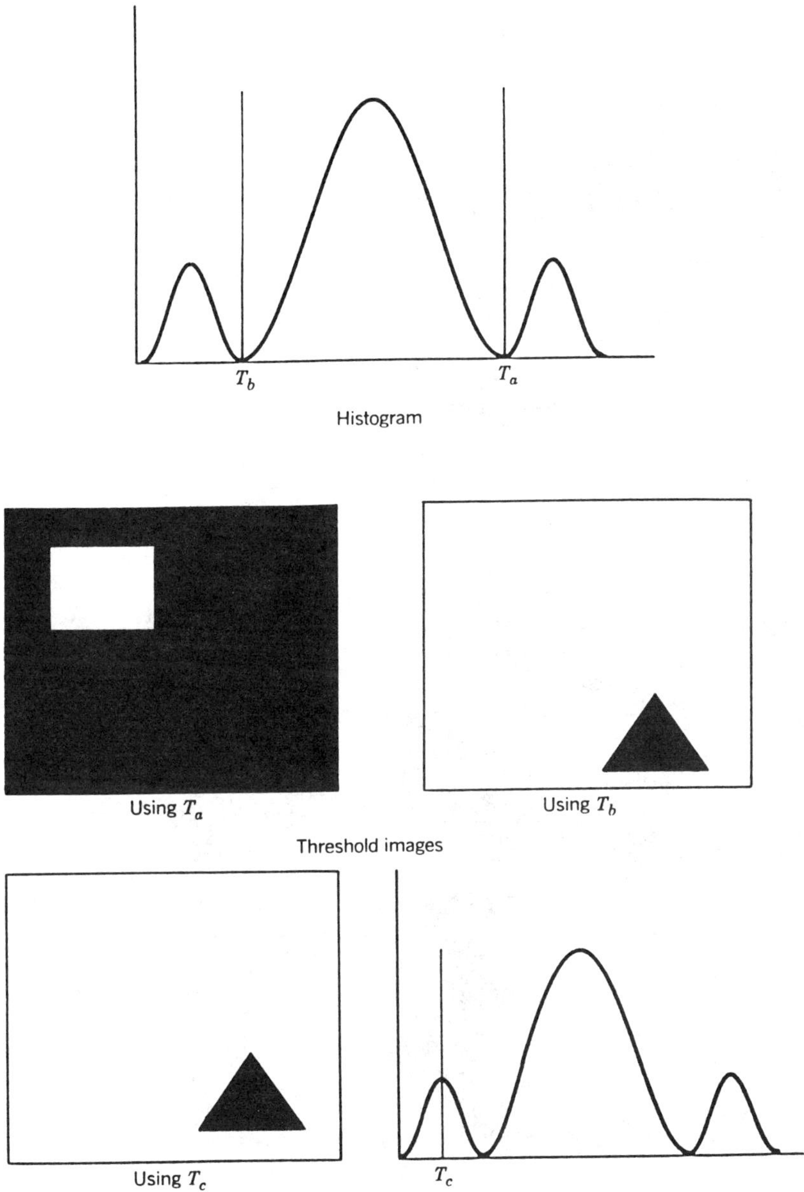

Figure 8.11 - Histogram and threshold images for an eight-bit system.

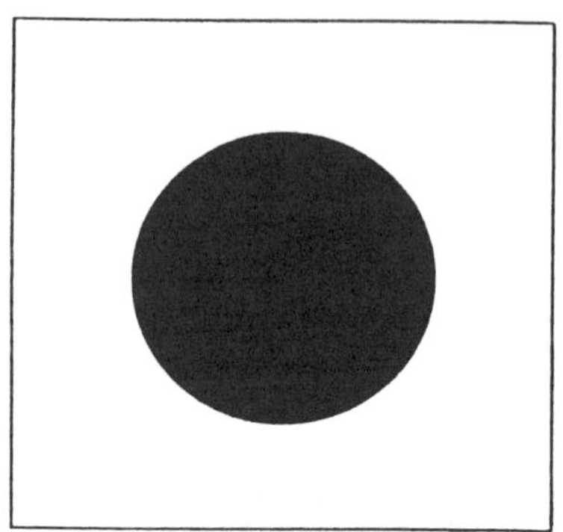
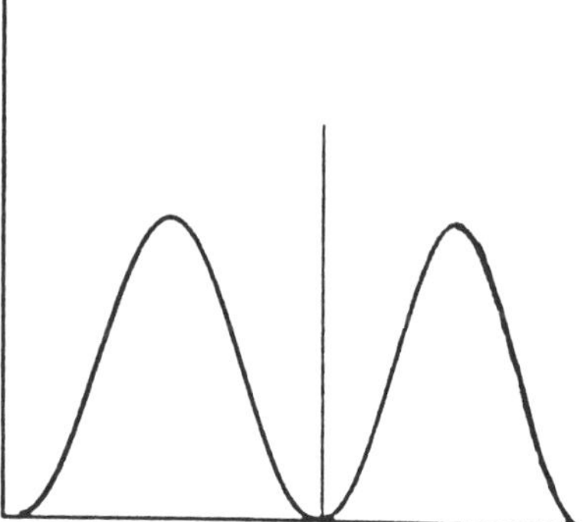

Figure 8.12 - Bimodal histogram.

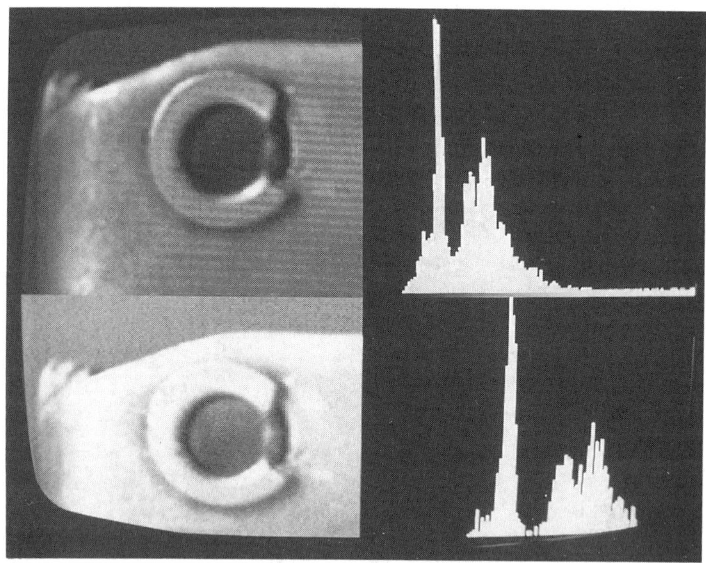

Figure 8.13 - Image shows how threshold chosen at two different places in histogram give two different binary images.

If this becomes extreme, one may not be able to threshold at all because no single value will discriminate objects from the background.

Depending on execution, since histogramming requires the examination of every pixel in the window, it may run slower for larger windows. However, the window should not be too small, or the histogram information will not represent the entire image accurately. Some implementations do not build the totals over several scenes. The first may be less accurate in representing the image; the second may not respond to short-term variations.

The procedure to examine the histogram and choose a threshold is not trivial. It must find how many lobes there are and possibly check this number, find the lobes of interest and then select the threshold value. This may be done by averaging the center points of two peaks, or an offset from a center peak, weighted average of the entire histogram, or many other methods. Histogramming is another feature extraction process, and its data can be used for many other things, as well as adaptive thresholding.

Note in the image of Figure 8.13 how thresholds chosen at two different places in the histograms give two very different binary images. In the upper frames, the threshold is at 110 (out of 256), while in the lower frames, it is 40.

The upper image would serve well to detect the wedge-shaped missing "chunk" in the bushing, but the part's top edge is weak. The lower image should enable us to find the edge of the part, but does not show the outside of the bushing. Which threshold is desired and, thus, how the threshold selection algorithm worked, would depend on which features one is interested in.

Localized Thresholding

So far, global thresholding has been discussed. As already noted, some images may not be separable using the same threshold(s) for the entire image.

However, different thresholds for different areas of the image can be chosen. If the image is brighter in the upper left corner, it may be due to brighter illumination, so both the features and the background are brighter. A higher threshold in that area will pick the feature from the background. This technique is called "localized" thresholding. Its drawback is that it requires prior knowledge; either the relative illumination (or brightness) throughout the image, or what features are being looked for, or their expected histograms must be known.

In the first case, fix the thresholds for the different regions according to their brightness; in the second, perform adaptive thresholding in each region by the techniques described above. As a corollary to this, the entire image need not be thresholded. If windows can be placed reliably, different thresholds can be assigned to each window (either fixed or adaptive).

8.3.3 Edge Segmentation

Features can also be extracted based on edges. Again edges can be obtained from a binary image based on transition locations in a gray scale image. In the case of the latter, points of rapid change characterize an edge in gray level intensity. While sensitive to changes in pixel intensity of a single pixel, edge detection is not related to the individual intensities within patterns. Analysis of the edge intensity within a single pixel results in sub-pixel calculations of the location of an edge. Some claim an ability to locate an edge to 1/64 of a pixel.

Many edge-segmenting systems are based on detecting patterns of increasing and decreasing intensities or gradients generally found at the edges of objects. Since they are based on gradients, they are less sensitive to illumination variations and can handle lower contrast scenes.

Neighborhood processing techniques have evolved which are generally employed in conjunction with edge-segmentation systems. These techniques involve evaluating each individual pixel according to its relation with its nearest neighbor pixels using a template or an array designed to detect some invariant regional property to perform the convolution. Figure 8.14 depicts a point template. The idea is to get a template response at every pixel location by centering the mask over each pixel in the image.

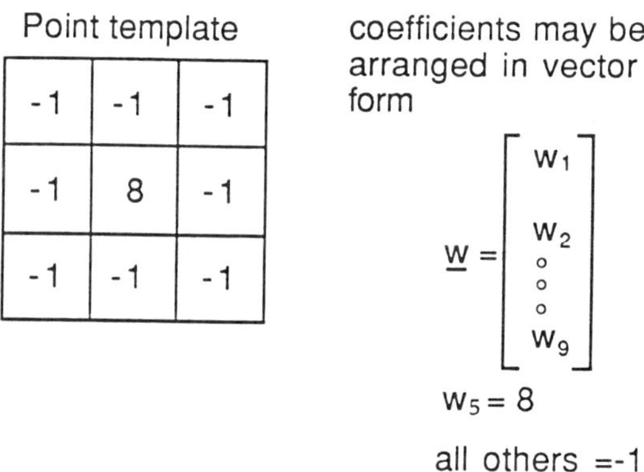

Figure 8.14 - Point template.

8.3.4 Morphology

Another approach to segmentation of regions or edges involves Boolean logical operations on images using set theory concepts adapted to images and is known as mathematical morphology. The following treatment of morphology is based largely on papers on the subject coming out of the Environmental Institute of Michigan. (See Becher, 1982; Lougheed and McCubbrey, 1980; Sternberg, 1980 and 1981.)

Mathematical morphology gets its name from the studies of shape analysis. It is based on investigating the association between shapes or structures contained within the image and a shape that is dictated as significant by the application. It treats images as sets of points in space that can only take on one of two states, active or inactive, that is, binary sets. Active points represent the foreground set and inactive the background set. Gray scale images are represented as binary images in a three-dimensional space, brightness being the third dimension (Figure 8.15).

Systems based on mathematical morphology efficiently perform operations that involve treating each pixel in a set identically resulting in a new or transformed image. These image transformations fall into three categories: unary (one image in, one image out), dyadic (two images in, one image out), and information extraction (image in, numbers out). Within each of these categories, the operations on binary images are either geometric or logical. In other words, in addition to the image processing, analysis itself is based on transformations of the image to other data structures, and the ordering of pixels is the basis of the decision. These systems need to be able to store multiple images and perform arithmetic and logical operations swiftly.

Unary operations include complement (not), reflection, and translation (shift in a given direction) (Figure 8.16). Complement, a logical operation, changes all

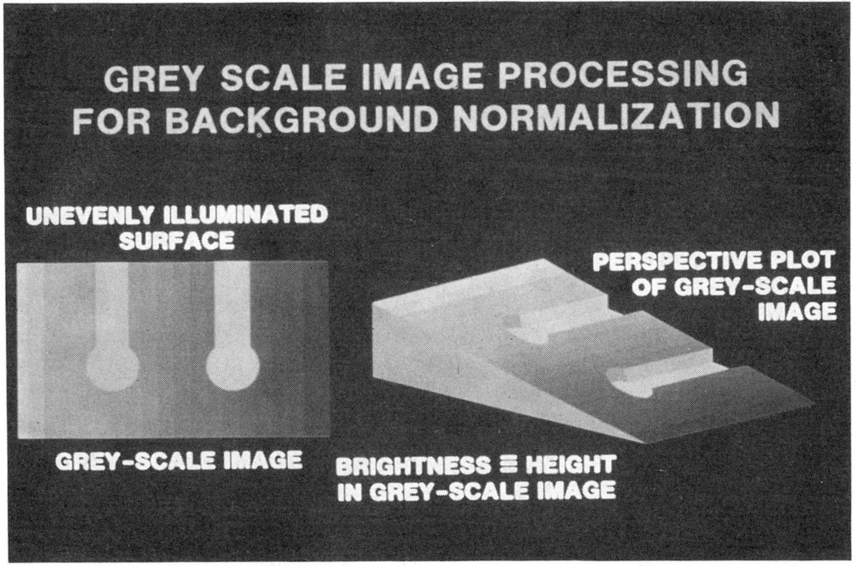

Figure 8.15 - Depiction of gray scale as third dimension (courtesy of General Scanning/SVS).

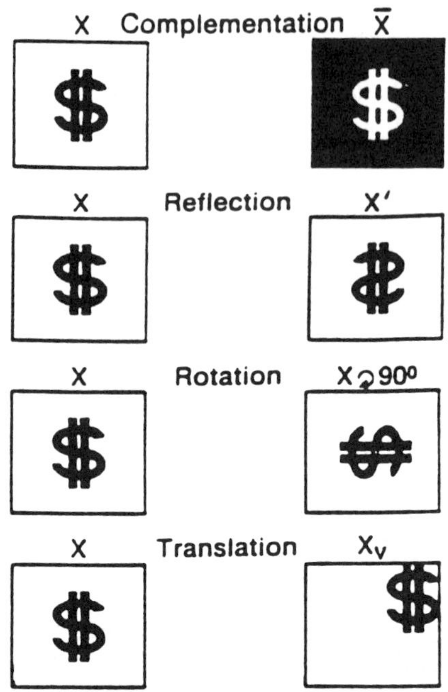

Figure 8.16 - Unary operations on binary images (courtesy of ERIM)

pixels that are active to inactive and vice versa. Translation shifts all pixels in a given direction a specified distance. Reflection assumes an origin for the image and coordinates for each point. Multiplying by -1, the foreground points are reflected across the "origin." For two-dimensional images this is analogous to rotating the image 180 degrees.

Dyadic operations combine two images into one. Given two sets, for example, from set theory the logical combinations of two images are union, intersection, and difference (Figure 8.17).

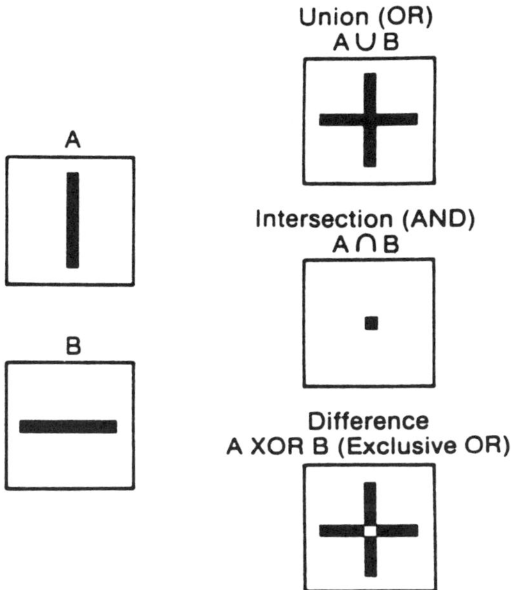

Figure 8.17 - Dyadic operations on binary images (courtesy of ERIM).

Two key operations are dilation and erosion. The dilation operation (Figure 8.18) between two sets A and B involves transforming each individual pixel in the A image by each pixel in the B image. In one definition of dilation, the transformed image that results is characterized as the outermost image made up of the center point of all the B images (typically the structured element) added to the A image. Erosion (Figure 8.19) is the opposite of dilation and is essentially a containment test. The erosion operation between two sets A and B (typically the structured element) results in a transformed image that is the universe of all center points of set B, where set B is fully contained in set A. Combinations of these operations are also possible.

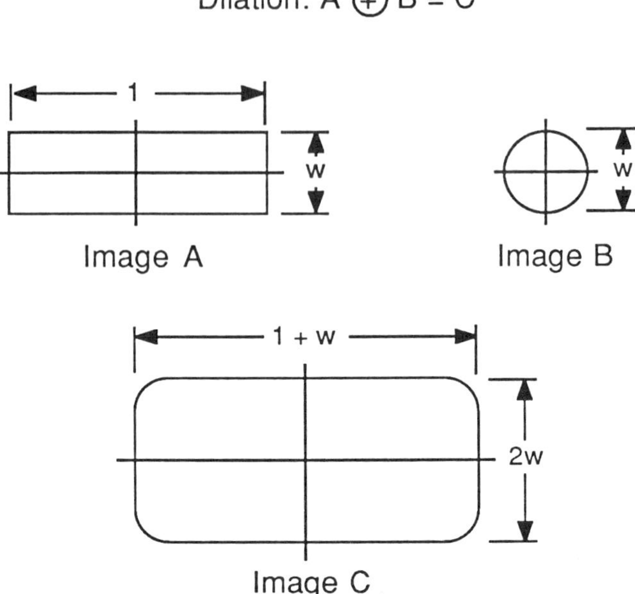

Figure 8.18 - Dilation operation (courtesy of ERIM).

The dilation and erosion operations are called duals because the dilation of the foreground is equivalent to the erosion of the background (Figure 8.20). With the ability to complement an image, patterns and shapes that can be generated can also be detected with similar sequences of operations.

Basic operations can be combined. For example, an erosion followed by a dilation with the same image will remove all of the pixels in a region that are too small to contain the structured element. This sequence is referred to as opening. If a disk-shaped structured element is used, all of the regions of pixels smaller than the disk will be eliminated. This forms a filter that suppresses positive spatial details. The opposite sequence, a dilation followed by an erosion, fills in holes smaller than the structured element. Figure 8.21 gives an example of combining operations to detect defects in traces on printed circuit boards.

Systems based on mathematical morphology have been used successfully to evaluate the cosmetic properties of objects. Figures 8.22(a)-8.22(e) depict a sequence. As suggested, treating the gray scale image as a three-dimensional surface where brightness is altitude, one first normalizes the background and the effect due to the nonuniformity of the illumination. The effect of the nonuniform lighting is depicted in A by the analog trace of a single line scan. The signal stemming from the nonuniformity of illumination is greater than the signal difference between the scratch and grinding marks on the automotive intake manifold.

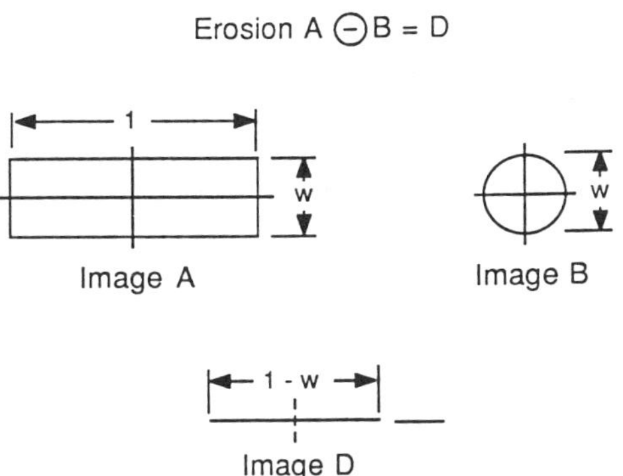

Figure 8.19 - Erosion operation (courtesy of ERIM).

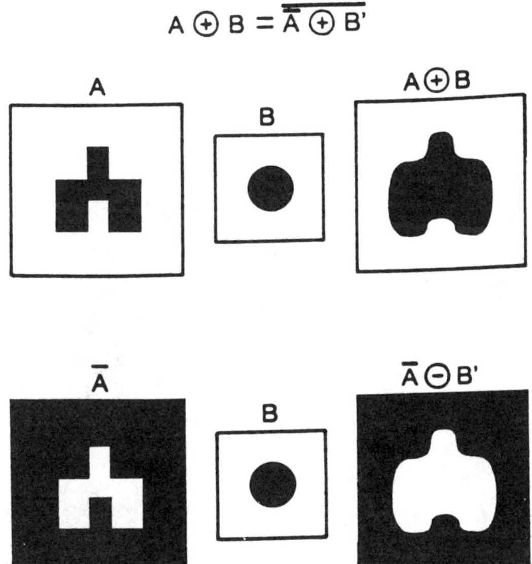

Figure 8.20 - Duality of dilation and erosion (courtesy of ERIM).

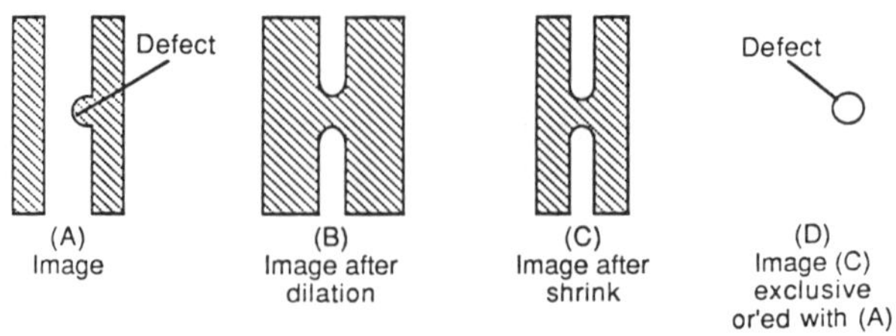

Figure 8.21 - Mathematical morphology in processing of PC board images (courtesy of General Scanning/SVS).

The results of the first operation-closing are depicted in Figure 8.22(b). A difference image is shown in Figure 8.22(c). A threshold is taken on the difference image [Figure 8.22(d)]. Filtering based on shape takes place to distinguish the noise from the scratch; that is, the detail that can fit in a structured element in the shape of a ring is identified in Figure 8.22(e) and subtracted [Figure 8.22(f)]. Figure 8.22(g) depicts the segmented scratch displayed on the original image.

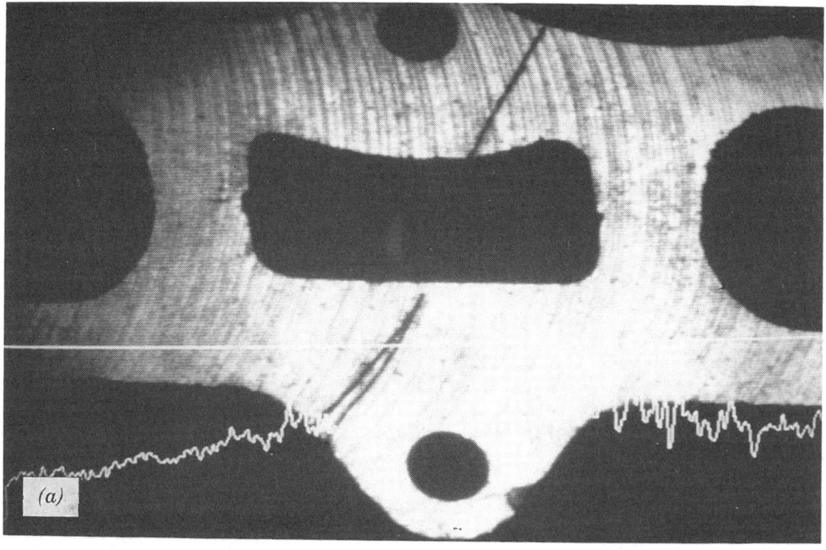

Figure 8.22a-g - Morphology operating on a scratch on manifold (courtesy of Machine Vision International).

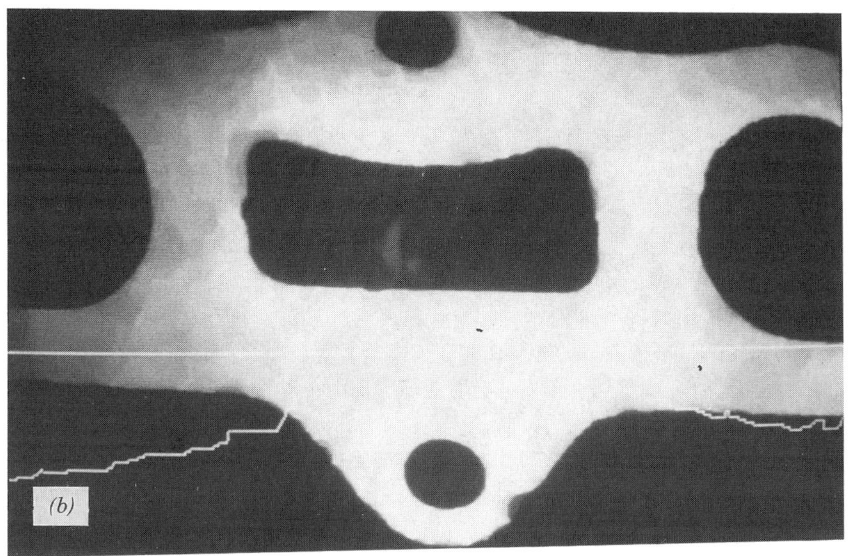

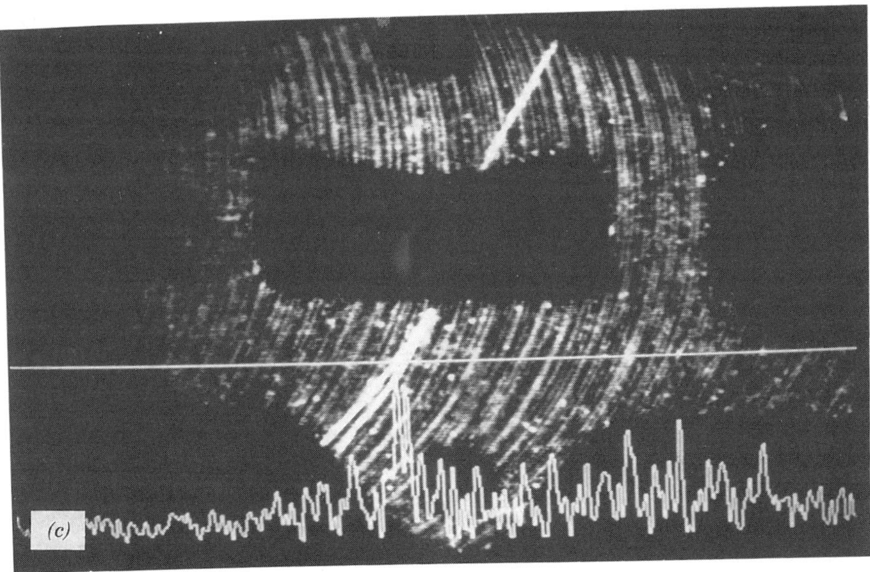

Figure 8.22 Continued.

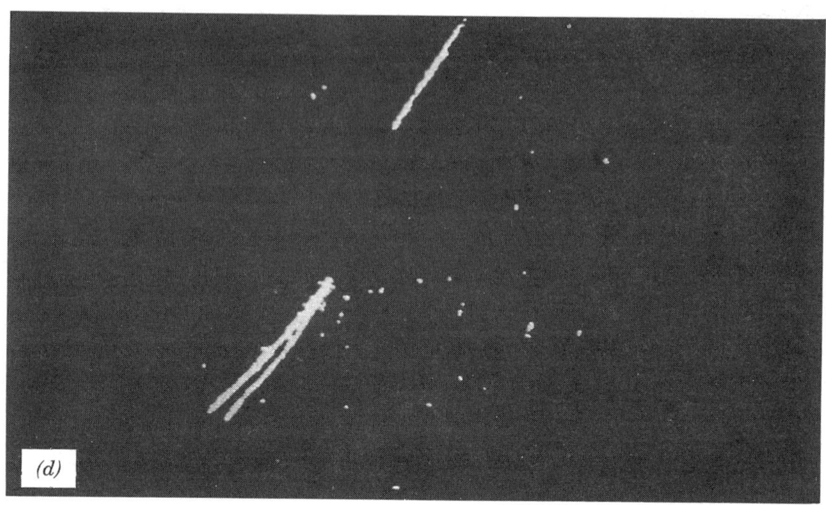

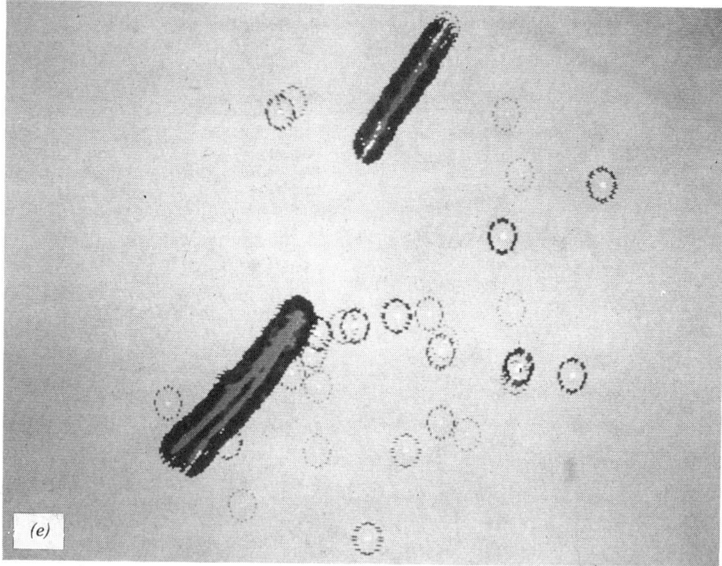

Figure 8.22 Continued.

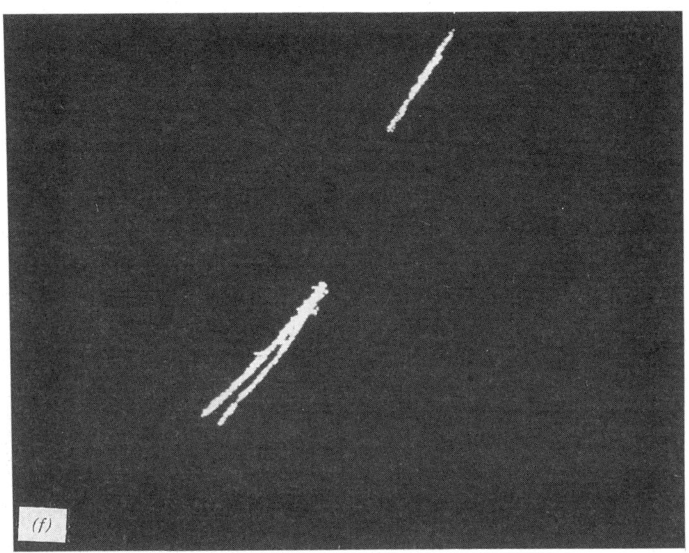

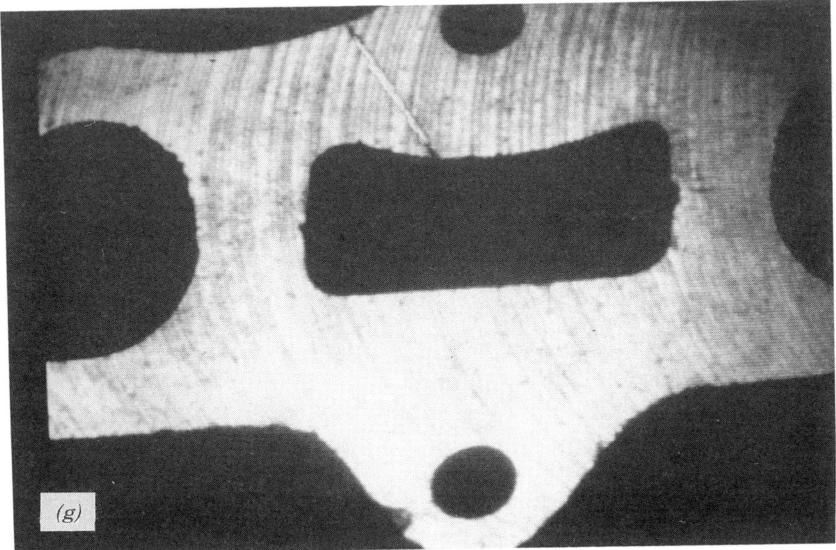

Figure 8.22 Continued.

8.4 CODING/FEATURE EXTRACTION

Feature extraction is the process of deriving some values from the enhanced and/or segmented image. These values, the features, are usually dimensional but may be other types such as intensity, shape, etc. Some feature extraction methods require a binary image, while others operate on gray scale intensity or gray scale edge-enhanced images. The methods described below are grouped into three sections: miscellaneous scalar features, including dimensional and gray level values; shape features; and pattern matching extraction.

8.4.1 Miscellaneous Scalar Features

Pixel Counting. For simple applications, especially part identification and assembly verification, the number of white pixels in a given window is enough to derive the desired result. This operation, finding the number of pixels above a threshold within a window, is called "pixel counting." It is a very widely used technique and runs very quickly on most systems.

Often pixel counting is used for tasks other than the main application problem such as threshold selection (as already described), checking part location, verifying the image, etc.

Edge Finding. Finding the location of edges in the image is basic to the majority of the image-processing algorithms in use. This can be one of two types: binary or enhanced edge. Binary-edge-finding methods examine a black-and-white image and find the X-Y location of certain edges. These edges are white to black, or black-to-white transitions. One technique requires the user to position scan lines, or tools ("gates") in the image (Fig. 8.23). The system then finds all edges along the tools, and reports their X-Y coordinates.

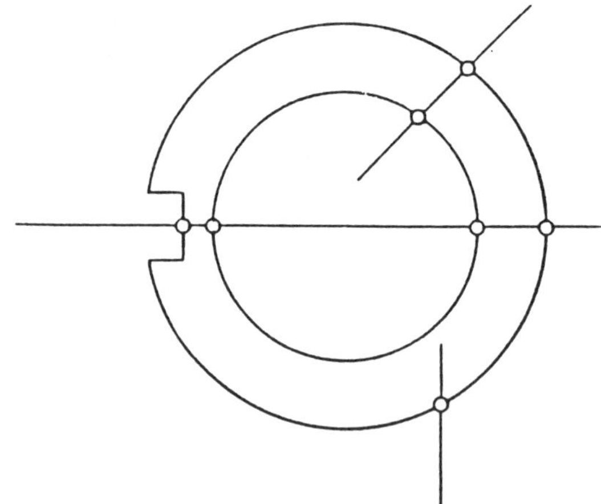

Figure 8.23 - Edge finding, circles are edge locations found.

These gates operate like a one-dimensional window, starting at one end and recording coordinates of any transitions. Many times they can be programmed to respond to only one polarity of edge, or only edges "n" pixels apart, or other qualifiers. Like windows, some systems allow orientation at angles, while others do not.

Because they are one-dimensional, and also because the video is binarized, any edge data collected using gates should be verified. This verification can be done by using many different gates and combining the results by averaging, throwing out erratic results, etc. Other features are often used to verify binary edge data such as pixel counting.

Gray-scale edge finding is very closely tied with gray-scale edge enhancement; in fact, usually the two are combined into one operation and are not available as two separate steps. Generally, the edge-enhanced (gradient) picture is analyzed to find the location of the maximum gradient, or slope.

This is identified as the "edge." The set of coordinates of these edge points are taken as features and used as inputs for classification. Sometimes, the edge picture is thresholded to create a binary image so that the edges are "thick." A "thinning" or "skeletonizing" algorithm is then used to give single pixel wide edges. The first method, finding gradient, gives the true edge but usually requires knowing the direction of the edge (at least approximately). The second method, thresholding the gradient and thinning, may not find the true edge location if the edge is not uniform in slope. The error is usually less than one pixel, depending on the image capture setup; and, thus, thresholding the gradient image is a very common gray-scale edge locating method.

There are systems that use algorithms that produce edge location coordinates directly from the gray scale data, using algorithms that combine filters, gradient approximations, and neighborhood operations into one step. These are usually tool-based (or window-based) systems. Many even 'learn" edges by storing characteristics such as the strengths of the edge, its shape and the intensity pattern in the neighborhood. These characteristics can be used to ensure that the desired edge is being found during run-time.

8.4.2 Shape Features

Some computationally more intensive image analysis systems are based on extracting geometric features. One such approach (developed at Stanford Research Institute International) involves performing global feature analysis (GFA) on a binary picture. In this case, the features are geometric: centroid, area, perimeter, and so on. In GFA, no inferences are made about the spatial relationships between features, and generally the parts are isolated.

Generally, operations are performed on the "raw" (unprocessed) video data (filtering) and preprocessed images (run length encoding) (Figure 8.24). Decision-making and control is the function of the microprocessor.

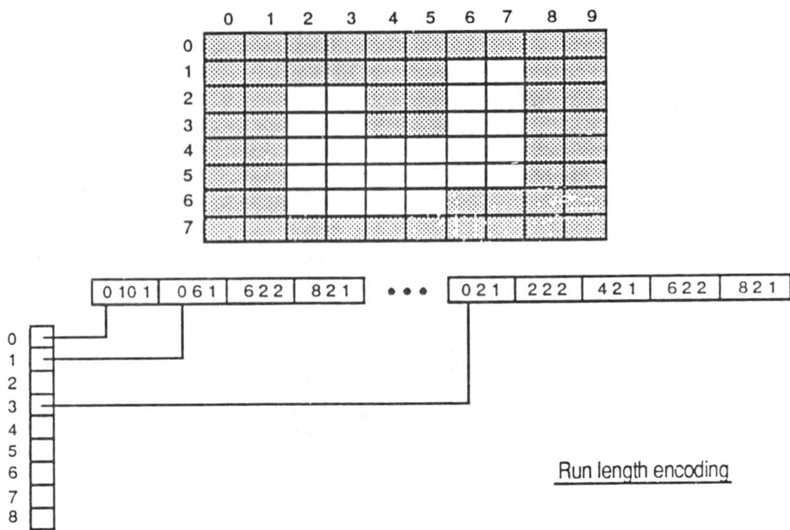

Figure 8.24 - Run length encoded image.

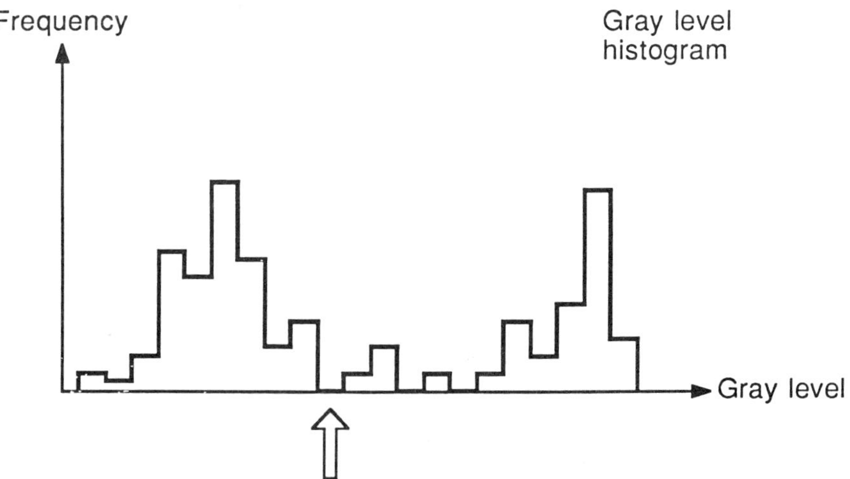

Figure 8.25 - Thresholded segmentation.

An enhancement of this approach involves segmentation based on either threshold gray scale or edges. In thresholded gray scale, a threshold (Figure 8.25) is set, and if the pixel gray level exceeds the threshold, it is assigned the value 1. If it is less than the threshold, it is assigned the value 0. An operator during training can establish the threshold by observing the effect of different thresholds on the image of the object and the data in the image sought. The threshold itself is a hardware or software setting.

A pixel gray value histogram (Table 4. 1) analysis display can provide the operator with some guidance in setting the threshold. In some systems, the threshold is adaptive; the operator sets its relative level, but the setting for the specific analysis is adjusted based on the pixel gray scale histogram of the scene itself.

Once thresholded, processing and analysis is based on a binary image. An alternative to thresholded segmentation is that based on regions, areas in an image whose pixels share a common set of properties; for example, all gray scale values 0-25 are characterized as one region, 25-30, 30-40, and so on, as others.

TABLE 8.1 Uses for Histograms

Binary threshold setting
Multiple-threshold setting
Automatic iris control
Histogram equalization (display)
Signature analysis
Exclusion of high and low pixels
Texture analysis (local intensity spectra)

SRI analysis is a popular set of shape feature extraction algorithms. They operate on a binary image by identifying "blobs" in the image and generating geometrical features for each blob. Blobs can be nested (as a part with a hole; the part is a blob, and the hole is a blob also). SRI analysis has several distinct advantages: the features it generates are appropriate for many applications; most features are derived independent of part location or orientation, and it lends itself well to a "teach by show" approach (teach the system by showing it a part).

SRI, or "connectivity" analysis as it is often called, requires a binary image. However, it only deals with edge points, so often the image is "run-length" encoded prior to analysis. Starting at the top left of the screen (Figure 8.26), and moving across the line, pixels are counted until the first edge is reached. This count is stored, along with the "color" of the pixels (B or W), and the counter is reset to zero. At the end of the line, one should have a set of 'lengths" that add up to the image size (shown below), 0 = black, 1 = white.

00000000	111111111	000	11	0000000000
8B	9W	3B	2W	10B

These set of like pixels are called runs, so we have encoded this line of binary video encoded by the length of the runs; hence, "run-length encoding." Run-length encoding is a simple function to perform with a hardware circuit, or can be done fairly quickly in software.

Figure 8.26 - Shape features.

Each line in the image is run-length encoded, and the runs are all stored. The runs explain about the objects in the scene. The image in Figure 8.26 may appear as:

-20W 15B 65W
-20W 17B 65W
-20W 10B 1W 5B 65W
-20W 10B 3W 5B 63W
-20W 10B 5W 5B 26W 5B 30W
-20W 10B 7W 5B 22W 9B 28W
-20W 10B 9W 5B 19W 11B 27W etc.

Note how the left "blob" split into two vertical sections. Similarly, as one works down the image, some blobs may "combine" into one. The keys to SRI are the algorithms to keep track of blobs and sub-blobs (holes), and to generate features from the run lengths associated with each blob. From these codes many features can be derived. The area of a blob is the total of all runs in that blob.

By similar operations, the algorithms derive:

Maximum X, Y
Minimum X, Y

Centroid X, Y
Length
Second moments of inertia
Minimum circumscribed rectangle
Maximum and minimum radii, etc
Orientation angle
Width
Eccentricity ("roundness")
Major and minor axes

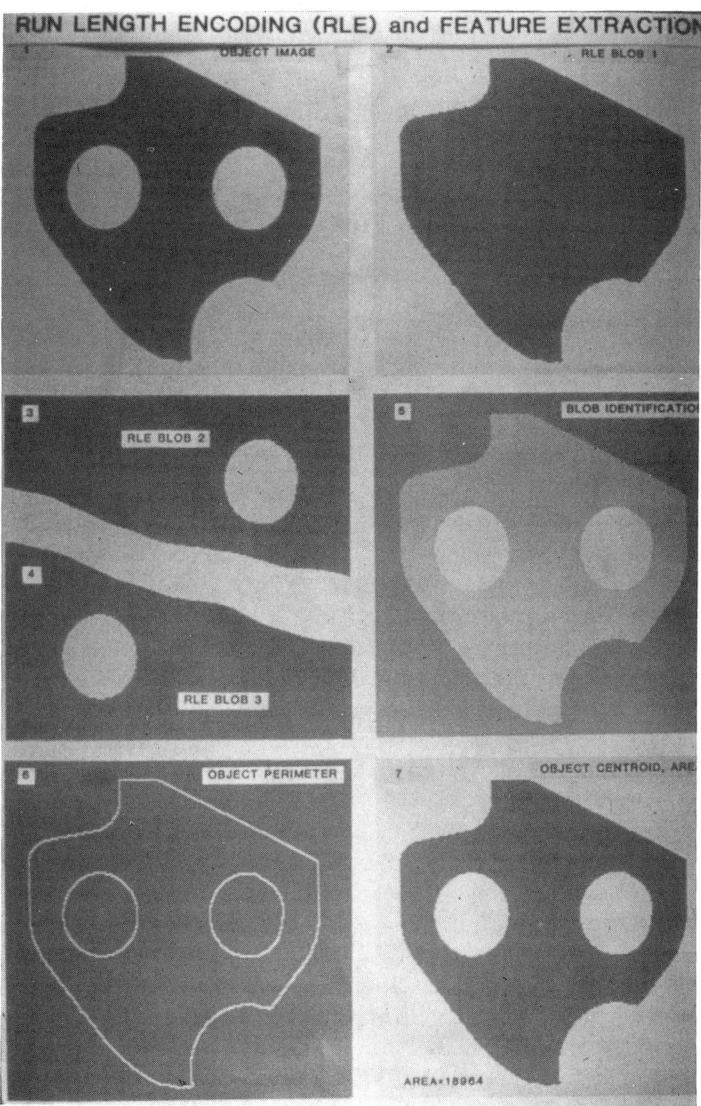

Figure 8.27 - Connectivity/blob analysis (courtesy of International Robotmation Intelligence).

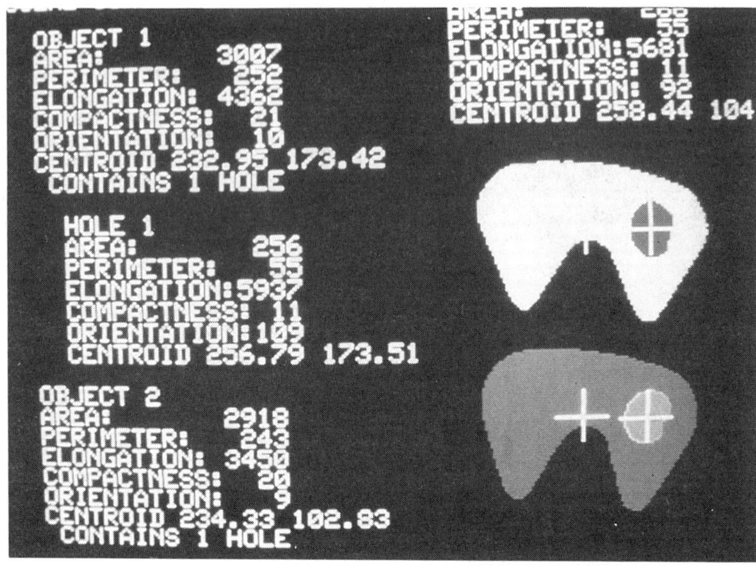

Figure 8.28 - Geometric features used to sort chain link (courtesy of Octek).

Figures 8.27 and 8.28 are examples of the SRI approach. It is apparent that knowing all these features for every blob in the scene would be enough to satisfy most applications. For this reason, SRI analysis is a powerful tool. However, it has two important drawbacks that must be understood.

1. Binary image - the algorithms only operate on a binary image. The binarization process must be designed to produce the best possible binary image under the worst possible circumstances. Additionally, extra attention needs to be paid to the image-verifying task. It is easy for the SRI analysis to produce spurious data because of a poor binary image.

2. Speed - because it (typically) examines the whole image, and because so many features are calculated during analysis, the algorithms tend to take more time than some other methods. This can be solved by windowing to restrict processing, and by only selecting the features necessary to be calculated. Most SRI-based systems allow unnecessary features to be disabled via a software switch, cutting processing time.

Some systems using SRI have a "teach by show" approach. In the teach mode, a part is imaged and processed. In interactive mode, the desired features are stored. At run time, these features are used to discriminate part types, to reject non-conforming parts, or to find part position. The advantage is that the features are found directly from a sample part, without additional operation interaction.

8.4.3 Pattern Matching

Pattern matching is also called "correlation, pattern recognition, or template matching" - a mathematical process for identifying the region in an image that "looks most like" a given reference subimage. The reference subimage, or "template," is overlaid on the image at many different locations. At each, goodness of match is evaluated. The location with the best is recorded, and the process is complete. Notice that process is inherently robust since it uses all the information in the image. Note, however, that a "match" operation, involving the template and part of the subimage, must be performed for each location in the image; this is a very time-consuming task. Many considerations discussed below involve ways of reducing this time.

Binary. For the case of a binary image and template, the match process is fairly simple. The total number of different pixels is evaluated; the smaller this number, the better the match. If one XORs the image and the template, white results where the pixel values are different, black where they were same (see Fig. 8.29).

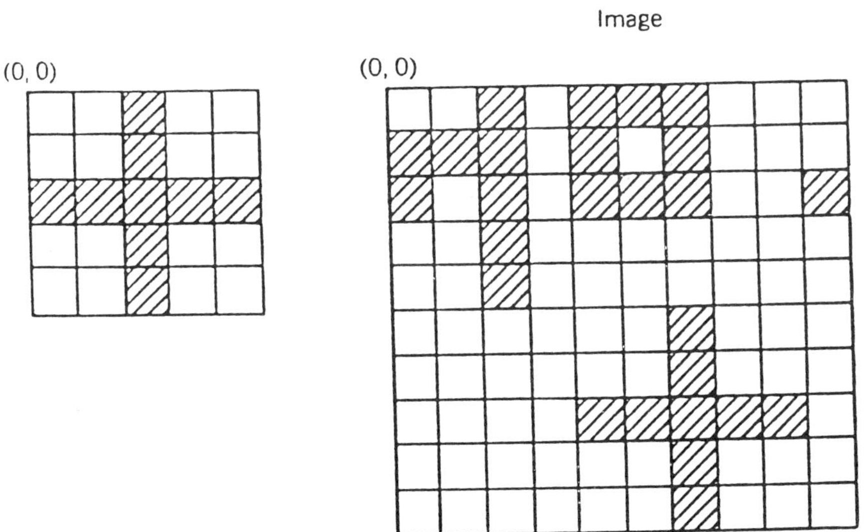

Figure 8.29 - Binary image and template.

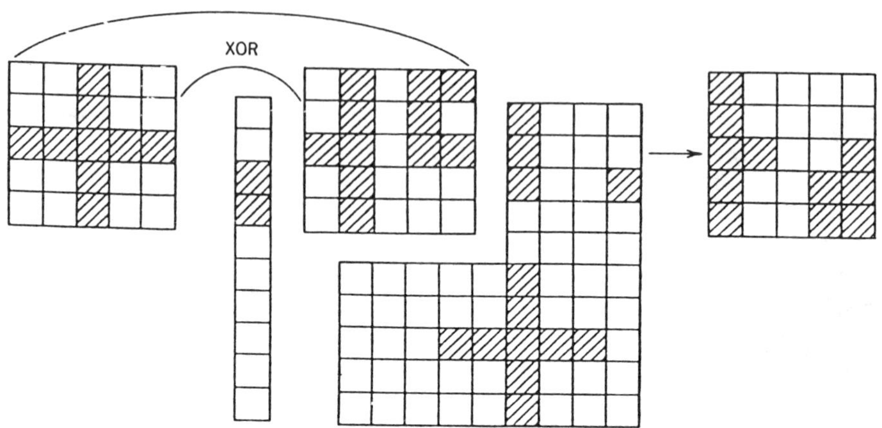

Figure 8.30 - Location 1. Total number of pixels = 25, number of black pixels = 21, match % = 21/25 x 100% = 84%.

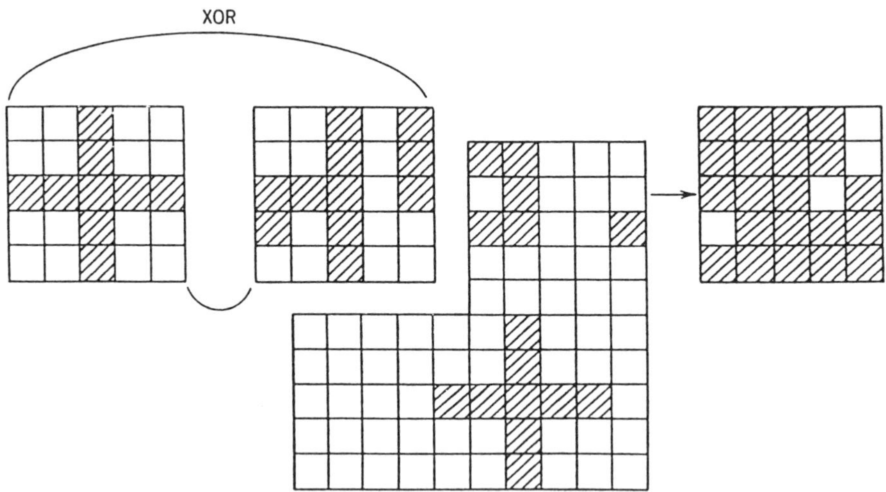

Figure 8.31 - Location 2. Number of black pixels = 13, match % = 13/25 X 100% = 52% match.

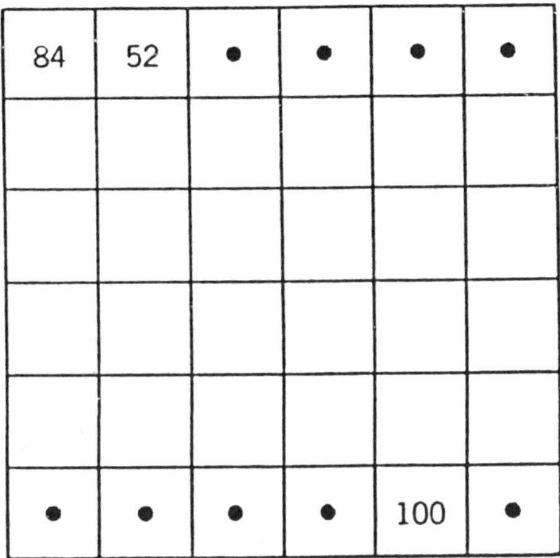

Figure 8.32 - Evaluation of percent match at each location and matrix formed.

Evaluate the match at several different locations in the recording each with its location. For location #1: template pixel (0, 0) at image pixel (0, 0) (see Fig. 8.30).

Now shift the template one pixel to the right. For location #2: template pixel (0, 0) (see Fig. 8.31). Notice that one can only shift the reference four more pixels to the right. The number of different displacements in one dimension is equal to the size of the image, minus the size of the template in that dimension, plus one. Mathematically, the number of different locations possible for an m X n image and an i X j template is: $(md-i+l) * (n-j+1)$.

If this process is continued exhaustively, evaluating the percent match at each location and forming a matrix of the results, Figure 8.32 is obtained.

From this, the best match is at (4, 5) in the image. This is the pixel in the image that corresponds to pixel (0, 0) in the plate. Therefore, this is also an offset that can be added at any pixel coordinates in the template to find their location in the image. This simple example brings out several points:

A full XOR and pixel counting operation must be done at each offset. If a match is tried at every possible location (exhaustive search), the time requirements are considerable.

100% match occurs only for an exact match. The closer the match, the higher the percentage.

Two unrelated patterns (random noise) will match to about 50%, using this quality calculation (XOR and count). The only way to get 0% match would be to have a reverse image (for example, a white cross on a black field). What useful information is obtained from the pattern matching process?

Most importantly:

1. Location of the template in the image.
2. The quality of match between the image and the template.

The location can be used for several things. In robot guidance applications, the part's X-Y coordinates are often the answer to the application problem. For most inspection systems, the material handling does not hold part position closely enough. Pattern matching can be used to find the part's location, which is then used to direct other processing steps. For instance, often the location found by pattern matching will relocate a set of windows so that they fall over certain areas of the part.

The quality of match can be used to verify the image. If the quality is lower than expected, something may be wrong with the part (mislocated, damaged, etc.) or the image (bad lighting or threshold). Also, more than one pattern matching process may occur, each using a reference corresponding to a different number. The reference that gives the best match is the same part number as the image; this can be used to sort mixed parts.

In reality, parts can not only translate (X and Y), but rotate (0). Therefore, a third variable must be introduced. At each location (X and Y), the template must be rotated through 360 degrees, and the match at each angle evaluated. This gives the system the ability to find parts at any orientation, as well as any position. However, the computational speed is quite substantial for an exhaustive search approach. Assuming that the match is evaluated at each location (every pixel) and at every angle (1 degree increments), we must make [(assume 256 X 256 image, 64 X 64 template) = (256 - 64 + 1) (256 - 64 /1) 360] = 13.4 million match operations, each one 64 X 64!

The most common approach for solving this problem is to use a "rough search" first, followed by a "fine search." The "rough search" may only examine every tenth pixel in X and Y, and every 30 degrees in 0. This would call for

$$(256 - 64 + 1)/10 *(256 - 64 + 1)/10 * 360/30 = 4470 \text{ matches}$$

Then, starting at the "most promising" point(s) found by the rough search, a fine search is conducted over a limited area. This search uses increments of 1 pixel and the smallest angle increment. The best match thus found is the final result. For most images used, this procedure works very well. However, this provides a guide to selecting a reference. To aid the rough search, the template should contain some large features (of course, they should be unambiguous - so they are not confused with undesired features). For the fine search, some small, sharp details should be in the template also.

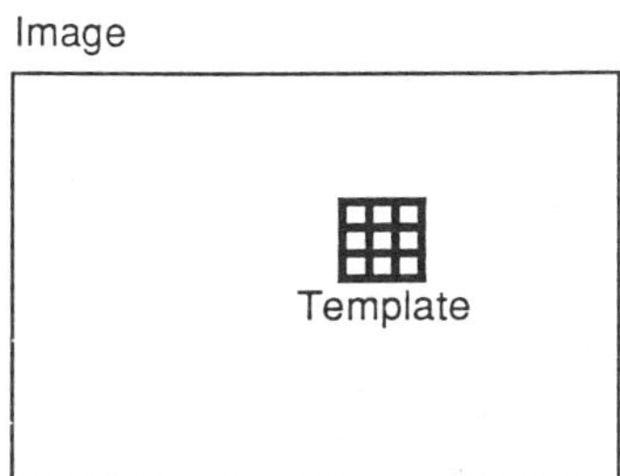

The idea is to get template response at every pixel location by centering the mask at each pixel in the image.

Figure 8.33 - Running template through image.

The search is an interesting problem, since it calls for some interpolation to lay the rotated template (Fig. 8.33). An often-used technique is binary pattern matching of edge pictures. In this case, both the template and the image have been edge enhanced and thresholded (but usually not skeletonized). This allows pattern recognition to be used on scones that cannot be binarized simply. Normalization of the gray scale images or auto-thresholding of the edge may be necessary to get consistent edge pictures

In Figure 8.33 the template was given. In practice the template is loaded during a "teaching phase". A part is imaged by the system and a nominal (0,0) point chosen. The appropriate size subimage is extracted from the image (or its image) and stored in memory as the template. Usually the teach phase allows matching with the part to check the template, loading of multiple references for part-sorting applications, and viewing of the reference for further verification. Sometimes the system allows "masking" of irrelevant portions of the template. For instance, if the part may or may not have a hole in a certain location, and one does not want its presence or absence to affect the quality of the match, it can be "masked" out; pattern matching will then consider only the rest of the template.

Gray-Scale Pattern Matching. This operation is similar to binary pattern matching, except that the part and the template contain gray-scale values.

The measure of goodness of fit is then based on the difference between pixel values. A "square root of the sum of squares" quantity is appropriate, for instance:

$$fit(m,n) = \sqrt{\frac{\sum_i \sum_j \left[image(m+i,n+j) - template(i,j) \right]^2}{\left[\sum_i \sum_j image(m+i,n+j) \right] \left[\sum_i \sum_j template(i,j) \right]}}$$

The denominator term is included to "normalize" the result. Without this, the index would be the lowest where the image had the lowest gray levels, not where the pattern match was the best.

A quantity of this type involves a lot of computation. When the rotation variable 0 is included, it becomes challenging to use this technique in industrial image processing for "full frame" searches. Gray scale pattern matching is used by several systems, however, in the following manner. At "teach time," several small windows are laid on the image (typically 7 X 7). Around each of these, a larger window is defined. The contents of each small window are stored as references. At run time, the system will search within each large window for its appropriate reference pattern. Then when each has been found, the offsets from nominal are combined to yield the part's true position (offset from nominal at teach time). If the search window is kept fairly small (40 X 40), the run time is not excessive.

8.5 IMAGE ANALYSIS/CLASSIFICATION/INTERPRETATION

For some applications, the features, as extracted from the image, are all that is required. Most of the time, however, one more step must be taken; classified interpretation.

The most important interpretation method is conversion of units. Rarely will dimensions in "pixels" or "gray levels" be appropriate for an industrial application. As part of the software, a calibration procedure will define the conversion factors between vision system units and real world units. Most of the time, conversion simply requires scaling by these factors. Occasionally, for high accuracy systems, different parts of the image may have slightly different calibrations (the parts may be at an angle, etc.). In any case, the system should have separate calibration factors in X and Y. Sometimes, especially in process control applications, the dimensional output must be converted again into units of control signal (stepper motor pulses, pulse rate, control voltage, etc.).

Reference points and other important quantities are occasionally not visible on the part, but must be derived from measurable features. For instance, a reference point may be defined by the intersection of two lines (as the "bend point" of a tube, defined by the axes of the tube on either side of the bend). To derive the location of this point, enough points must be measured to define the two lines (Fig. 8.23) and find their intersection by geometry.

Another common indirect measurement is to locate the center of a circle by finding points on its perimeter. Most systems have fast methods for doing this. In fact, indirect measurement calculations should present no problem to an experienced applications engineer.

An almost trivial classification step is a good/bad test. This consists of some logical combination of the measured dimensions and some preset limits. For instance, if the measured length is between 3.3 and 3.5, AND the diameter is no more than 0.92, the part is good; otherwise, it is bad. Any system that performs a good/bad check of this type should also make the measured dimensions available. During system setup and debug, it will be necessary to see the quantities. It is extremely difficult to verify the image processing performance of a system that says only "good" or "bad."

Error checking, or image verification, is a vital process. By closely examining the features found, or extracting additional features, test the image itself to verify that it is suited to the processing being done. Since features are being checked, it can be considered a classification or interpretation step. Without this, features could have incorrect values because the part is mislocated, upside down or missing, because a light has burned out, because the lens is dirty, etc. A philosophy of "fail-safe" programming should be adopted; that is, any uncertainty about the validity of the image or the processing should either reject parts or shut down the process. This is vital in inspection applications, where errors could lead to bad parts being passed. This is imperative in the process control, process verification, and robot guidance, where safety is at risk. Unfortunately, error-checking procedures are usually specific to a certain type of image; general procedures are not available. However, here are some possibilities:

Pattern matching - too low of a quality index suggests that something has changed.

Pixel counting windows - one on a known bright spot to check lights; one on a known area of the part to check presence.

Histogramming - smoothing of the histogram may indicate poor focus; shift will show changes in light level; distortion or smoothing may be due to dirt on lens or sensor malfunction.

Redundant feature extraction - if two quantities should be similar and are not, something is not right.

Edge counts - a tool of window that sees too few or too many edges may indicate part movement, dirty part or lighting problems.

8.6 DECISION-MAKING

Decision-making, in conjunction with classification and interpretation, is characterized as heuristic, decision theoretic, syntactic or edge tracking.

8.6.1 Heuristic

In this case, the basis of the machine vision decision emulates how humans might characterize the image:

Intensity histogram
Black-white/black-white transition count
Pixel counts (Figure 8.34, 8.35, and 8.36)
Background/foreground pixel maps
Background/foreground pixel counts
Average intensity value

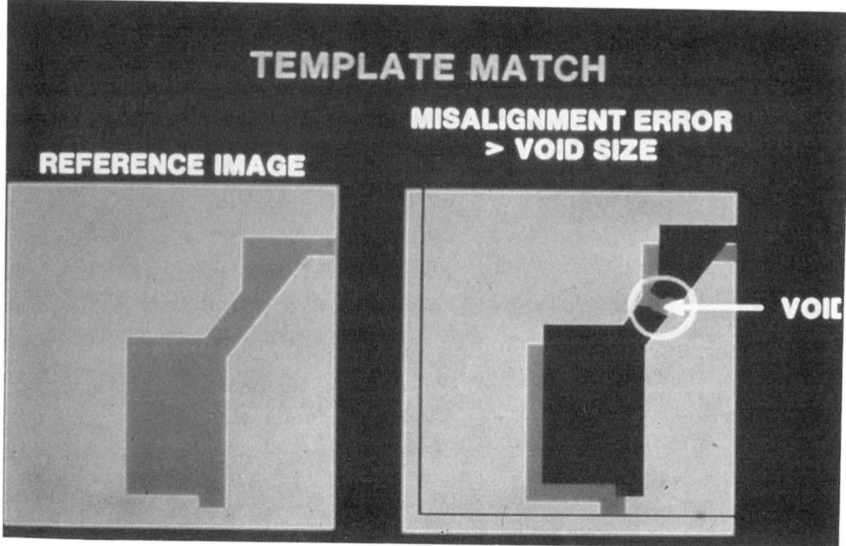

Figure 8.34 - Pixel map (courtesy of General scanning/SVS).

Figure 8.35 - Pixel counting measuring technique (courtesy of Automated Vision Systems).

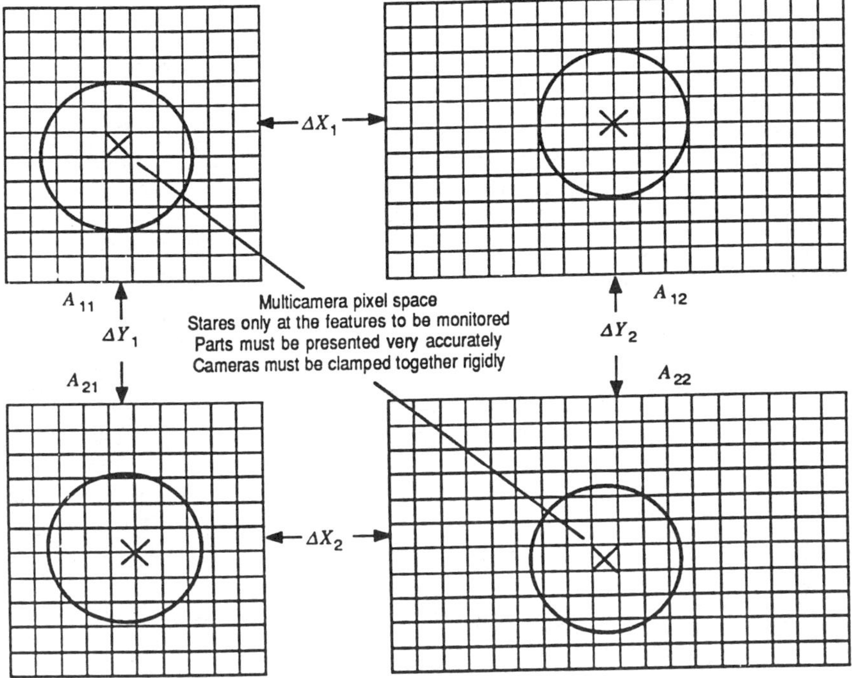

Figure 8.36 - Using multiple camera based vision system to gage large objects (courtesy of General Motors).

Delta or normalized image intensity pixel maps

X number of data points, each representing the integration of intensity over some area in the picture row/column totals.

Often companies that offer these products refer to the representation so established as a "fingerprint" or template. Some companies have executed these programs in hardware and, consequently, can handle some decision-making at throughputs as high as 3000 per minute.

These systems typically operate in a "train by showing" technique. During training (sometimes called learning), a range of acceptable representative products is shown to the system, and the representation, which is to serve as a standard, is established. The representation may be based on a single object or on the average of the images from many objects, or may include a family of known good samples, each creating a representation standard to reflect the acceptable variables.

In the operating mode, decision-making is based on how close the representation from the present object being examined compares to the original or standard representation(s). A "goodness-of-fit" criterion is established during training to reflect the range of acceptable appearances the system should be tolerant of. If the difference between the representation established from the object under test and the standard exceeds the "goodness-of-fit" criteria, it is considered a reject. Significantly, the decision may be based on a combination of criteria (pixel counts and transition count, for example). The goodness-of-fit criteria then becomes based on statistical analysis of the combination of each of the "fit" criteria.

Decision-making, in conjunction with these approaches, can be either deterministic or probabilistic. Deterministic means that given some state or set of conditions, the outcome of a function or process is fully determined with 100% probability of the same outcome. Probabilistic means that a particular outcome has some probability of occurrence (100%), given some initial set.

A major reason these techniques work in imagery stems from the fact that imagery is highly redundant. Changes in images as a function of spatial coordinates is generally slow, and more often than not, neighboring pixels look very much like each other.

Some of these systems have an ability to compensate for translation and rotation errors to account for positional uncertainty. Some will employ simple timing analysis-start processing when transition is first detected, for example. This can compensate for both horizontal and vertical translation, but not rotation.

Translation, as well as rotation compensation, is generally obtained using correlation techniques as described in the pattern matching section (Figures 8.37 and 8.38). Significantly, in some applications, it may be the objective to determine the extent of rotation to be able to feed the information back to the machine to compensate accordingly for the next operation. Such is the case with wire bonders and die slicers in microelectronics manufacturing. In this instance, X, Y and theta data are fed back to the machine to make corrections before operating on the semiconductor chip or silicon wafer.

Although systems exist capable of providing some translation and rotational compensation before representation extraction to eliminate position as a variable, such systems are bound by requiring that the object always remain in the field of view. By expanding the field of view to make the system tolerant of positional uncertainty, one does sacrifice the size detail such a system can reliably detect, or as the basis of a decision.

Many of these type systems base their representation on threshold images. Recognizing that the encoded value of each pixel is based on the average intensity value across the pixel, because of sampling and part variables, there is inherent uncertainty associated with the decision of which region to identify a pixel with, especially along the boundaries. Consequently, these systems generally have some uncertainty related to what constitutes "good."

The "goodness-of-fit" criterion must be tolerant of this scene-to-scene interpretation variation. The fit criteria can be established by experimenting with comparing the range of representations established from routine production items. One inherent weakness of these systems may be that to be forgiving of acceptable vari-

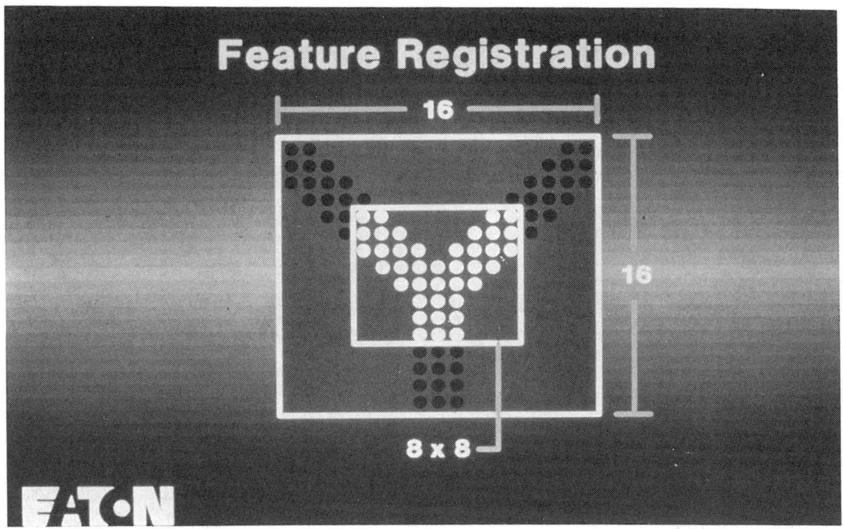

Figure 8.37 - One tactic for correcting for translation positional uncertainty (courtesy of Inex Vision Systems).

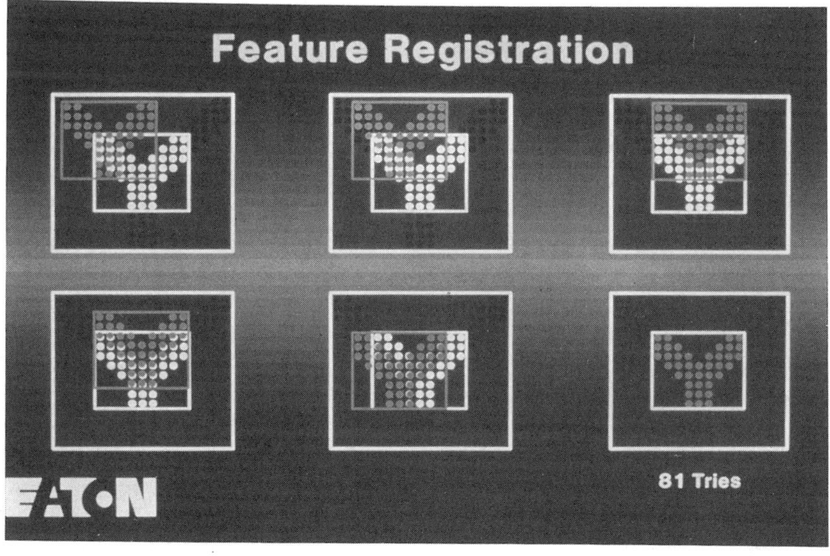

Figure 8.38 - Correlation routine used to compensate for rotation and translation positional uncertainty (courtesy of Inex Vision Systems).

ables, the "goodness-of-fit" criterion becomes too lenient, and the escape rate for defective products becomes excessive. Alternatively, by establishing the fit too tightly, the incidence of false rejects will increase, a condition that may be equally intolerable in a production environment. As a general rule, systems based on heuristic decision analysis techniques are most successful where significant contrast changes represent the basis of the decision. The gray scale shade difference should exceed 10% and, ideally, 20%. Where backlighting is possible, these systems can be very effective. Where contrast can be achieved by inferring data from the distortion due to structured lighting, these techniques can also be effective. In these instances, the contrast stems from the use of filtering techniques to isolate the decision on the structured light-HeNe laser light (632.8 nanometers), for example.

8.6.2 Decision Theoretic

Decision theoretic analysis is frequently associated with the SRI set of techniques. Decisions are made based on comparisons of the feature (Table 8.2) vector created from the specific geometric features selected upon which to base the decision during training, for example area and perimeter (Table 8.2). In these "decision-theoretic" approaches, objects are represented by N features (Figure 8.28) or an N-dimensional feature vector, and the decision is based on a distance measure in vector space (Figure 8.39). These techniques are especially well suited to recognition, verification, and location analysis.

8.6.3 Syntactic Analysis

The ability to make decisions based on pieces of an object is usually based upon "syntactic" analysis, unlike the decision theoretic approach. In this case, the object is represented as a string, a tree, or a graph of pattern primitives and their

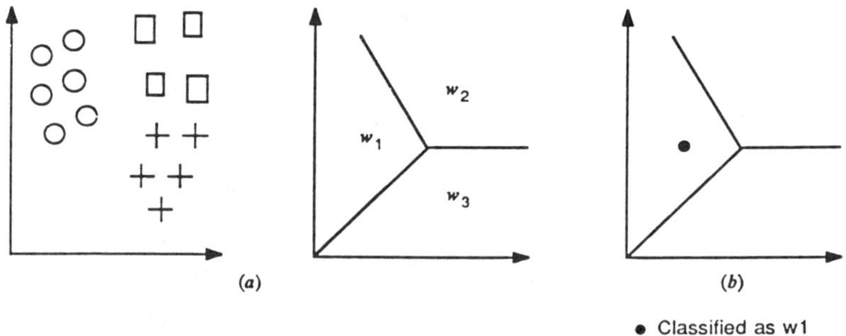

Figure 8.39 - Decision vector and pattern recognition paradigm: nearest neighbor method.

TABLE 8.2 Prototype Development

Moments

General moment Y-1 ,- j X, N, Y, M, 1, J = 0,1,2, Where each (X, Y) pixel is either black (1) or white (0)

Zero Moment Area = M_{00} = N =Number of black cells

First Moments X centroid = X = $M_{01}/M_{00} = \sum_i X_i$

Y centroid = Y = $M_{01}/M_{00} = \sum_j Y_j$

Second moments $M20 = (\sum_i X^2)Y^2$, $M_{02} = (\sum_j Y^2)X^2$, $M_{11} = (\sum_i \sum_j XY) - XY$

Angle = $1/2 + \tan^{-1}(2M_{11}/(M_{02}-M_{20}))$

Magnitude = $1/\{2(M_{20} + M_{02}) + [(M_{02} - M_{20})^2 - 4M_{11}^2]^{1/2}\}$

Sixth moments M_{06}, M_{15}, M_{24}, M_{60}

Area Algorithm

Binary threshold area = number of pixels that are white (or black)

$Area = N = \sum P(X, Y)D_l$ $D_l = 1$ if pixel is white

$D_l = 0$ if pixel is black

Whenever a white pixel is met, the counter is incremented by 1. Alternatively, the counter increments every pixel between threshold edges. If gray scale is used, D_l *becomes the gray scale value itself. This is sometimes called the zero moment.*

Max-min Algorithm

Once the centroid is known, the object perimeter is scanned and each edge pixel position is subtracted from the centroid value. The first such value is stored in two counters (max and min). Each subsequent perimeter value is compared to these counts and if it is larger, it replaces the max count. If it is smaller, it replaces the min count.

Centroid Algorithm

The two centroids X and Y are first moments and are sometimes referred to as the mean. Similarly, second, third, and fourth moments are analogous to standard deviation, skew, and kurtosis.

$X = \sum_l X_l/A$ $Y = \sum_j Y_j/A$

Each time a white pixel is encountered, its X coordinate and its Y coordinate are added to the centroid count. After the count is complete, it is divided by the area.

Min R Max for Orientation Calculation

This feature is like the semimajor axis feature, but it is completely general rather than specific to ellipses. It may also be used to calculate part orientation. It is assumed that all parts can be stored or referenced with their max R direction parallel to the Y axis. The max R is known by its two end points (centroid and perimeter). These numbers can be used to calculate the relative orientation.

Min R Max for Determination of Handedness

(*continues*)

When the direction of max R is calculated and aligned to the Y axis and the direction of min R is calculated and if the min R directions of both the image and the standard are in the same direction, the object is right handed. If the directions are opposed, the object is left-handed. Thus, the attributes of only one of an enantiomorphic pair need to be stored.

---

relationships. Decision-making is based on a parsing procedure. Another way to view this is as local features analysis (LFA) - a collection of local features with specified spatial relationships between various combinations thereof. Again, these primitives can be derived from binary or gray scale images thresholded or edge processed.

For example, three types of shape primitives include curve, angle and line that together can be used to describe a region. Image analysis involves decomposing the object into its primitives, and the relationships of primitives results in recognition. The primitive decision-making can be performed using decision theoretic or statistical techniques.

An easy example involves finding a square in an image. It is known that a square has four corner points equidistant from each other. A corner point in a square has two contiguous points that are at right angles to it. A system based on syntactic analysis would first find all the points satisfying the definition of a corner point [Figure 8.40*(a)*] and then find all the points whose relation to each other satisfies the equidistant syntax [Figures 8.40 *(b)* and 8.40 *(c)]*.

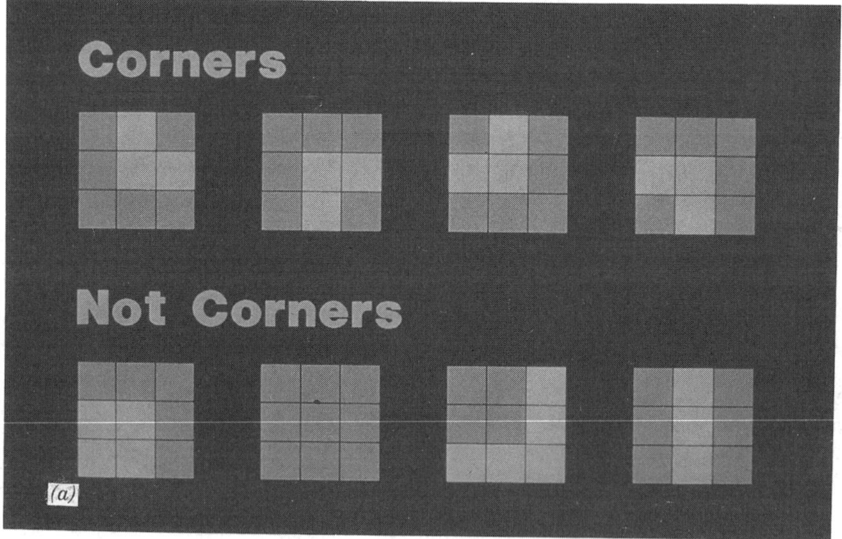

Figure 8.40a-c. Example of syntactic analysis (courtesy of RVSI/Itran).

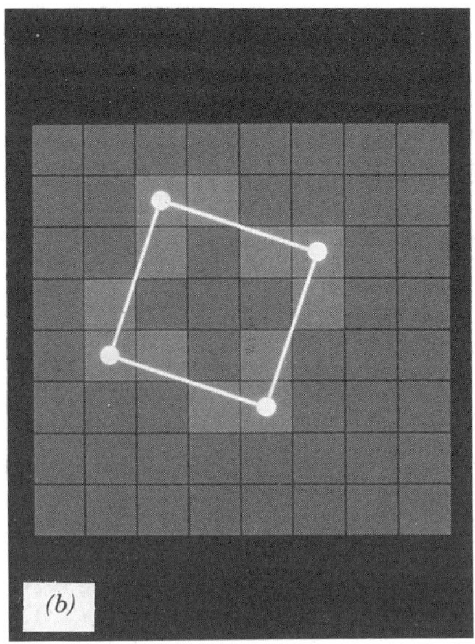

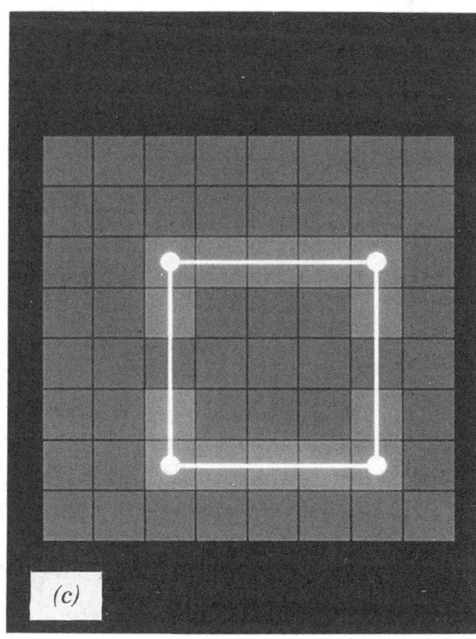

Figure 8.40 Continued.

8.6.4 Edge Tracking

In addition to geometric feature extraction of boundary images, image analysis can be conducted by edge tracking; when the edge is detected, it is stored as a link of edge points. Alternatively, line encoding and connectivity analysis can be conducted. That is, the location of edge points detected is stored and line fitting is performed.

Decision-making is then based on comparison of line segments directly or based on probability theory. Line segment descriptions of objects are called structural descriptions. The process of comparing them to models to find the most similar model is called structural pattern recognition.

8.7 A WORD ABOUT GRAY SCALE

Gray scale processing is a much overused term in machine vision. Gray levels are physical measurements that represent the integration of many properties: hue, saturation, lightness, texture, illumination, shadows, viewpoint, surface reflectance, surface orientation, filter properties, sensor sensitivity nonuniformity, and so on. All companies that offer products that encode pixel sites into a digital value corresponding to brightness refer to their systems as "gray scale" systems. In many instances, the next operation involves using an established threshold to assign all brightness values above a certain level to one region and all those below to another. In other words, binary picture processing is what actually takes place. If based on adaptive thresholding, it may be appropriate to consider such a system a gray scale system; otherwise it is actually a binary system.

Some gray scale systems establish a representation based on the gray scale content (the average gray value in the scene or in the window), a histogram of the gray values, Fourier coefficients based on the gray values, and so on. Deviations from the standard representation based on contrast changes are detected.

Some products develop edges based on gray scale gradients. The edge segmentation that ensues becomes the basis upon which to model the scene; geometric features can be extracted on models based on, for example, line vectors and can be the basis of the representation stored. In these systems each of the eight (in the case of a 256-shade system) arrays of bits is stored in its own individual RAM memory bit plane (Figure 8.41).

In addition to video noise (present in virtually all imaging sensors) contributing to a vision system's inability to repeat precisely its data when a system is presented the same scene, is a phenomenon often referred to as spatial quantization. This stems from the fact that scene edges are not precisely merged with the array pixel pattern of the sensor. Subtle subpixel variations in the position of the scene can cause major variations in the gray levels of the pixels located along the edges. Keeping this in mind, where edges are by definition places where the image gray levels cross a binarization threshold, thresholding amplifies edge variations, causing uncertainty in the perimeter of an object. Gray scale processing

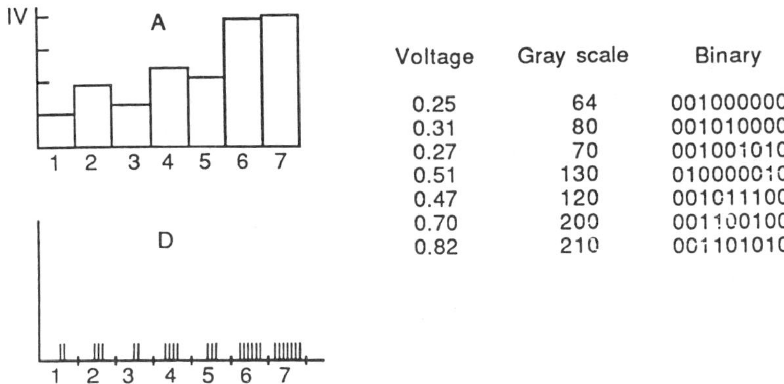

Voltage	Gray scale	Binary
0.25	64	001000000
0.31	80	001010000
0.27	70	001001010
0.51	130	010000010
0.47	120	001011100
0.70	200	001100100
0.82	210	001101010

If the gray scale is to be used throughout the vision system, each of the 8 arrays of bits must be stored on its own individual RAM memory bit plane.

Figure 8.41 - Analog-to-digital conversion for gray scale.

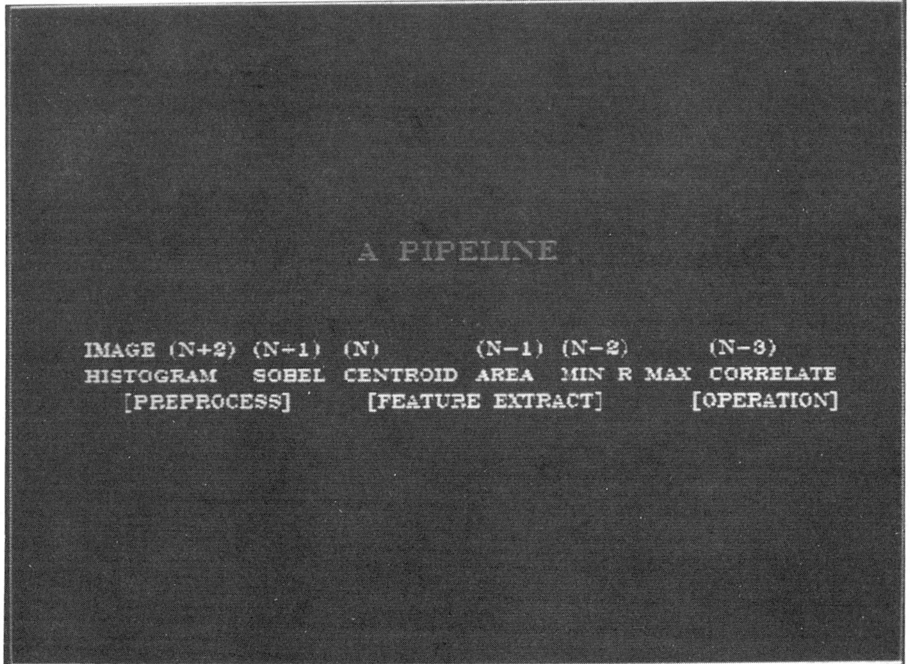

Figure 8.42 - Pipeline architecture.

algorithms, while also sensitive to edge variations, tend to normalize them or at least not amplify variations.

While systems based on gray scale analysis are computationally more expensive, they are more tolerant of contrast variations within an object and from object to object as well as those due to shadows, lighting, and so on. Gray scale boundary descriptive (edge) techniques also lend themselves to analyzing touching and overlapping parts as well as flexible objects or objects with several stable positional states as long as the necessary structured elements or pieces of the object can be recognized.

Each approach offers advantages in certain applications. Where contrast is available upon which to base a decision, techniques involving, for example, binary representations and average gray values, can be effective. Typically, assembly verification can be performed with these approaches. Where the decision must be based on low contrast and the representation based on edges is essential, gray scale gradient techniques will be required. Gaging with reasonable accuracy requirements usually benefit from systems that employ these approaches.

8.8 SUMMARY

In summary, image-processing procedures include image enhancement, feature extraction and image analysis/classification. Enhancement is generally performed to remove unwanted clutter or noise. It may involve a variety of techniques from simple image frame averaging to spatial, temporal, median, or other filtering. As suggested, feature extraction may involve histograms, segmentation, masking (structuring an element), line thinning, and so on. Image analysis/classification may involve region labeling, line segment labeling, histogram, pixel counts, and so on.

The steps associated with image acquisition and image enhancement as described herein are called the "iconic" processing steps as they deal with pictorial data, or at least a gray value digital representation. Image analysis and image coding or classification are sometimes referred to as the symbolic processing stage since they deal with a representation of the features extracted from an image rather than with the image itself. Symbolic phases include recognition of objects and object attributes, and interpretation of the relationship between attributes.

Systems that are more tolerant of contrast and part registration generally perform more computationally expensive preprocessing, processing, and analysis. While much of this could be done in software, the time would be prohibitive for many shop-floor applications. Consequently, many commercial systems are based on creative approaches in the following:

Custom hardware (edge, correlation, convolution, histogram, etc.)
Algorithm-driven architectures using floating-point processors
Memory-driven architectures that minimize need to access data from memory

Tailored instruction sets (edge detection, histogram, etc.)
Data flow computers with multitasking capabilities
Parallel computer architectures using array processors or multiprocessors

These can obtain gains of 1000-100,000 in operations per pixel.

Pipeline architectures (Figure 8.42) are those in which the image moves into the processor and the various operations are performed sequentially. After each operation is performed, it is moved to the next operator, and the next image moves into the original operator. A processed image comes out of the pipeline every frame (once the first cycle of operators is completed), but the time lapse between entering and leaving the pipeline can be tens of seconds.

A parallel-pipeline architecture is one in which the image enters the pipeline, and many different operators work on the image simultaneously.

REFERENCES

IMAGE PROCESSING

Alekeander, I., Artificial Vision for Robots, Methuen, London, 1984.

Ankeney, L. A., "On a Mathematical Structure for an Image Algebra," National Technical Information Service, Document AD-AI50228.

Ballard, D. H., and Brown, C. M., Computer Vision, Prentice-Hall, Englewood Cliffs, NJ, 1982.

Barrow, H. G., and Tenenbaum, J. M., "Computational Vision," Proceedings of the Institute of Electric and Electronics Engineers, May, 1981.

Batchelor, B. G., et al., Automated Visual Inspection, IFS Publications, Bedford, England, 1984.

Baxes, G. A., "Vision and the Computer: An Overview," Robotics Age, March 1985.

Becher, W. D., "Cytocomputer, A General Purpose Image Processor," ASEE 1982, North Central Section Conference, April 1982.

Brady, M., "Computational Approaches to Image Understanding," National Technical Information Service, Document AD-AI08191.

Brady, M., "Seeing Machines: Current Industrial Applications, " Mechanical Engineering, November, 1981.

Cambier, J. L., et al., "Advanced Pattern Recognition," National Technical Information Service, Document AD-AI32229.

Casasent, D. P., and Hall, E. P., "Rovisec 3 Conference Proceedings," SPIE, November 1983.

Chen, M., et al., "Artificial Intelligence in Vision Systems for Future Factories," Test and Measurement World, December 1985.

Cobb, J., “Machine Vision: Solving Automation Problems,” Laser Focus/ElectroOptics, March 1985.

Corby, N. R., Jr., “Machine Vision for Robotics,” IEEE Transactions on Industrial Electronics, Vol. IE-30, No. 3, August 1983.

Crowley, J. L., “A Computational Paradigm for Three-Dimensional Scene Analysis,” Workshop on Computer Vision: Representation and Control, IEEE Computer Society, April 1984.

Crowley, J. L., “Machine Vision: Three Generations of Commercial Systems,” The Robotics Institute, Carnegie-Mellon University, January 25, 1984.

Eggleston, Peter, "Exploring Image Processing Software Techniques," Vision Systems Design, May, 1998.

Eggleston, Peter, "Understanding Image Enhancement," Vision Systems Design, July, 1998.

Eggleston, Peter, "Understanding Image Enhancement, Part 2," Vision Systems Design, August, 1998.

Faugeras, O. D., Ed., Fundamentals in Computer Vision, Cambridge University Press, 1983.

Fu, K. S., “The Theoretical Background of Pattern Recognition as Applicable to Industrial Control,” Learning Systems and Pattern Recognition in Industrial Control, Proceedings of the Ninth Annual Advanced Control Conference, Sponsored by Control Engineering and the Purdue Laboratory for Applied Industrial Control, September 19-21, 1983.

Fu, K. S., “Robot Vision for Machine Part Recognition,” SPIE Robotics and Robot Sensing Systems Conference, August 1983.

Fu, K. S., Ed., Digital Pattern Recognition, Springer-Verlag, 1976.

Funk, J. L., “The Potential Societal Benefits From Developing Flexible Assembly Technologies,” Dissertation, Carnegie-Mellon University, December 1984.

Gevarter, W. B., "Machine Vision: A Report on the State of the Art," Computers in Mechanical Engineering, April, 1983.

Gonzalez, R. C., “Visual Sensing for Robot Control,” Conference on Robotics and Robot Control, National Technical Information Service, Document AD-A134852.

Gonzalez, R. C., et al., “Digital Image Processing: An Introduction,” Digital Design, March 25, 1986.

Grimson, W. E. L., From Images to Surfaces, A Computational Study of the Human Early Visual System, MIT Press, Cambridge, MA, 1981.

Grogan, T. A., and Mitchell, 0. R., “Shape Recognition and Description: A Comparative Study,” National Technical Information Service, Document ADA132842.

Heiginbotham, W. B., “Machine Vision: I See, Said The Robot,” Assembly Engineering, October 1983.

Holderby, W., “Approaches to Computerized Vision,” Computer Design, December 1981.

Hollingum, J., “Machine Vision: The Eyes of Automation, A Manager’s Practical Guide,” IFS Publications, Bedford, England, Springer-Verlag, 1984.

Jackson, C., “Array Processors Usher in High Speed Image Processing,” Photomethods, January 1985.

Kanade, R., "Visual Sensing and Interpretation: The Image Understanding Point of View," Computers in Mechanical Engineering, April, 1983.

Kent, E. W., and Schneier, M. O., “Eyes for Automation,” IEEE Spectrum, March 1986.

Kinnucan, P., “Machines That See,” High Technology, April 1983.

Krueger, R. P., “A Technical and Economic Assessment of Computer Vision for Industrial Inspection and Robotic Assembly,” Proceedings of the Institute of Electrical and Electronics Engineers, December 1981.

Lapidus, S. N., “Advanced Gray Scale Techniques Improve Machine Vision Inspection,” Robotics Engineering, June 1986.

Lapidus, S. N., “Advanced Gray Scale Techniques Improve Machine Vision Inspection,” Robotics Engineering, June 1986.

Lapidus, S. N., and Englander, A. C., “Understandings How Images Are Digitized,” Vision 85 Conference Proceedings, Machine Vision Association of the Society of Manufacturing Engineers, March 25-28, 1985.

Lerner, E. J., "Computer Vision Research Looks to the Brain," High Technology, May 1980.

Lougheed, R. M. and McCubbrey, D. L., “The Cytocomputer: A Practical Pipelined Image Processor,” Proceedings of the 7th Annual International Symposium on Computer Architecture, 1980.

Marr, D. "Vision - A Computational Investigation into the Human Representation and Processing of Visual information," W. H. Freeman & Co., New York, 1982.

Mayo, W. T., Jr., “On-Line Analyzers Help Machines See,” Instruments and Control Systems, August 1982.

McFarland, W. D., “Problems in Three-Dimensional Imaging,” SPIE Rovisec 3, November 1983.

Murray, L. A., “Intelligent Vision Systems: Today and Tomorrow,” Test and Measurement World, February 1985.

Nevatia, R., Machine Perception, Prentice-Hall, Englewood Cliffs, NJ, 1982.

Newman, T., " A Survey of Automated Visual Inspection," Computer Vision and Image Understanding, Vol. 61, No. 2, March, 1995.

Novini, A, "Before You Buy a Vision System," Manufacturing Engineering, March, 1985.

Pryor, T. R., and North, W., Ed., Applying Automated Inspection, Society of Manufacturing Engineers, Dearborn, MI, 1985.

Pugh, A., Robot Vision, IFS Publications, Bedford, England, 1983.

Rosenfeld, A., "Machine Vision for Industry: Concepts and Techniques," Robotics Today, December 1985.

Rutledge, G. J., "An Introduction to Gray Scale Vision Machine Vision," Vision 85 Conference Proceedings, Machine Vision Association of the Society of Manufacturing Engineers, March 25-28, 1985.

Sanderson, R. J., "A Survey of the Robotics Vision Industry," Robotics World, February 1983.

Schaeffer, G., "Machine Vision: A Sense for CIM," American Machinist, June 1984.

Serra, J., Image Analysis and Mathematical Morphology, Academic, New York, 1982.

Silver, W. M., "True Gray Level Processing Provides Superior Performance in Practical Machine Vision Systems, " Electronic Imaging Conference, Morgan Grampian, 1984.

Sternberg, S. R., "Language and Architecture for Parallel Image Processing," Proceedings of the Conference on Pattern Recognition in Practice, Amsterdam, The Netherlands, May 21-30, North-Holland Publishing Company, 1980.

Sternberg, S. R., "Architectures for Neighborhood Processing," IEEE Pattern Recognition and Image Processing Conference, August 3-5, 1981.

Strand, T. C., "Optics for Machine Vision," SPIE Proceedings Optical Computing, Vol. 456, January 1984.

Warring, R. H., Robots and Robotology, TAB Books Inc., Blue Ridge Summit, PA, 1984.

Wells, R. D., "Image Filtering with Boolean and Statistical Operations," National Technical Information Service, Document AD-AI38421.

West, P., "Overview of Machine Vision," Seminar Notes Associated with SME/ MVA Clinics.

9

Three-Dimensional Machine Vision Techniques

A scene is a three-dimensional setting composed of physical objects. Modeling a three-dimensional scene is a process of constructing a description for the surfaces of the objects of which the scene is composed. The overall problem is to develop algorithms and data structures that enable a program to locate, identify, and/or otherwise operate on the physical objects in a scene from two-dimensional images that have a gray scale character.

What are the approaches to three-dimensional machine vision available commercially today? The following represent some "brute-force" approaches: (1) two-dimensional plus autofocusing used in off-line dimensional machine vision systems; (2) 2Dx2Dx2D, that is, multiple cameras each viewing a separate two-dimensional plane; (3) laser pointer profile probes and triangulation techniques; and (4) acoustics.

Several approaches have emerged, and these are sometimes classified based on triangulation calculations:

1. Stereoscopy
    - A. Passive
        1. Binary
            - a. Point
            - b. Edge

c. Area
2. Gray scale
a. Point
b. Edge
c. Area
d. Template
3. Color

B. Active, using projected bars and processing techniques associated with A.1 and A.2

C. Passive/active, based on laser scanner techniques, sometimes referred to as based on signal processing

2. Controlled illumination

A. Structured light
1. Sheet
2. Bar
3. Other

B. Photometric stereo

Another class of approaches emerging and largely fostered by projects affiliated with the autonomous guided vehicle programs of the military are based on time of flight: (a) time of arrival and (b) phase shift.

In addition, research is being conducted into three-dimensional systems based on shape from texture, shading, and motion as well as laser holography. At this time, however, the three most popular methods of acquiring the third dimension of data are (1) stereo views, (2) range images, and (3) structured light projections (Figure 9. 1).

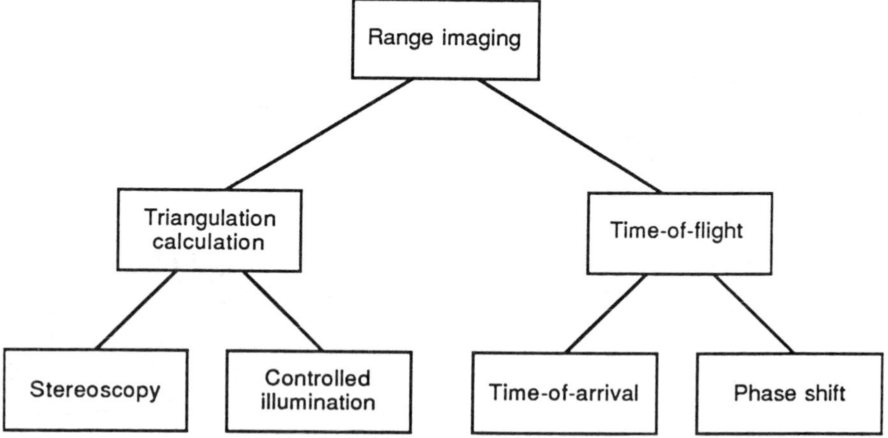

Figure 9.1 - Various approaches to obtaining three-dimensional data.

Methods 1 and 2 rely on triangulation principles, as may some ranging techniques. These systems can be further classified as active or passive systems. In active systems, data derived from a camera(s) are based on the reflection of the light source off the scene. Most ranging techniques are active. Passive systems utilize the available lighting of the scene.

It has been suggested that the most complicated and costly three-dimensional image acquisition system is the active nontriangulation type, but the computer system itself for such a system may be the simplest and least costly. On the other hand, the simplest image acquisition system, passive nontriangulation (monocular), requires the most complex computer processing of the image data to obtain the equivalent three-dimensional information.

9.1 STEREO

An example of a passive stereo triangulation technique, depicted in Figure 9.2, is the Partracking system developed by Automatix (now part of RVSI). They overcome the massive correspondence dilemma by restricting examination to areas with specific features. Essentially, two cameras are focused to view the same feature (or features) on an object from two angles (Figure 9.3). Trigonometrically the feature is located in space.

The algorithm assumes a "pinhole" model of the camera optics; that is, all rays reaching the camera focal plane have traveled through a common point referred to as the optics pinhole. Hence, a focal plane location together with the pinhole location determines a unique line in space. A point imaged by a pair of

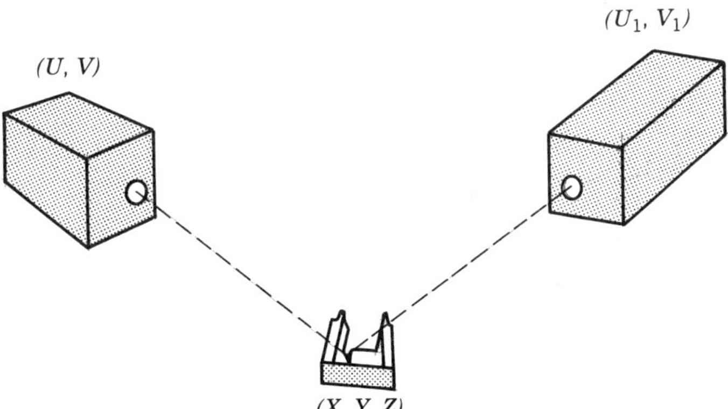

Figure 9.2 - Triangulation from object position as practiced in Automatix partracking system.

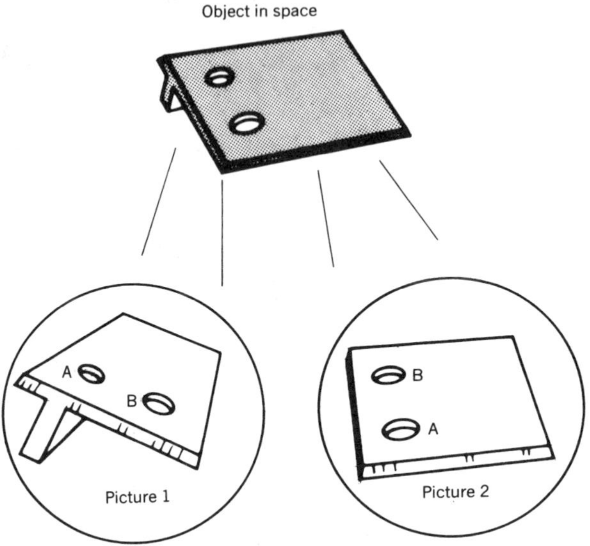

Figure 9.3 - Stereo views of object in space.

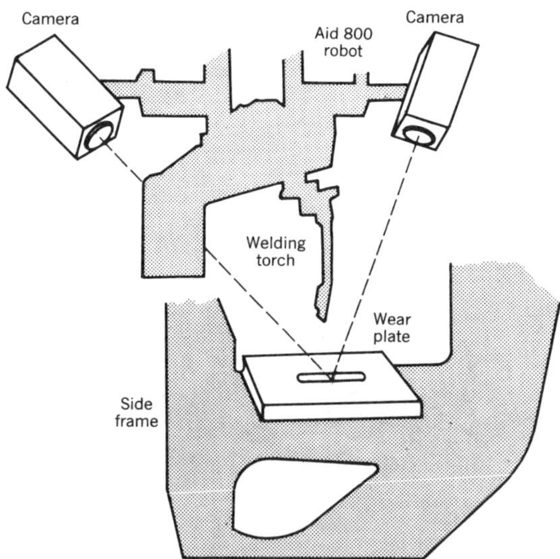

Figure 9.4 - Use of stereo vision in welding. Vision locates edges of slot and welding robot arc welds wear plate to larger assembly (train wheel).

cameras determines a pair of lines in space, which intersect in space at the original object point. Figure 9.4 depicts this triangulation of the object's position. To compensate for noise and deviations from pinhole optics, point location is done in a least-squares sense - the point is chosen that minimizes the sum of the squares of the normal distances to each of the triangulation lines.

The further apart are the two cameras, the more accurate the disparity depth calculation, but the more likely it is to miss the feature and the smaller the field of view overlap. The displacement between the two cameras is inversely proportional to depth. This displacement in the image plane of both cameras is measured with respect to the central axis; if focal length and the distance between cameras are fixed, the distance to the feature can be calculated.

In general, the difficulty with this approach is that in order to calculate the distance from the image plane to the points in the scene accurately, a large correspondence or matching process must be achieved. Points in one image must be matched with the corresponding points in the other image. This problem is complicated because certain surfaces visible from one camera could be occluded to the second camera. Also, lighting effects as viewed from different angles may result in the same surface having different image characteristics in the two views. Furthermore, a shadow present in one view may not be present in the other. Moreover, the process of correspondence must logically be limited to the overlapping area of the two fields of view. Another problem is the trade-off of the accuracy of the disparity range measurement (depends on camera separation) and the size of the overlap (smaller areas of overlap with which to work).

As shown in Figure 9.4, a pair of images are processed for the features of interest. Features can be based on edges, gray scale, or shape. Ideally the region examined for the features to be matched should be "busy." The use of edges generally fulfills the criteria for visual busyness for reliable correlation matching and at the same time generally requires the least in computational cost. The actual features are application dependent and require the writing of application-specific code. The image pair may be generated by two rigidly mounted cameras, by two cameras mounted on a robot arm, or by a single camera mounted on a robot arm and moved to two positions. Presenting the data to the robot (in the cases where interaction with a robot takes place) in usable form is done during setup. During production operation, offset data can be calculated and fed back to a robot for correction of a previously taught action path. The Automatix Partracker is shown in Figure 9.4.

A key limitation to this approach is the accuracy of the image coordinates used in the calculation; this accuracy is affected in two ways: (1) by the inherent resolution of the image sensor and (2) by the accuracy with which a point can be uniquely identified in the two stereoscopic images. The latter constraint is the key element.

9.2 STEREOPSIS

A paper given by Automatic Vision Corporation at the Third Annual Conference on Robot Vision and Sensory Controls *(SPIE,* Vol. 449) described an extension of photogrammetric techniques to stereo viewing suitable for feedback control of robots.

The approach is based on essential differences in shape between the images of a stereo pair arising out of their different points of view. The process is simplified when two images are scanned exactly synchronized and in a direction precisely parallel to the base line. Under these conditions the distance to any point visible in the workspace is uniquely determined by the time difference *-dt* - between the scanning of homologous image points in the left and right cameras.

Unlike outline processing, stereopsis depends upon the detailed low contrast surface irregularities of tone that constitute the input data for the process. All the point pairs in the image are located as a set, and the corresponding XYZ coordinates of the entire scene are made available continuously. The function required to transform the images into congruence is the Z dimension matrix of all points in the workspace visible to the local scaling of the XY scanning signals.

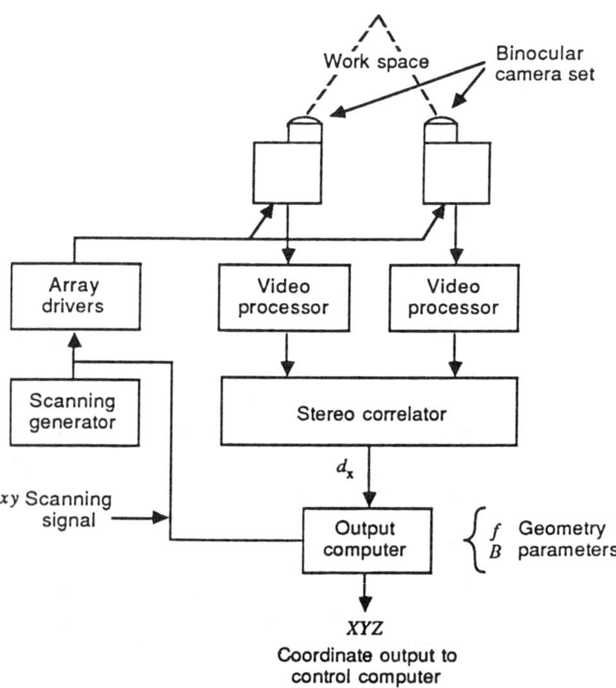

Figure 9.5 - Robot stereopsis system (courtesy of Automatic Vision, Inc.).

A block diagram of the system is shown in Figure 9.5. The XY signals for the synchronous scanning of the two images are produced by scanning generator and delivered to the camera array drivers simultaneously. The video processors contain A/D converters and contrast enhancement circuits. The stereo correlator receives image data in the form of two processed video signals and delivers dimensional data in the form of the *dx* signal that is proportional to 1/Z. The output computer converts the *dx* signal into *XYZ* coordinates of the model space.

This approach relies on a change from the domain of brightness to the domain of time, which in turn becomes the domain of length in image space.

9.3 ACTIVE IMAGING

Active imaging involves active interaction with the environment, a projection and a camera system. This technique is often referred to as structured illumination. A pattern of light is projected on the surface of the object. Many different patterns (pencils, planes, or grid patterns) can be used. The camera system operates on the effect of the object on the projected pattern (a computationally less complex problem), and the system performs the necessary calculations to interpret the image for analysis. The intersections of the light with the part surface, when viewed from specific perspectives, produces two-dimensional images that can be processed to retrieve the underlying surface shape in three dimensions (Figure 9.6).

The Consight system developed by General Motors is one such system. It uses a linear array camera and two projected light lines (Figure 9.7) focused as one line on a conveyor belt. The camera detects and tracks silhouettes of passing objects by displacing the line on the belt. The displacements along a line are proportional to depth. A kink indicates a change of plane, and a discontinuity in the line indicates a physical gap between surfaces.

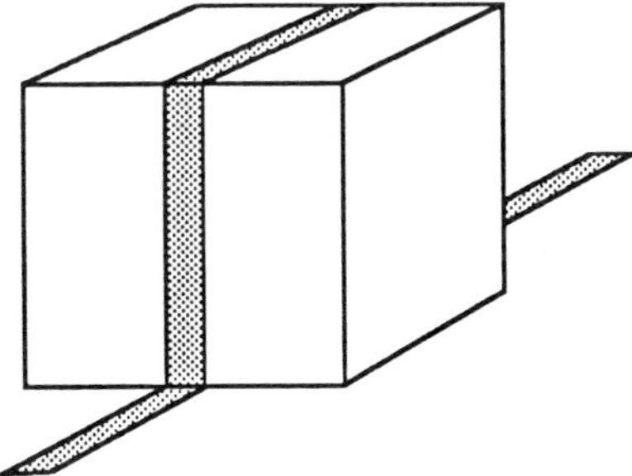

Figure 9.6 - Light stripe technique. Distortion of image of straight line projected onto three-dimensional scene provides range data.

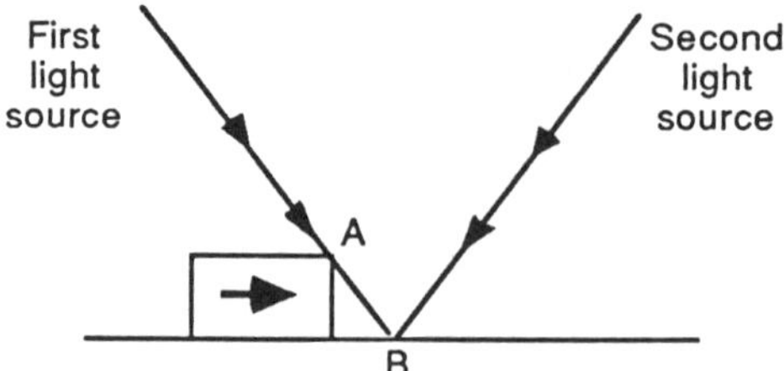

Figure 9.7 - The General Motors Consight uses two planes of light to determine bounding contour of object to finesse shadowing problem depicted. If only the first light source is available, light plane is intercepted by the object position A. Program interpreting scan line will conclude incorrectly that there is an object at position B.

The National Institute of Standards and Technology also developed a system that used a line of light to determine the position and orientation of a part on a table. By scanning this line of light across an object, surface points as well as edges can be detected.

When a rectangular grid pattern is projected onto a curved surface from one angle and viewed from another direction, the grid pattern appears as a distorted image. The geometric distortion of the grid pattern characterizes the shape of the surface. By analyzing changes in this pattern (compared to its appearance without an object in the field), a three-dimensional profile of the object is obtained. Sharp discontinuities in the grid indicate object edges. Location and orientation data can be obtained with this approach.

Another active imaging approach relies on optical interference phenomena. Moire interference fringes can be caused to occur on the surfaces of three-dimensional objects. Specifically, structured illumination sources when paired with suitably structured sensors cause surface energy patterns that vary with local gradient. The fringes that occur represent contours of constant range on the object. The fringe spacing is proportional to the gradient of the surface. The challenge of this method is processing the contour fringe centerline data into nonambiguous contour lines in an automatic manner. Figure 9.8a depicts a Moire fringe pattern generated by an EOIS scanner.

9.4 SIMPLE TRIANGULATION RANGE FINDING

9.4.1 Range from Focusing

This technique senses the relative position of the plane of focus by analyzing the image phase shift that occurs when a picture is out of focus. Knowledge of the focal length and focal plane to image plane distances permits evaluation of focal

(a)

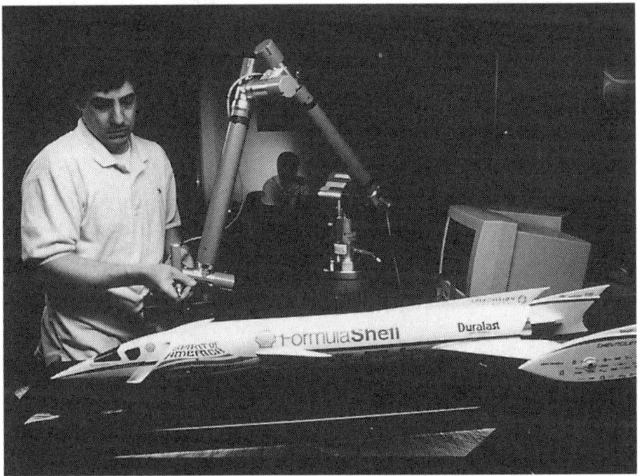

(b)

Figure 9.8 (a) Fringe pattern generated by an EOIS miniscanner. (b) EOIS miniscanner mounted on Faro Technology arm to capture 3D data.

plane to object distance (range) for components in a three-dimensional scene in sharp focus. The sharpness of focus needs to be measured on windows on the image over a range of lens positions to determine the range of corresponding components in the scene. Analysis of the light intensity in these windows allows a microcomputer to calculate the distance to an object. Such a technique can be used to detect an object's edge, for example, and feed that data back to a robot previously trained to follow a procedure based on the edge as a starting point.

9.4.2 Active Triangulation Range Finder

A brute-force approach for absolute range finding is to use simple, one-spot-at-a-time triangulation. This does not use image analysis techniques. Rather, the image of a small spot of light is focused onto a light detector. A narrow beam of light (displaced from the detector) can be swept in one direction or even in two dimensions over the scene. The known directions associated with source and detector orientation at the instant the detector senses the light spot on the scene are sufficient to recover range if displacement between the detector and the source is fixed and known. This approach costs time to scan an entire object. It is suitable, however, for making measurements at selected points on an object, especially where the head can be delivered by a robot to a family of regions on an object where such measurements must be made. This is described further in Chapter 14.

9.4.3 Time-of-Flight Range Finders

Direct ranging can be accomplished by means of collinear sources and detectors to directly measure the time it takes a signal to propagate from source to target and back. The main two approaches are based on acoustic or laser techniques: speed of sound or speed of light. No image analysis is involved with these approaches and assumptions concerning the planar or other properties of the objects in the scene are not relevant.

The time-of-flight approach can be accomplished in two ways: (1) time of flight is directly obtained as an elapsed time when an impulse source is used (Figures 9.9 and 9.10) and (2) a continuous-wave (CW) source signal is modulated and the return signal is matched against the source to measure phase differences (Figures 9.11 and 9.12). These phase differences are interpreted as range measurements.

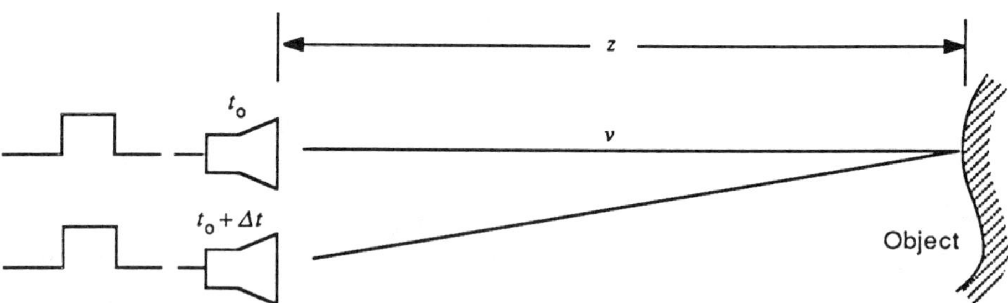

Figure 9.9 - Time-of-flight principle. Pulse traveling with known velocity v is transmitted from detector plane, travels to object, and returns to detector. Distance is determined from elapsed time, Δt.

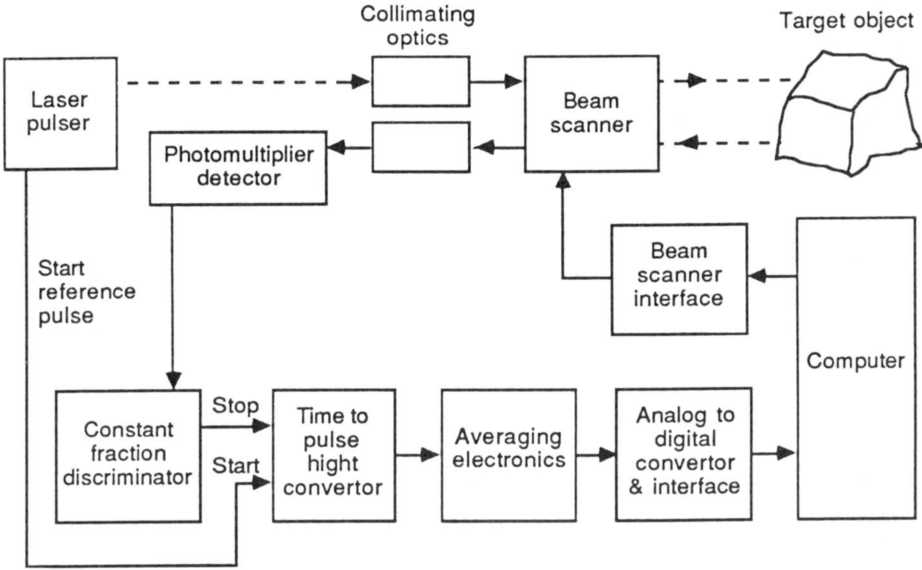

Figure 9.10 - Time-of-flight laser range finder schematic.

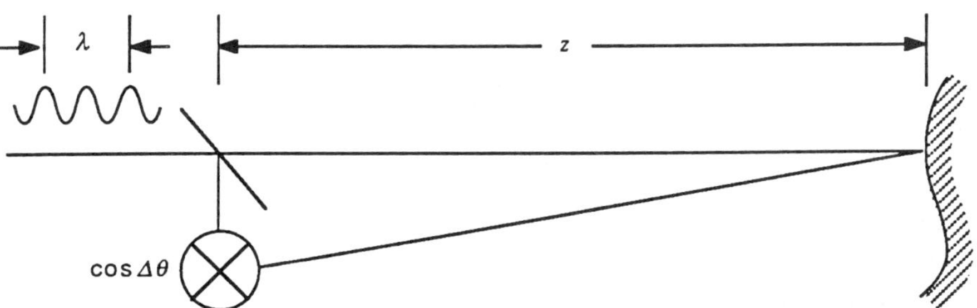

Figure 9.11 - Interferometry principle. Signal beam with known wavelength is reflected off object and allowed to interface with reference beam or local oscillator.

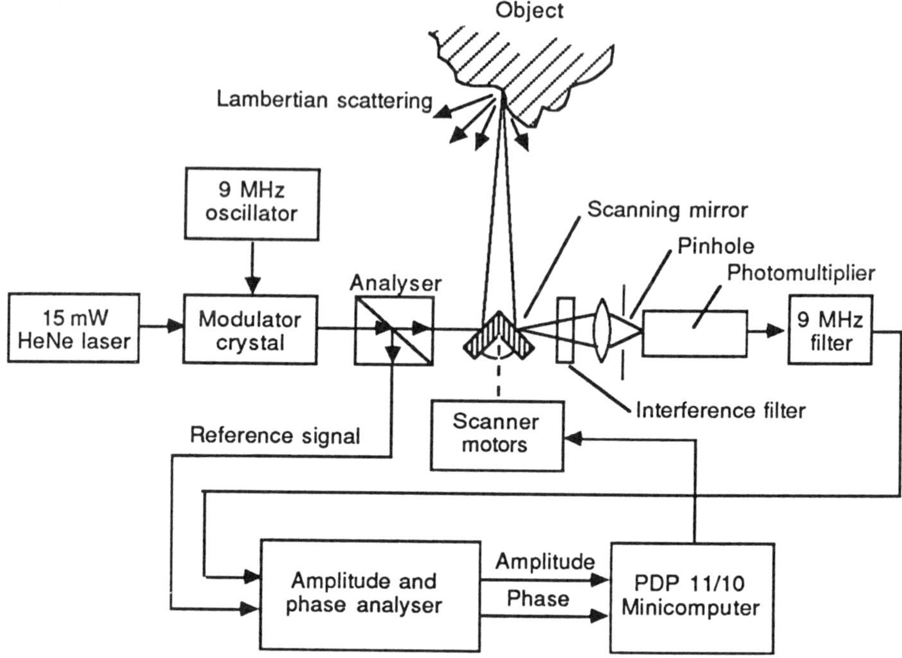

Figure 9.12 - Phase detection laser range finder block diagram.

While acoustic approaches represent an approach, they do not have the resolution required for scenes containing hard objects with surfaces whose normals are in arbitrary directions. Most work conducted with this approach has been based on laser techniques. In these imaging laser scanner approaches, a laser source is used in a pulsed or CW mode to illuminate the object. In the pulsed mode, time-of-flight range gating is employed; in the CW mode, phase modulation with heterodyne detection is used for ranging. While the CW technique is more difficult to implement, it is potentially capable of greater range resolution.

Several advantages cited for this type of system are as follows:

1. No triangulation calculations are required; there is no correspondence problem.

2. No missing-points problem exists when two scenes differ due to occlusion or nonoverlapping areas.

3. The error does not increase with range.

4. Range images are easily generated.

5. There are no shadows (transmitter and receiver are coaxial).

6. Registered intensity images are also available.

On the other hand, some problems include the following:

1. It is not practical for mirror-like surfaces.
2. The system must accommodate a large dynamic range of the reflected beam over the surfaces.
3. Reflectance properties can be a challenge.
4. Noise effects exist.
5. There is inherent variation of the path length and scanning velocity as the scanning mirror rotates. This causes a problem obtaining range values equally spaced along each line of the raster.

9.5 SURFACE MEASUREMENT USING SHADING DATA

Three-dimensional data can be obtained by inferring the surface shape properties based on the analysis of a luminance image. Physical shapes and shape changes usually manifest themselves as luminance changes. This technique assigns an intensity or color to every picture element in the image that accurately simulates the viewing situation. The shading of a surface point depends upon the surface reflection characteristics, surface geometry, and lighting conditions, each of which must be considered in the development of a surface-shading model.

Surface geometry can be depicted as the surface normal vector at the point of interest. Surface reflection is the composite of spectral reflectance (how surface reflects light of specific wavelengths), surface texture (determines the diffusivity and specularity components), and surface transparency (determines amount of light that is refracted by the surface rather than reflected). Various models have been developed for simulating these properties and are utilized to obtain three-dimensional information that represents surface shape by examining the observed intensity of a set of surface points. The intensity of a surface point identifies the solution space for the normal vector to the surface at that point if the viewing geometry and lighting conditions are known.

The ability to perform this task with sufficient accuracy depends on the ability to select an appropriate shading model (models that combine specular and diffuse models) that closely approximate the reflection characteristics of the surface material. One problem with this approach is the requirement for a new set of calibration parameters for every surface to be examined.

If this technique can be effectively implemented, its advantage is that it does not require a high-resolution image to obtain a sufficient number of sample points for surface modeling. A shortcoming is that only an object's shape can be inferred. Other types of analysis and procedures must be applied to derive absolute size, orientation, and location.

A similar approach that employs active lighting uses photometric stereo and reflectance mapping techniques to compute the surface orientations of Lambertian workpieces. If three images of an object are obtained using a single stationary camera, by varying the direction of illumination, the local surface gradients can be determined from the point of intersection of reflectance maps corresponding to those light sources. The limitations of this technique are the assumptions of Lambertian reflection and point source illumination.

9.6 DEPTH FROM TEXTURE GRADIENT

Texture gradient refers to the increasing fuzziness of visual texture with depth observed when viewing a two-dimensional image of a three-dimensional scene containing approximately uniformly textured planes or objects. Changes in texture "coarseness" provides cues about surface slant and relative range; the direction in which coarseness is changing most rapidly corresponds, for a uniformly textured surface, to the slant direction, while an abrupt change in coarseness indicates the possibility of an occluding edge.

Some challenges of these techniques are the following:

1. The regions of the image over which texture features are to be extracted must be uniformly textured in the three-dimensional sense.
2. Prior segmentation is required.
3. Application is restricted to highly textured scenes.
4. Computational cost is high.

Most of the work in this field has dealt with texture as an image property and has been primarily concerned with uniformly textured regions, such as might arise from nonperspective views of uniformly textured surfaces. Practical applications of this technique require a good deal more research.

9.7 APPLICATIONS

One of the earliest industries adopting 3-D machine vision techniques was the wood products industry. Three-dimensional volume measurements are in widespread use in sawmills today. The measurements are used to adjust sawing to obtain the most yield and/or value out of the wood. Applications of these optimizing techniques include bucking, primary breakdown, cant, edger and trimmer operations.

In the semiconductor industry the single biggest application for a 3D ASMV system has been for co-planarity measurements on leaded IC packages. **Co-planarity** is an issue with multi-leaded active SMD's. Both 2-D and 3-D measurements should be made - co-planarity (3-D) and lead alignment (2-D). Co-planarity is also a consideration associated with solder bumps on wafers where high density interconnects are required. Accuracies on the order of +/- 0.00025" are required.

In the electronic industry stand-alone three-dimensional **solder paste inspection** systems are in widespread use to measure the volume of the solder paste after screen-printing and before component placement. Three-dimensional systems are also being used to measure board warpage as solder joint quality can be impacted where components with high density interconnects are used.

In the automotive industry 3-D systems are used for gap and flushness measurements of sheet metal assemblies. Another 3-D ASMV system senses part position in spot welding fixtures to provide positional feedback information to optimize welding spot location.

In the food industry 3-D-based ASMV systems have emerged to support the needs of **water-jet-based portioning systems**. Based on the volume of a fish fillet, chicken fillet, beef or pork, these systems are able to determine where to cut the product for a given weight portion.

Structured light machine vision techniques are used to find and track the weld seam slightly in advance of the arc. The environment is brutal. There are many variations on the same theme with optional technique a function of application specifics: MIG, TIG, etc.; thin, thick, butt, lap, etc.; corners, curves, etc. Requirement can also call for determination of volume of seam.

In addition to these application-specific implementations of 3-D-machine vision systems there are systems of a more general-purpose nature. These are used to provide input to CAD systems for surface rendering and reverse engineering applications or even for comparison to actual dimensional data.

REFERENCES

Barrow, H. G., and Tenenbaum, J. M., "Computational Vision," *Proceedings of the IEEE,* Vol. 69, *No. 5,* May 1981, pp. 572-595.

Braggins, D., "3-D Inspection and Measurement: Solid Choices for Industrial Vision," Advanced Imaging, October, 1994.

Boissonat, J. D., and Germain, T., "A New Approach to the Problem of Acquiring Randomly Oriented Workpieces in a Bin," *Proceedings IJCAI-81,* August 1981, pp. 796-802.

Brady, M., "Seeing Machines: Current Industrial Applications," *Mechanical Engineering,* November 1981, pp. 52-59.

Corby, N. R. Jr., "Machine Vision for Robotics," *IEEE Transaction on Industrial Electronics,* Vol. IE-30, No. 3, August 1983.

Edson, D., "Bin-Picking Robots Punch In," *High Technology,* June 1984, pp. 57-61.

Geo-Centers, "A Review of Three Dimensional Vision for Robotics," ARPA Contract No. DNA-001-79-C-0208, May 1982.

Harding, Kevin, "Improved Optical Design for Light Stripe Gages," SME Sensors Conference, 1986.

Henderson, T. C., "Efficient 3-D Object Representations for Industrial Vision Systems," *IEEE PAMI, Vol.* 5, No. 6, November 1983, pp. 609-618.

Jarvis, R. A., "A Perspective on Range Finding Techniques for Computer Vision," *IEEE PAMI, Vol.* 5, No. 2, March 1983, pp. 122-139.

Kanade, T., "Visual Sensing and Interpretation: The Image Understanding Point of View," *Computers in Mechanical Engineering,* April 1983, pp. 59-69.

Lees, D. E. B., and Trepagnier, P., "Stereo Vision Guided Robotics," *Electronic Imaging,* February 1984, pp. 61-64.

Ray, R., "Practical Determination of Surface Orientation by Radiometry," Society of Manufacturing Engineers, MS82-181, Report from Applied Machine Vision Conference, April 1982.

Rosenfeld, A., "Computer Vision," DARPA Report DAAG-53-76C-0138, April 1982.

Strand, T. C., "Optics for Machine Vision," *SPIE Proceedings-Optical Computing-Critical Review of Technology,* Vol. 456, January 1984.

PAPERS FROM THIRD INTERNATIONAL CONFERENCE ON ROBOT VISION AND SENSORY CONTROLS, NOVEMBER 1983, SPIE PROCEEDINGS, VOL. 449.

Band, M., "A Computer Vision Data Base for the Industrial Bin of Parts Problem," General Motors Research Publication, GMR-2502, August 1977.

Chiang, Min Ching, Tio, James B. K., and Hall, Ernest L., "Robot Vision Using a Projection Method."

Hobrough, T., and Hobrough, G., "Stereopsis for Robots by Iterative Stereo Matching.

McFarland, W. D., and McLaren, R. W., "Problems in 3-D Imaging."

McPherson, C. A., "Three-Dimensional Robot Vision."

10

Applications of Machine Vision in Leading User Industries

10.1 SEMICONDUCTOR INDUSTRY

The semiconductor manufacturing process is inspection intensive. Inspection is performed not only as a means of sorting reject conditions but also to provide feedback to process performance. The goal is to optimize yield. In some fab houses a 1% improvement in yield can add $250,000 per month to the bottom line.

Today much instrumentation exists to: assist inspectors in performing their tasks; automate repetitive tasks, especially manual tasks such as material handling; enhance the performance of the inspector; replace the inspector entirely; or perform a task not possible by a person. Some of this instrumentation incorporates machine vision technology either to enhance performance or replace the inspector.

In addition to inspection, machine vision is embodied in much of the production equipment used in the manufacture of semiconductors to provide visual servoing - feedback of offset position data for motion control.

The process starts with the design of the IC. This is done generally by CAD systems with the aid of CAE systems that incorporate design rule checking techniques to convert design to layered geometries. The digitized design geometries are stored on tape, creating a master database for producing reticles.

The information from the tape is converted to a geometrical image and transferred by electron beam to glass reticle. 1X, 5X or 10X reticle for each design layer is produced for image transfer to photomask or directly to the wafer. The photomask or reticle is compared with the design data with equipment that compares each ½ micron square on the plate to a reformatted set of design data. Simultaneously, die-to-die comparisons are made. Every difference is highlighted and recorded. The coordinates of the difference on the photomask or reticle are then sent along to a repair system, where the plate is made perfect.

The optical stepper process reduces each reticle to 1X and repeats the pattern 100 or more times to create a master photomask. A die-to-die comparison of the photomask is performed at this point. Defects are recorded and the photomask is sent to the repair station for correction before working photomasks are made. After repair, the "perfect" photomask is ready for image transfer to the wafer.

Also leading up to the next step in the manufacturing process is the process leading up to the production of the wafer. These steps include crystal growth using either Czochralski or float zone methods followed by grinding to specific diameters which are typically measured using laser gaging techniques.

The crystal next is sawed into thin wafers using diamond-edged blades. Thickness and flatness over a "Flatness Quality Area" (FQA) are checked, in some cases using machine-vision-based interferometry. Roll off from the manufacturing process at the edge varies, and most flatness-measuring systems provide a means for excluding measurements at the very edge. Typically up to 5 mm can be excluded in 1-mm increments. However, some instruments optionally permit measurements to the very edge. This provides insight to the true character of the wafer, which at its actual circumference can be of interest to the wafer manufacturer. FQA is defined as a SEMI standard. The smaller the FQA, the more repeatable the measurements become as they are made when the edge discontinuities are excluded.

Flatness is generally defined as the deviation of the wafer surface relative to a reference plane. Because virtually all advanced technology mask aligners use a vacuum chuck to hold the wafer during exposure, image surface flatness is usually specified in the "clamped" condition relative to either front or backside referencing. Projection aligners generally employ a "global" (i.e., with respect to the entire wafer surface) front side reference, for which the wafer is gimbaled and tilted so as to remove the effects of linear thickness variation and to minimize wafer surface deviation above and below the plane of the aligner. The exposure cycle is either one time in a single cycle or scanned in smaller sectional areas. Projection aligners usually specify image plane tolerances relative to a global focal plane. Focal depths are on the order of +/- 4 or 6 microns.

Applications for flatness measurements include:

1. As-cut saw areas: saw set-up, saw qualification, SPC of saws
2. Lap and etch: equipment qualification, SPC
3. Reclaim: material removal monitor, QC of final thickness, sampling
4. Grind and polish: equipment qualification, SPC

Some systems targeted at incoming receiving inspection include the ability to measure resistivity as well as thickness and flatness.

Three approaches to making flatness measurements include acoustical, capacitive and interferometric. Such instruments can also measure thickness at the same time.

Wafer dimensions may also be checked with machine vision and the border grind area inspected for scratches, etc., as well as measured. During these operations, machine vision may also be used to detect the location of a small orientation notch on the wafer. Roundness as well as notch detection has also been performed using laser triangulation techniques.

Different processes are used to perform the deposition process (chemical, physical and epitaxial). A film of silicon oxide may be required as a preliminary to a diffusion or implant operation or to cover layers previously placed on the surface. In any event, where films are involved, instruments exist to measure film thickness using either capacitive or electro-optical techniques (ellipsometer).

The unpatterned wafer (before and after film deposition steps) is also checked for geometric defects - pitting, scratches, particulates, etc. Again, either capacitive or electro-optical techniques are used. The electro-optical techniques are either based on laser scanning and light scattering or machine vision and dark field illumination.

This is essentially a 3-D application. It requires the ability to detect: particulates, pits, scratches, haze, grain uniformity, mounds, dimples, saw marks, grooves, fractures, slip, epi-spikes, small particles on large grained surfaces, gel slugs, particles buried in or under film, and film thickness non uniformities. Particles and artifacts as small as 0.1 microns must be detected. This suggests a z-axis resolution that should 3–10 times better. Systems that use flying spot scanning approaches are often used for this application.

While implantation and diffusion steps can also be monitored based on electrically measured properties, most of the following key lithographic or patterning processes (which include spin/bake, align, expose, and etch) can only be monitored by optical techniques.

To form patterns on a silicon wafer, a photographic process of great precision is used. A thin metal layer on a glass or quartz plate (the photomask) contains an image of the desired pattern. A layer of photosensitive material (resist) is spread on the surface of the silicon wafer and dried. It is exposed to ultraviolet light through the mask to cause chemical changes in certain areas of the resist. The pattern is developed chemically, leaving areas of the wafer covered with resist while other areas are clear.

For a diffusion or implantation process, the wafer may then be etched to remove silicon dioxide in the clear areas, following which the remaining resist is stripped away. Doping during diffusion or implantation reaches the silicon only through the windows in the oxide, but not elsewhere. Or, for implantation of and for metal deposition processes, the resist may be left in place. After the resist is stripped off, it leaves a pattern of doping from the implant or a pattern of metal from the deposition.

Frequently it is necessary to provide an inorganic coating on the semiconductor wafer to protect it from exterior influences. This is called a passivation step. The last step involves metallization - evaporation in vacuum of thin metal films onto selected areas to provide the interconnections required. Non-contact film thickness measuring instruments are used. These include techniques based on UV refection spectrophotometry and ellipsometers. In addition to thickness, information based on optical properties and reflectivities enable tighter critical dimension (CD) control as well as optimization of stepper exposure time.

Etch engineers want to be able to detect thicknesses before and after etch to better control etch rate. Diffusion engineers map thickness of unpatterned oxide layers as a means to evaluate gate oxide integrity. CVD engineers are interested in monitoring the refractive index and thickness of nitride and oxynitride films to evaluate film quality. UV or visible or combination spectrophotometers are used at this point as well as instruments based on ellipsometry. X-Ray fluorescence-based instruments are used to measure thickness of metal layers. Some of these instruments are designed as post process inspection devices and some as in-situ instrumentation.

Sample inspection is generally performed after the develop-and-bake cycle and after the etch cycle and before the diffusion stage. This usually involves a die-to-die and die-to-reference image comparison. Such systems basically check for both pattern defects and particles, though not all do both.

Also, some can only handle single layers and some are geared for on-line operation versus off-line. Those based on light scattering techniques are generally in-line and only suitable for finding particles or geometric problems.

Off-line inspection is suitable for statistical quality control and engineering analysis, including verification of reticle quality and stepper set up. Instruments inspect single layers on specially prepared wafers. The objective is qualification of a photolithographic process via the inspection of a resist image on a patterned test wafer that goes through a process prior to processing product wafers. Significantly, on-line does not necessarily mean a 100% inspection - it, too, may be on a sample basis especially where imaging is involved. Scanning electron microscopes with review stations are also used to perform these inspections.

Applications for patterned wafer defect inspection fall into five categories. Image qualification refers to verifying that exposure equipment images are defect-free. Partitioning/engineering analysis is utilized for process characterization and the elimination of defect-causing mechanisms. Foreign material monitoring examines contamination levels in process equipment and process segments. Sector-limited yield uses process monitoring to detect defect density excursions and to monitor yield in specific portions (sectors) of a process cycle. Develop inspection refers to verification of photoresist pattern integrity prior to wafer etching or implantation.

Where defects are detected, that information is fed to a review station where an operator revisits each site to verify and classify the condition. The goal is to understand types, locations and distribution of defects and their effects on yield. Some work has lead to using neural nets to automate the classification step to eliminate inspector subjectivity.

In addition to pattern matching applications there are also instruments that perform critical dimension (CD) measurements and check for overlay registration. In some cases these capabilities are built into one instrument. Machine vision techniques have been applied to performing these inspections, too, although in many cases an operator is very much in the loop, interactively establishing points between which measurements are to be made. Various products have emerged, such as automatic cursors and edge finders, to make the operator more reliable. More and more because of the diffraction limits of optics, secondary electron microscopes are being used.

Significantly, CD measurements are only made on a sampling basis on any given wafer. It is also noted that CD SEM equipment is not particularly well suited for overlay measurements. A trend, however, is the emergence of optical, dedicated overlay tools to complement the CD SEM equipment. The ultimate in automated metrology will see measurement tools completely integrated into the lithography line. But this is not yet the case.

Until now, this application has been addressed with 2-D image processing based systems. Given that the conductor paths are actually 3-D, it may make some sense to address this application with 3-D-based machine vision. That way detail, such as the line width at the base vs. line width at the top can also be analyzed.

Up until now we have described the process end of semiconductor manufacturing, or the front end. The back end, or packaging side, of the process starts with the electrical prober that automatically aligns and tests each IC on the wafer electrically. Ink dots mark the failed circuits. In each tester, machine-vision-based pattern recognition systems are used to align the wafer.

The next step is the dicing operation that saws the wafer into individual dies. Again machine vision systems are used to align the wafer before sawing. In some cases they also have the capability to inspect the IC after sawing for chips, cracks, and evidence of blade wear.

Immediately before die bonding there is a requirement for machine vision to: determine if an ink dot is present (indicates an IC that is a reject and should not be bonded), and inspect for: metallization issues, saw damage, probe mark damage, scratch, smears on die surface or other blemishes. The machine vision system used in die bonding will generally have the ability to inspect for the presence of the die mark and maybe some gross problems as it also provides alignment data.

At this point there may be other requirements out of a machine vision system besides alignment. In some cases the die are packaged into "waffle" packs before bonding. Die bonding operation would require more complex 'find' capability (Figure 10.1). Post die bonding there should be a need for a machine vision system to check for: smears on die, eutectic or epoxy contamination, wetting, lifted die, missing die, preform melt on die surface, die orientation after bonding. Individual semiconductor companies are probably adapting general-purpose machine vision platforms for this application as there is no known turnkey solution.

Figure 10.1 - Cognex machine vision system verifies presence/absence and precise alignment of semiconductor dies bonded to leadframes (courtesy of IBM Microelectronics).

The next operation is wire bonding. Machine vision pattern recognition systems are integrated into these systems to provide: chip alignment; outer lead location on chip carriers, leadframes, CERDIPS, hybrid headers, pin alignment and jumper chip alignment.

The next operation entails packaging. Before packaging it has been suggested that there is a requirement for 'pre-cap' or '3rd op' inspection. This involves checking for: bond location, bond size, bond shape, wire path/position, wire loop/height, lifted wires/non-stick on die or package, wire separation, wire tail length, wire diameter, crack/tear at bond, and crossed wires.

Following packaging there is a requirement to check out the package itself (Figure 10.2). This involves inspecting the package itself, inspecting the markings on the package, and measuring the coplanarity of the leads. Machine vision is used

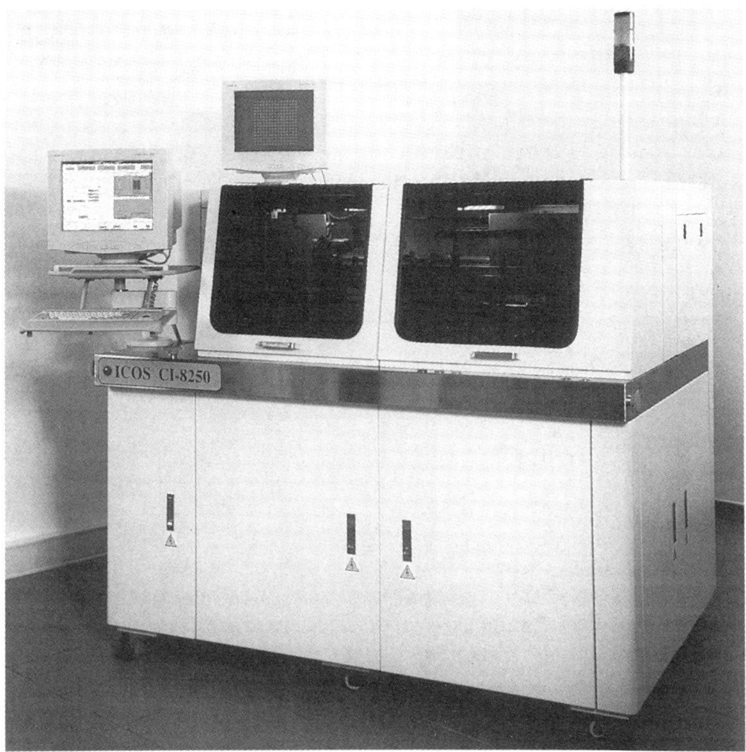

Figure 10.2 - ICOS machine vision system to inspect packaged integrated circuits for marking, mold, and lead concerns.

to inspect markings and some cosmetic properties (which include things like: chip-outs, cracks, discolorations, etc.). The markings are verified as correct for the product and checked for print quality and cosmetics. In some cases, the marking equipment suppliers offer an optional vision system on an OEM basis. Ultrasonic scanners are used for internal plastic IC package inspection for delaminations, cracks and voids.

Following electrical test the packages may be checked for lead straightness and coplanarity before placing them in shipping containers. Where the lead density is relatively low, these devices are often placed on tape-based carriers for subsequent automatic assembly onto printed circuit boards. Consequently, inspection for lead co-planarity takes place immediately before mounting onto the tape carrier. In these cases the manufacturers of these tape/reel arrangements often offer a machine vision system to check co-planarity. These systems typically employ multiple camera arrangements to infer 3-D properties of the leads. This measurement approach is referred to as "implied co-planarity."

The IC packages with higher lead counts (generally the more expensive components) are generally mounted on PCB boards manually or with robots. Hence, these are delivered in nested trays. The IC manufacturers want to measure the true 3-D co-planarity of these leads. This inspection includes: lead count, lead spacing, finishing, lead finish, twisted/bent leads, debris and coplanarity. Both 2-D and 3-D measurements should be made - coplanarity (3-D) and lead alignment (2-D).

The following are the requirements associated with true 3D co-planarity measurements:

	Accuracy	Repeatability
Co-planarity (regression plane) 0.0002”	0.0002”	0.0004”
True position error	0.0003”	0.0003”
Pitch	0.0005”	0.0003”
Ball diameter (in the case of BGA pkg)	0.0010”	0.0003”
Board warpage	0.0003”	0.0003”
Height resolution	0.0000625”	

These are three sigma-based measurements. The ideal approach should be able to handle all types of IC packages, of which there are many.

Significantly, as noted throughout the production processes, there is a requirement for alignment. Pattern recognition systems used are generally purchased under OEM agreements by the equipment builders and embedded in their machines.

Another major generic application in semiconductor manufacturing is monitoring work in process inventory. This is done based on OCR or bar-code-reading techniques. In the case of OCR, again general-purpose machine vision systems are used.

10.2 ELECTRONIC MANUFACTURING

As in semiconductors, the process begins with the creation of the circuit design using CAD systems. Using the CAD-generated files, the artwork masters are created by computer-driven photoplotters that design and lay out the conductor patterns.

Generally, the artwork masters are silver halide film transparencies representing an unmagnified picture of the conductor line pattern. In some instances, glass is used in place of film where ultra high quality is required. This practice is consistent with the use of glass reticles in IC manufacture.

The artwork master is inspected to assure it satisfies the design rules, matches the referenced image, is defect-free and that the dimensions of the conductor paths are correct. Defects would include: mousebites, line breaks, pinholes, or other blemishes on the conductor pattern that would be reproduced on phototools or production boards. Dust, dirt and scratches on the artwork will produce rejects.

Measuring the dimensions of the artwork is sometimes done with machine-vision-based off-line gauging stations. The phototools or the working artwork films are then produced from the master artwork. The phototools are used to actually transfer the conductor line circuit print to the PC board. A photographic exposure process makes the transfer. The phototools are silver halide or diazo transparencies that are prepared by a standard contact printing process from the artwork master.

Inspection of these phototools is just as extensive and detailed as that of the master artwork. It is of prime importance because the next step in the process is to transfer the print to the actual board. If defects are made in this sequence, they must be found before the inner layers are laminated together in the case of multilayer boards. The same equipment that is used to inspect the master artwork is used to inspect the phototools, again for pattern correctness, dimensions and defects.

Along a parallel production path the base laminate material is being prepared. The inner layers are actually discrete sheets of laminate material, the substrate being plastics such as polymide, epoxy/glass or teflon-based laminates, usually 0.003", or greater. Any number of products known as web scanners can inspect the substrate materials at their point of manufacture.

The laminate is generally a continuous coil stock so the first operation is a shearing one to produce the discrete layer. In some cases, holes are drilled into the substrate and machine-vision-based hole inspection can be done to verify presence, location and completeness of holes.

A deburring or desmearing operation might follow. Electroless copper is then deposited. The holes are inspected to assure plating coverage, generally using a microsectioning on a sample basis or magnifying glass.

The next production process is exposure and development. The conductor pattern is transferred to the inner layers. Once again the pattern should be inspected closely to determine if there are any defects that can be touched up and repaired.

The substrate may be laminated to a thin copper sheet. The copper clad laminate is overcoated with photoresist and has the conductor pattern exposed on it. Again, inspection is important at this stage because many defects which may have been caused by defects that were in the photoresist pattern can be corrected. The same equipment used to inspect the phototools can be used to perform this inspection as well.

Copper and tin plating operations may be performed at this time and the resist stripped to get rid of excess material. At this point in the process, the board is acid-etched to create the conductor pattern in the copper. This is a critical inspection point, because after this inspection the layers will be laminated together and defect correction is difficult if not impossible.

Some time ago it was analyzed that to repair the artwork only costs $.25; to repair the inner layer, about $2.50; to repair a multilayer board for an inner layer defect, about $25; to repair a populated board for an inner layer defect, about $250.

Bare board inspection (Figure 10.3) is made for conditions such as: line breaks/opens could be caused by flaking photoresist or by a scratch in the artwork; short, often caused by underetch or a blob of excess copper; pinhole, often shows up in a pad and can be caused by a thin area in the copper cladding; overetch, can create narrow conductor lines; underetch, can create thick conductor lines and leave excess copper; excess copper, can be attached to the circuit pattern where it can affect the performance of the board, or be separate from the pattern (both should be removed); mousebite or gouge out of a line, a thin area of line.

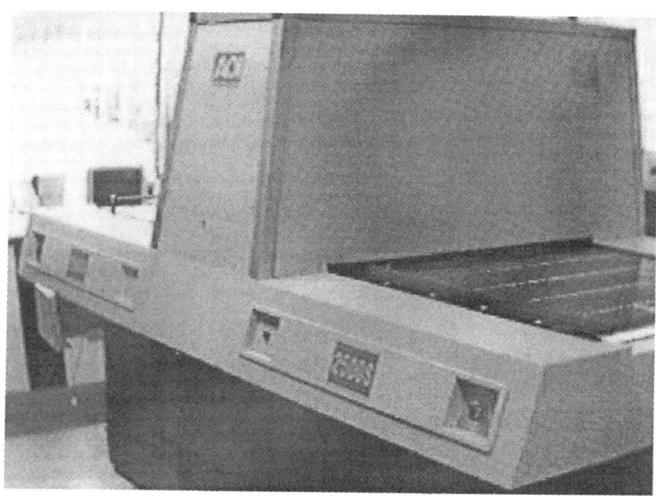

Figure 10.3 - Bareboard inspection system offered by AOI Systems.

At this stage the inner layers are laminated together (if multilayer). In the case of multilayered boards, the outer layer of the board is then processed. The same steps as enumerated above are followed: exposure and development etch and strip. The multilayered boards have pad holes drilled in them so leads of components can be inserted into the boards. These drilling machines may use X-ray-based imaging systems and machine-vision-based pattern recognition systems to align the boards to assure correct hole drilling.

It is noted that in some cases X-ray inspection equipment is used in the artwork compensation process, typically when hole sizes exceed pad size tolerances. This is done on a sample at the beginning of a run. Based on registration measurements obtained from the 'first piece', a program can be generated that compensates the original coordinates with dimensional offsets which are used during subsequent hole drilling.

The holes are through-plated to assure perfect contacts. Again visual inspection is performed and, on a sample basis, microsectioning is performed to verify reliability of the plating process. Connector fingers may also be plated at this time. Special machine-vision-based systems have been built to perform some of this inspection.

It is noted that wherever plating is performed, coating thickness can be measured using any number of techniques: X-ray fluorescence, eddy current, magnetic induction, Beta backscatter, and microresistance.

At this stage, the inspection process is complicated by the presence of holes drilled in the boards, which leads to a potential defect called "breakout." When holes are drilled in the pads at the end of a conductor line, they leave an annular ring around the hole. If the line is slightly out of alignment, or slightly short or long, the drill will be off center and the hole will cut the annular ring causing "breakout."

X-ray systems permit the detection and measurement of: inner-layer shifts, pad registration, breakout measurement, drill offset, contamination and shorting. Real-time X-ray systems are those that have replaced film with a display viewed by an operator to provide immediate decision. In some cases the systems can do some image processing to enhance the images.

At this time, solder reflow and solder mask operations are performed. The solder mask is generally applied using screen-printing techniques. Next, there is a final inspection and most likely an electrical test. Machine-vision-based inspection of solder mask has also been offered; however, there may currently be limits to the effectiveness of the system given the complexity of the appearance variables.

Today automatic optical inspection (AOI) techniques employing machine-vision-based approaches are in widespread use throughout the bare board manufacturing cycle, especially in the manufacture of high value, multilayered boards. Specifically, AOI is used after the following operations:

Artwork masters and production phototools
Inner and outer layers after development
Inner layers after etch and strip
Outer layers after reflow

Significantly, after reflow the application is especially challenging for three reasons: 1) the holes make it necessary to inspect the annular rings for width and breakout; 2) the reflective characteristics of reflowed solder are quite different from those of bare copper and 3) whereas the copper surface is flat, the reflowed solder surface is curved, making it difficult to properly gage the width of the conductor using specular reflection. It is noted that after reflow is the last point at which repairs can be made to the conductor pattern.

A final inspection after routing and solder masking may be performed for cosmetic defects primarily. Defects in the conductor pattern are difficult to find because the solder mask obscures the conductors. Significantly, automating this visual inspection is difficult because of the wide variations in appearance in the solder mask itself, due to all the "background noise" stemming from the patterns on the board - conductors and markings.

Preloaded Board Inspection (warpage) - Some companies have suggested a need to determine board warpage and verify hole presence and completeness. Speeds on the order of 9 square feet/minute are desired. This appears to be more of a diagnostic tool to assess the effects of board materials and board designs by checking board warpage on a populated printed circuit board before and after soldering operations. They suggest that this approach permits local evaluation of board warpage versus global evaluation. This way they can adjust tolerances to be part specific. For example, a BGA may require a warp spec of 0.5% across the interconnect area. However, to ensure good interconnection otherwise a 0.5% tolerance may be required across the entire board, whether such flatness is necessary beyond the BGA or not. With their system a 0.5% tolerance can be set in the BGA area only.

The **population or assembly of the printed circuit board** is generally performed at a completely different facility from the one that manufactures the bare board. The requirements for inspection depend on the board design itself - lead through hole (LTH) components, surface mount device (SMD) components, or mixed (Figure 10.4).

Where companies are producers of lower volumes but high product mixes, robotic-based assembly machines with their inherent flexibility are used. These generally use machine-vision-based motion servoing.

In the case of LTH, the first inspection is performed after the board has been populated and before soldering. Inspection is for: presence of lead in correct hole, verification of clinch, and in some cases, lead length.

After soldering, inspection can also be performed by equipment from these same companies or from companies that use X-ray based approaches for solder joint inspection with automated image processing.

In fabricating SMD boards, solder paste or epoxy must first be applied to the pads. Solder paste is generally applied using screen-printing techniques. The more automated screen printers employ machine vision servoing based on monitoring fiducials. In some cases this is an aid in set up. In other cases the capability is embodied in the screen printer for ongoing correction for screen stretch, etc. While the products that are usually offered for fiducial finding are used on an OEM basis by the screen printers, a leading supplier of such systems for the screen printer niche is ORS automation.

Immediately after screen printing, either on-line or off-line, an inspection is performed to verify presence, placement and, in some cases, volume. While this may be an on-line operation, it is unlikely to be done on a 100% basis because of

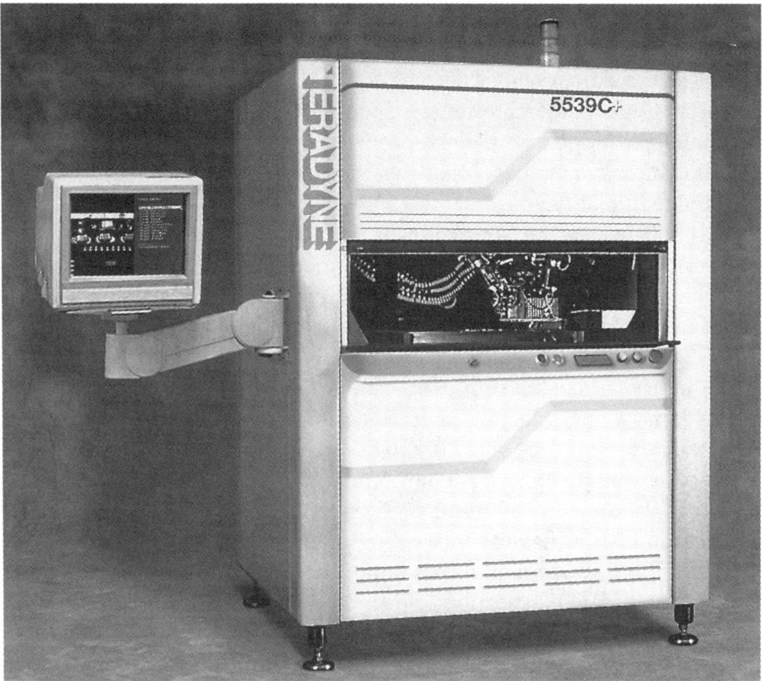

Figure 10.4 - Automated Optical Inspection system from Teradyne to inspect populated printed circuit boards before or after different soldering stages.

the speed involved. Rather, even on-line it is likely to be a sample inspection, generally on a rotating basis so that after 'X' number of prints all the solder paste pads will have been inspected. As the pitch of the components decline it is generally conceded that volume-based measurements are critical.

While in general it is the responsibility of the semiconductor component supplier to do co-planarity checking before shipping his product, it is possible that some board assemblers have invested in coplanarity measuring equipment to perform this check at in-coming receiving.

In the case of SMD or mixed designs the production equipment used to assemble SMDS will be a function of volumes produced. "Chipshooters" can apply 15,000-20,000 passive type components per hour using multiple vacuum nozzles and fast X-Y positioning tables to position the PC board. More flexible placement systems usually employ overhead X-Y gantry systems into which a pick-and-place head is integrated. Placement speeds in these machines range between 2500-4000 components per hour. These can generally handle active, multi-pin devices. In all these cases, a machine vision system is used for positional feedback.

Many of these systems embody machine vision to provide board offset correction. In the case of high pitch components, machine vision is being used to look specifically at the component leads themselves, the pad pattern on the PC board and provide a precision locate for the specific component to assure that all leads are physically positioned on their appropriate pad. It is noted that some of the placement machine companies offer machine vision value adders that are based on their own technology.

Because of the critical requirements of fine pitch component placement, many of these placement machines are now beginning to incorporate machine-vision-based techniques to assess pin co-planarity immediately before positioning the component onto the PC board.

In many SMD boards the passive components are found on the bottom of the board. In the case of mixed designs, the leads of the LTH components can be observed on the same side. Machine vision systems can perform presolder inspection on this side of the board. This involves verifying presence of the passive components, that the components have not 'tombstoned' and the presence and clinch of the leads. In the case of an all SMD board, these same systems might also be used for assembly verification.

Significantly, there is ongoing debate about the merits of post placement inspection/pre-solder automated inspection. Some studies have shown that the SMD placement equipment is very reliable and consequently there are very few problems at this point. This opinion varies, however, from company to company, perhaps based on the individual company's specific experience or the bias of individuals within the company.

The post solder inspection of SMD or mixed boards can be performed by either X-ray or optical/machine vision techniques by the same companies cited above.

Significantly, another sensor modality that may be used in conjunction with the assembly of boards is thermal imaging. In some cases these systems embody image-processing techniques to enhance images being viewed by an inspector. Most often this is used as a diagnostic tool to debug designs and production processes.

10.3 AUTOMOTIVE INDUSTRY

There has been a misconception in the machine vision market that the automotive industry is a dominant and perhaps the leading user of machine vision technology. That is definitely not the case. In the early 1980's the automotive industry was definitely an early adopter of the technology. Significantly, virtually all of the applications in automotives in the early days were unique. The result, in fact, was the development of a lot of projects but very few products.

Because they were projects, they had a heavy engineering content to them. In many cases the cost of the project might have even gone up over .5 to 1 million dollars. The dollars associated with such projects resulted in a distortion in the size of the machine vision market concentrated in automotives. Significantly, over the years the amount of money that the automotive industry spends on machine vision has remained relatively constant.

The typical machine vision system installed in the automotive industry is one that is used to address a generic application such as coarse gaging, assembly verification, flaw detection, etc. As noted, there is a major application engineering component to virtually all vision applications in the automotive industry even today. Few, if any, are opportunities for major multiple sales. Consequently, in recent years the application of machine vision into the automotive industry has fallen into the domain of the merchant systems integrator rather than the merchant vision company.

These merchant integrators, depending on how conversant they are with computer technology, will either integrate image processing board level products or vision computers. The main distinction between the two is that the vision computer will typically include some overhead software designed to communicate to the user in machine vision terms as opposed to in computer languages.

In the early 1980s General Motors conducted a rather exhaustive analysis of all of their manufacturing facilities to determine the importance of machine vision to their manufacturing processes. The much publicized survey suggested that there were over 44,000 potential applications. Pretty much as a consequence of that study and the analysis of applications, GM made investments in four machine vision companies. Applied Intelligent Systems Inc. was to become their lead company with respect to cosmetic inspection and assembly verification types of applications.

Diffracto, a company out of Canada, was to become a supplier of sheet metal gauging systems. Robot Vision Systems Inc. was to become a supplier of vision-guided robotic systems such as sealant-applying systems. View Engineering was invested in because of its expertise in metrology. Among other things they visualized a role for View in potentially integrating machine-vision-based metrology systems with machining operations.

GM almost simultaneously developed a subsidiary called GMF Robotics in conjunction with Fanuc out of Japan. While aimed at robotics applications, this operation nevertheless developed their own machine vision capability. For the most part, the applications were related to robotic guidance.

10.3.1 Taxonomy Of Machine Vision Applications In The Auto Industry

The applications of machine vision in the automotive industry can be characterized as falling into three broad classes: inspection, identification, and location analysis or guidance and control. Inspection itself can be further classified as either: verification, dimensional analysis or cosmetic/flaw detection. In the case of verification, the typical application is to make sure that an assembly is complete or that in the course of an assembly operation that a part is present, is correctly oriented and is the correct part. This is frequently referred to as a part presence/part absence type of application. Significantly, virtually every type of pattern recognition algorithm employed in machine vision can be used to perform a verification task.

Dimensional measurement applications can be further refined as those that involve low tolerances, that is, tolerances that are greater than 20 mils; high tolerances, tolerances between 20 mils and one mil; and very high tolerances, those below one mil. In addition, one can differentiate dimensional measurements as two-dimensional vs. three-dimensional, and three-dimensional also include being able to inspect surface contours. Flaw detection involves the use of machine vision to detect both surface anomalies or surface conditions that are three-dimensional in nature such as porosity, or two-dimensional in nature such as stains.

Part identification refers to the use of geometric or photometric features associated with objects to recognize those objects for the purposes of counting or sorting them. It may also include the use of markings on those objects. A separate class of applications are those that use alphanumeric markings or optical-character-recognition applications. The automotive industry may use machine-vision-based OCR systems or 2-D symbol readers, especially in view of the federal regulations regarding monitoring correctness of all of the components associated with the fuel emissions standards.

Guidance or location analysis applications basically again can involve either two-dimensional or three-dimensional vision techniques. In addition to alignment operations that are widely used throughout the electronic assembly areas, in the automotive industry vision guidance/vision servoing is used for robotics as well as for motion control. Applications include things like palletizing, depalletizing, machine loading, vehicle body finding in final assembly operations or part finding in component (e.g., carburetors) assembly operations, and seam tracking in welding operations.

10.3.2 Specific Applications Of Machine Vision In The Automotive Industry

In the final assembly of cars, one will find three-dimensional visual servoing techniques in conjunction with applications that involve finding the car body in three dimensional space for purposes of performing something on that car body; for example, windshield insertion or rear window insertion, applying sealant in critical locations in the underbody of the car, finding the car body to bring down mechanization to hydropierce the body for trim holes, etc.

Another application that is in widespread use involves inspecting sheet metal assemblies, including the final car assembly. In this case, machine vision is used for flushness and gap measurements. A typical installation might include over 50 specific machine vision sensors to make the measurements. Some sensors might be specifically designed to detect features on the part such as holes.

General Motors has publicized several developments that they have made in the area of machine vision. Back in the early 1980's they had developed a system called Sight One that was used at Delco Electronics largely for aligning operations as understood. They also developed Consight which was a system that used structured light techniques to identify objects of a common family passing by a conveyor. The structured light technique involved the use of a line of light projected on a conveyor and as the part passed under it observing the shifting of the pattern of light which relates to the geometry of the object. Specifically, this was used to sort castings where they would ultimately be palletized by type.

The third system that was developed was Keysite. This system looked at engine valve assemblies to verify that the retainer keys were present and properly seated.

Specific applications of machine vision in the automotive industry have included:

1. Inspecting of speedometers for calibration purposes and to verify properties - bounce, smoothness,
2. Inspecting LED/LCD instrument clusters,
3. Inspecting instrument assembly itself,
4. Using color-based machine vision techniques, verify that the color of different objects in an assembly are the same,
5. Using color based image processing techniques, look at the painted surface for DOI, gloss, and orange-peeling properties,
6. Looking at radiator assemblies to make sure none of the holes are clogged with excess solder,
7. Inspecting sheet metal parts after they are stamped to make sure that all of the features (i.e., holes, cutouts, etc. are present and also that there are no splits in the deep drawn areas),
8. Looking at a crankshaft gear to verify proper alignment,
9. Looking at gears to make sure that the teeth are all present as well as other features such as holes,
10. Looking at fasteners to verify they are the correct length and that all the features are present such as threads, head, etc.
11. Machine vision is also being used in conjunction with machining operations to lead to untended machining. In these cases the vision system is used to monitor the operation on the part itself, for example, monitor dimensions and/or monitor the cutting tool property.
12. In conjunction with electrical assemblies, machine vision has been used to verify assembly is complete and, using color techniques, to verify that all of the wires are properly connected and make sure the wires are properly stripped of their insulator.
13. Machine vision techniques have also been adapted to surface inspection of sheet metal assemblies as well as painted cars. These systems detect dimples, dirt pimples, and other types of surface conditions.

14. Applications in foundry operations include verifying the properties of a casting, and examining a casting such as a connecting rod for cracks using fluorescent penetrate imaging techniques.
15. In forging operations, vision systems have been used to verify dimensions and presence of features such as holes.
16. In the case of air conditioning assemblies, machine vision systems have been used to identify different assemblies as they come down a line based on a complement of components on an assembly.
17. Machine vision has been used to verify the completeness of a MacPherson Strut Assembly and to make sure that the threads are correct.
18. In the case of spark plug manufacturing, machine vision has been used to measure the gap as well as to look at the ceramic to make sure there are no cracks or chip-outs.
19. In the case of ball bearing assemblies, machine vision has been used to verify the correct assembly of a ball bearing, make sure all the balls are present and to make sure the grease is present.
20. In conjunction with crankshaft manufacturing, machine vision systems have been used to measure the critical dimensions of a crankshaft.
21. Systems have been described which use vision integrated with robotics for automatically assembling of parts. One example involves utilizing vision to automatically assemble various components to the stator or support assembly used in automatic transmissions.
22. Vision systems with special optical front ends have been used to inspect boreholes such as piston holes for flaws.
23. Vision systems have been used for welding seam tracking. In this case several different types have been developed. There are those that are based on simply finding the seam and then the robot welds in accordance with a preprogrammed path. The next level of sophistication involves actually using vision integrated with the welder where it basically looks typically through the arc to provide visual feedback of direction of path to the robot for path correction. The next level of sophistication involves also monitoring the weld process itself to verify the integrity of the process. The former two techniques have been adapted by the automotive industry to a certain extent.

These are meant to provide examples of the generic machine vision applications in automotives. In terms of absolute numbers, outside of applications in the electronics part of the automotive business, the largest number of applications can be found in assembly operations. In these cases they are either used to verify the completion of an assembly task or as an integral part of the assembly task to verify something before assembly takes place or in combination with a robot for an automatic assembly workstation.

It is understood that the next largest number is used for robot guidance in some way, shape or form. This is followed by gauging applications including sheet metal gauging as well as small parts gauging.

In the electronic operations, vision can be found throughout the assembly operations providing visual servoing. It is also found in virtually all the generic machine vision applications in electronics manufacturing. These include: traces and spaces examinations on bare boards, solder paste verification, component placement verification, both before and after soldering, and solder joint inspection. The solder joint inspection actually uses x-ray imaging techniques.

It is also noted that in incoming inspection, as well as in some cases on shop floors, there is use of machine vision based off-line dimensional measuring systems. Some people use the analogy of optical coordinate-measuring machines or TV-based optical comparators. These systems typically employ machine vision in combination with motion control in order to provide precision measurement capability.

Laser gauging techniques that have typically been applied to extruded parts or cylindrical parts are also in widespread use in the automotive industry. Some of these can even be found alongside machine tools where they provide an immediate post-process inspection on cylindrically shaped objects. Electro-optical, machine-vision-based triangulation techniques are also finding use as a sensor input in combination with a coordinate-measuring machine. These are used not only to make non-contact measurements on parts or automobile models but also for purposes of reverse engineering - that is, capturing details of an object and feeding it into a CAD system.

It is also noted that in the major industries that supply a product to the automotive industry there is also reasonably widespread use of machine vision. For example, in the glass industry, machine vision is used to inspect the float glass for cracks and blemishes, etc. It is also used to inspect discrete glass parts such as side windows for holes and contour and edge surface finish, etc. In the case of the tire industry, machine vision has been adapted to measuring thickness, measuring thread properties, examining sidewall properties to make sure that the white wall, for example, is blemish free, etc. OCR techniques have also been applied.

In the case of steel suppliers, there is a growing use of machine vision techniques to inspect the galvanized metal to verify its properties. There is an entire set of suppliers that have concentrated on these types of applications. In the steel industry one might also find machine-vision-based techniques performing dimensional measurements on manufactured products such as billets, etc. One would also find vision-based web-guided equipment.

10.4 APPLICATION-SPECIFIC MACHINE VISION SYSTEMS IN THE CONTAINER MARKET

The container industry includes establishments that fabricate packaging containers for the food, beverage, detergent and other consumer product industries, such as pharmaceutical and personal products. Specifically these include: glass, plastic, metal (both steel and aluminum) and a variety of containers of aseptic design for beverage and microwave cooking compatibility. The different materials compete in the respective niche markets they serve. For the most part machine vision companies participating in the container industry applications have aligned themselves with one or the other basic container material - metal, for example.

Within each material class there are also two possible markets. One is at the container manufacturer and the other the container filler. In the latter case there is the additional requirement for post-filling inspection - label present and cosmetically correct, cap present and straight, absence of spill over, etc. For the most part, these latter systems are based on the application of general-purpose machine vision systems. The main reason why machine vision is being used by the container industry, whether for empty or filled containers is to sort rejects.

In the case of glass containers, machine vision is being used to read mold codes; do dimensional checks; verify shape at both the hot and cold end; check sidewalls for defects such as air lines, bubbles, blisters, etc.; check the mouth or finish to make sure there are no chips or cracks; check the neck area to make sure there is no evidence of weaknesses; make sure the threaded areas are correctly formed; to look inside the bottle to make sure that it is empty with no birdswing off the sidewalls or glass particulate in the bottom of the container; and, in combination with polariscopes, to assess strain. It is also used at the hot end of the process to eliminate "freaks."

The glass container niche is made of two distinct niches that correspond to different industries: primary manufacturer of glass container and the bottler. In the former market the objective is to assure the quality of the glass bottle/jar being shipped to the company that will be filling it with their proprietary product. At the filler, the requirement is to guarantee there are no problems with the bottle before filling. In the latter case, the big market is where returnable/reusable bottles are used - virtually every country outside of the U.S., including Canada. In advance of the filler, systems exist that: verify empty state, check finish of lip, check threads, check for excess scuffing on the outside sidewall and verify that it is the correct bottle based on shape/color/etc.

In metal containers, machine vision systems are designed to examine the can end (unconverted or converted) or can itself for cosmetic flaws that are either reflectance flaws or geometric flaws as well as for geometry. Speeds as high as 2200/minute might be encountered on lines that produce these products. Systems are also used to measure the score depth associated with the pull-tabs.

In the case of plastic bottles, machine vision is being used to detect similar defects to those found in glass and metal containers. Consequently, some of the same companies serving the glass and metal container market are pursuing the plastic container market. In addition many general purpose machine vision systems are used in plastic-container-related applications.

10.5 APPLICATIONS OF MACHINE VISION IN THE PHARMACEUTICAL INDUSTRY

The pharmaceutical industry is emerging as a leading adopter of machine vision technology. As a consequence, the machine vision industry has responded by developing a number of "canned" or application-specific machine vision systems. In all cases the objective is to improve the quality of the human vision functions to avoid user complaints and FDA scrutiny. Most of the applications in the pharmaceutical industry involve packaging.

Studies have shown that packaging and labeling errors are the reason for a major portion of drug recalls. These errors include: label mix-ups, product mix-ups, printing errors, label vs. container errors and wrong insert. Studies have also shown that these errors can be avoided by the use of machine vision, bar codes, and other intelligent sensors that operate on markings that differentiate labels, inserts, containers, cartons, etc. to verify correctness of product, label, and labeled container.

Because of the unique function performed by the pharmaceutical product and because of the strict regulations imposed by the FDA to control and monitor the preparation of the product there are certain "critical defects" that a bottled product cannot have. A critical defect will cause a production lot to be reworked in-house or recalled if it has been already shipped.

Examples of a critical defect are:

1. Wrong product or mixed product in a container,
2. Mislabeled or unlabeled bottles,
3. Missing or illegible lot or control number that would prevent the product from being traced back to its date of manufacture if required.

Other defects which might result in user complaints or FDA scrutiny include:

1. Defective container (off-color, bad neck, finish, etc.),
2. Product miscount,
3. Missing cotton,
4. Loose caps,
5. Loose label,
6. Torn carton, loose flap.

These items, besides detracting from the professional appearance of the finished package, can also adversely affect the line efficiency by causing jams or forcing the operator to stop the line to clear a situation. These may also be considerations in the cost calculation when justifying the purchase of equipment like machine vision.

Some specific packaging concerns include:

1. Printing devices on machinery (date/lot codes) must be monitored to assure conformity to specification.
2. The quantities of labels and printed materials issued, used and returned must be reconciled.
3. Procedures must assure that the correct label/packaging/caps, etc. are used and prevent mix-up.
4. Unique identification of a lot or control number of each batch is required.
5. Production must be monitored to assure that containers and packages in a lot have the correct label.
6. The integrity of the containers must be monitored to avoid possible future contamination.
7. The contents of a package/container/vial, etc. should be verified as contaminate free - free from foreign matter.
8. The seals of packaging/container/vial, etc. should be verified as sound to avoid possible future contaminants.
9. The presence and correctness of all components to a package should be verified: cap, tamper-proof seal, carton, and actual product in package.

10.5.1 Packaging/Product Integrity

Machine vision is in widespread use in order to perform packaging/product integrity verification. While perceived as a generic application, a system with configurability is essential in order to provide a comprehensive solution for any particular packaging line. Most machine vision applications inevitably involve more than one of the machine vision type functions. That is, one may need a system which has an ability to enable a find or location analysis routine before it can process the image data.

It may also require an ability to do gauging in order to verify the shape of a container or the correct position of a label and its registration. Inevitably it will also be required to do some form of pattern recognition in order to verify that it is the correct label. In addition, it will also be necessary to make certain that the package and label are aesthetically pleasing: there is no spill onto the outer packaging walls, that the label is not torn or wrinkled, or has no folded corners, etc.

Any machine vision systems that are used for these types of applications must be tolerant of a range of appearance variables in addition to position variables, both within a given product or across different products, that may be processed on a given packaging line. These might include different size or shape containers or packages, different colors, different labels with different colors and patterns, different shapes and colors of caps, etc.

In addition, one must be aware of the need for contrast as the means to separate conditions or patterns to serve as the basis of the inspection decision.

Another issue is that of resolution. When one is looking at a container it may require a multi-camera arrangement in order to have sufficient resolution to detect the level of detail required for an application. For example, it may necessitate that one camera view the shoulder and cap of the bottle and another the label area. If there are back and front labels it will necessitate a separate camera for each.

If the label is applied on a round bottle it may necessitate some form of material handling in order to capture the image of the label in a repeatable fashion. Any decision on which machine vision product to use for a packaging/product integrity machine vision application should be based on a detailed evaluation of the available "tools" in the vision product.

10.5.2 OCR/OCV

In the case of pharmaceutical labeling there are two distinct requirements associated with optical character recognition (OCR) and optical character verification (OCV)(Figure 10.5):

1. to recognize the alpha numeric character designation related to the label/product; and
2. to verify the presence and the integrity of the date and lot code as it is being imprinted on the label.

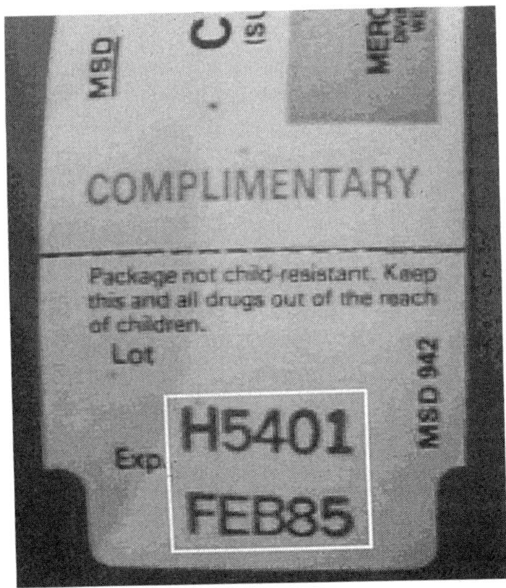

Figure 10.5 - Depiction of OCV application on label for pharmaceutical product.

While seemingly very similar, **verification** and **recognition** are actually two different applications. Verification reflects a condition where one knows what the character string is going to be and the requirement calls for making sure that the specific character exists at a specific location. Recognition, on the other hand, implies that one does not know which character is present at a specific location but that it is one of 26 characters or one of nine numerals.

The label/product identification code, often referred to as the National Drug Code (NDC), is printed when the label is printed. Often the font style associated with this printing will differ from the font style used in a date and lot code printer. Consequently, a machine vision system has to be able to handle both font styles.

In order for a single camera to have sufficient resolution applied across each of the characters, the NDC code should be imprinted on the label in the general area where the date and lot code is to be printed. Otherwise it may require two cameras which would lead to a somewhat more expensive machine vision system.

Both NDC code and date/lot code should be no smaller than 6-point type. Ideally the character style should be bold, and there should be no more than 16-18 characters in the string. Furthermore the print should be black and the background as light as possible with ideally at least a character height distance between the background and any neighboring patterns on the label.

In the case of date and lot code the application not only calls for the system to verify the characters are correct but that they are complete and not potentially subject to misinterpretation as a result of a printing error (contrast declining, character obliteration, etc.). Consequently, vision systems that are used for this application should generally have an ability to examine sub-features of the character such as specific lines and loops.

Machine vision systems used for such applications should have false reject rates that are lower than 1/10 of 1 percent and the capacity to operate at 300 containers per minute. Of course, no false accepts should be allowed. If ink jet printing is the technique for encoding the date/lot code, OCR/OCV systems should offer the capability to perform enhancements on the character image to make discretely segmented ink dots into continuous character strokes.

In some cases rather than verify the proper label/product based on the NDC code, it may be possible to use a bar code verifying system. Many labels exist that have adopted bar codes where real estate is available and aesthetics are not compromised.

In verifying characters there are basically three concerns:

1. Are the correct characters present?
2. Is the quality of the characters acceptable?
3. Is the contrast between the characters and the background sufficient?

Most machine vision systems do date and lot coding by virtue of a canned program that is configured for the specific date and lot code by a "train-by-showing" technique. Where a specific font style has already been trained, this operation only has to be performed once. Once trained on the font style, in general the system is then just trained on the specific date and lot code at the beginning of the batch run by the line operator.

In operation, because there are always some translation errors in the imprinting head, the vision system must first do a location analysis before it can actually do verification. Generally the entire code is first located and then each individual character is located. Once each character is located it can be checked for such features as: line width, hole fill in, breaks in the character and completeness of the character. Sensitivity of the system to these concerns is generally established ahead of time in training.

10.5.3 Glassware Inspection

The main requirements associated with glassware inspection are shape analysis and cosmetic analysis. Shape is generally performed by using a geometric type of approach. The pharmaceutical industry uses many different types of glassware for packaging, such as ampules and vials. These can have different sizes and to a certain extent different shapes as well.

While it is often left to the manufacturer to assure the quality of the glassware, because of transportation issues and general handling issues, it makes sense that the pharmaceutical manufacturer should also be using some of the same machine-vision-based "canned and uncanned" systems to inspect the glassware before filling. Certainly glassware defects are critical because they can affect the integrity of the container itself which could result in contaminating the contents.

A major concern is empty bottle inspection. That is, using a machine vision system to guarantee that no glass is attached to the bottom or any other contaminants for that matter. A typical empty glassware inspection system might include the following capabilities:

1. Detection of oversized openings
2. Detection of chips on the sealing surface
3. Detection of cracks on the sidewall of the tube glass
4. Detection of "air lines"
5. Detection of glass contamination in the base
6. Determination of oversize body dimensions
7. Assess height
8. Assess that the body and the openings are concentric
9. Determine if there are any dirt or opaque spots.

Such machine vision systems have the ability to handle 400 pieces of glassware per minute. As many as six cameras might be employed in order to provide input from all of the views that are necessary to examine the glassware comprehensively. The information that the system might also provide could include: total pieces of glassware presented to the inspection station, total pieces of glassware inspected, total pieces of glassware diverted and possibly also some statistics by generic defect condition detected.

10.5.4 List of Applications in Pharmaceuticals

The following represents a list of many applications that have been addressed by machine vision of one type or another at one time or another in the pharmaceutical industry. It includes some of the already mentioned applications in order to be complete.

- Date/lot code verification on labels and shelf cartons
- NDC code reading on labels, shelf cartons and case cartons
- Packaging conformance to appearance standards:
  - Correct shape
  - Label completeness, correct and appropriately positioned
  - Cap presence/straight
  - No content spill over
  - Outsert present
  - Tamper seal present
  - Blister pack/bubble pack - tablet/capsule/caplet - present, correct and complete
  - Slat counter - tablet/capsule/caplet - present, correct and complete
  - Glassware inspection - size, shape, flaws, empty state
  - Ampule/Vial package, powder - fill level, container/closure defects, missing caps/stoppers, foreign matter
  - Ampule/Vial package, solution - fill level, container/closure defects, missing caps/stoppers, foreign matter
  - Label/insert, proof reading and text verification
  - Label/insert/carton counting
  - Carton - squareness, content presence, sealed properly
  - Color code identification
  - Solid dosage stranger elimination
  - Solid dosage inspection
  - Spray coating analysis
  - Verify markings on solid dosage
  - Verify presence of capsule band
  - Vial /ampule counting

10.5.5 Validation

One challenge for machine vision in the pharmaceutical industry is the total cost of an installation. The demand for a turnkey system can drive the apparent cost up. In addition, there is a cost to validate the system and maintain a challenge procedure to verify the system continues to work correctly. Validation of all computer-based equipment used in pharmaceutical manufacturing operations as mandated by the FDA requires verifying the machine vision system performs as it is supposed to. It also requires verifying that the new practice based on a machine vision system yields results equal to or better than previous practices.

Validation is basically a structured documentation activity designed to prove that a machine vision system does what it purports to do. Validation of computerized systems, such as machine vision, that are used in the pharmaceutical/biomedical/medical device manufacturing industries receive a lot of attention from the FDA.

This can cost half again as much as the installed system to perform. Basically it involves both validating that the performance of the machine vision system is what it is supposed to be as well as verifying that the performance of the machine vision system is equal to or better than the previous procedures used to achieve the same functionality. Even when a vendor does validation, the pharmaceutical manufacturer must still maintain records related to the validation. A mere certification of suitability from the vendor is inadequate.

The tactics associated with validating computer systems such as machine vision systems in the pharmaceutical industry have evolved over the last decade. Along the way, the concept of life-cycle approach to computer system validation developed. This life cycle concept embodies four activities:

1. Define functional requirements of the system.
2. Assure structural and functional quality of software.
3. Provide adequate change control.
4. Provide clear and adequate documentation.

The functional requirements are distinguished from system specifications. Functional requirements are what the system is expected to do. They include the musts/wants, and the wants should be prioritized. The system spec, on the other hand, defines how - what type and what the system may be and what the system may be physically. The functional requirements detail the total parameters and characteristics of performance under which the system should operate. They define the output data that are necessary for an end user to complete the process. Functional testing is data-driven or input/output driven.

The system specification defines how the system will work and provides a physical description of its equipment, software, and standard operating procedures. It includes details such as controls, IOs, operator interface, engineering and communication interfaces, electrical specs, system staging, and system documentation, services/project management, engineering implementation, startup, training, and maintenance. A system spec describes what the system is and the functional requirements describe what it purports to do.

Appendix B includes a "Machine Vision Application Checklist" which can guide you through this process. It includes a series of questions related to the following:

Production process
Benefits of inspection
Application
Part to be inspected
Material handling
Operator interface
Machine interfaces
Environmental issues
System reliability/availability
Other issues/requirements
Acceptance Test/Buyoff procedure
Other responsibilities.

Validation requires:

Validation protocol: an all-inclusive document which describes the methodology used to create, execute, and document the results of the Installation Qualification (IQ), Operational Qualification (OQ) and Performance or Process Qualification (PQ). It defines team members and their responsibilities for approving, witnessing and documenting test results. It provides a detailed description of how to conduct and document each test, as well as define the overall acceptance criteria for the document.

Acceptance criteria for approval of the validation effort:

1. All essential documents have been adequately produced and approved
2. It has been demonstrated and documented that all user requirements have been satisfied
3. Each individual test to document these requirements will be considered successful when actual results meet the expected results or differences rationalized to an acceptable manner
4. Each installed sub component must comply with the engineering design and equipment data sheet/specification
5. The system must be installed according to the manufacturer's specifications
6. The system must perform reproducibly and consistently within its full application-specific range of operation.

Installation qualification: documentation verifying that the supplied, installed components (sensors, computers, software, PLCs, vision processors, etc.) conform to the specified design. This objective is achieved by conducting a series of tests and observations to confirm such items as wire continuity, environmental specifications (for each piece of equipment) and check that both hardware and software components were installed according to manufacturer's guidelines.

Operation qualification: Documentation verifying that the supplied, installed system and its components consistently operate as outlined in the specified design under test conditions. This objective is achieved by staging a series of field situations and observing the system's response. This includes testing of sensors and control elements, confirming feedback information from specific sensors, forcing alarm conditions to insure correct response, and full test on all normal modes of operation, as well as a test of system responses to abnormal modes of operation.

Performance qualification: Documentation verifying that the supplied, installed system and its component consistently operate as outlined in the specified design under actual production conditions. This objective is typically achieved by running three separate lots of a single product.

10.6 ANALYSIS OF THE SALE OF MACHINE VISION TO THE FOOD INDUSTRY

The food industry includes companies manufacturing or processing foods and beverages and includes: meat, poultry, dairy, canned and frozen fruits and vegetables, grain mill products, bakery products, oils and alcoholic and nonalcoholic beverages.

In addition to employing general-purpose machine vision systems mostly for package-line-related applications, the food industry has also become a major adopter of application-specific machine vision systems. This latter market is very segmented. Within each segment it is further segmented. For example, in the case of fruit and vegetable sorting and grading, first it is segmented by customer: food growers and/or their cooperatives and food processors, or whole product versus converted product. It is also segmented by type of food being inspected generically wet or dry. It is also segmented by material handling approach - channel or tube and belt or conveyor.

Of most interest is that in general the companies that provide systems for the growers do not provide systems for the processors. Systems for the growers have generally emerged from those providing the material handling usually at or for the cooperatives. Of interest is that these material-handling-system companies often specialize in the product for which they provide the handling, a situation that has emerged in response to the needs of a local producer; e.g., potatoes, apples, tomatoes, etc. Consequently, some have the capacity to handle near round product (apples, tomatoes, etc.) and no other shape, while others handle more oblong products (cucumbers, potatoes, etc.) and no other shape. In general the main requirement is to eliminate “freaks”.

At the processor side, again the suppliers of conveyors have generally emerged to provide machine-vision-based inspection systems, especially in the case of belt sorters. In the case of channel or tube sorters, it seems that these products mostly emerged in response to perceived market need and are supplied by merchants specifically of inspection or sorter systems. For the most part, these type sorters have not used machine vision techniques, although some have emerged in recent years. Generally they use bichromatic techniques to detect spectral signatures that differ from a norm. Consequently, they are application-specific to the product being sorted; e.g., systems that inspect rice have different spectral range sensitivities than those that inspect lentils, or peas, etc.

The objective in using sorters is to find foreign objects or conditions on a product indicative of an anomaly (Figure 10.6). In some cases, the ability has also been given to these sorters to sort based on size and grade. In the case of grade, the techniques are generally based on color image processing. At the process end the applications are more demanding, looking for finer defects, although there may be better control of the product movement.

Figure 10.6 - SRC Vision/AMVC food sorter system applied to raisins.

The market for these types of products has expanded dramatically in the last couple of years because: advances in the technology have improved the capacity of the systems to do the more difficult grading task; the adoption of color makes it possible for the same system to be reconfigured at the plant to handle a family of similar products that can be conveyed by the same material handling; the value perceived is a better return on investment based on recovery of good product and reduction of disposal costs and more consistent grading than humans can perform; and it is becoming more difficult to hire people for seasonal work.

As the larger growers continue to "add value" to their product by packaging semi-prepared portions, one can expect their requirements to increase. As the color technology continues to improve, these systems should be able to handle more applications satisfactorily. As more channel/tube sorters move to more machine vision processing, the overall market should increase as these systems will replace older versions with less functionality.

Another market within the food industry for machine vision involves applications with automatic portioners in the seafood, poultry, beef and pork industries especially. Several companies have developed water-jet cutters that employ three-dimensional based machine vision systems to calculate volume. The volume dictates where the cutting should take place in order to obtain the optimal yield from a piece. This works since in most such products the density is relatively consistent.

There are significant economic drivers to adopt this type of equipment. It is becoming more difficult to hire people to perform these cutting operations. Furthermore, when a company does so, it exposes itself to workman's compensation claims and worker liability claims stemming from carpal tunnel syndrome that many cutters eventually acquire.

10.7 APPLICATION-SPECIFIC MACHINE VISION SYSTEMS IN THE PRINTING INDUSTRY

The printing industry includes establishments engaged in printing by one or more of the following processes: letter press, lithography (including offset), gravure or screen. It includes companies printing books, newspapers, periodicals, greeting cards, wallpaper, etc. It includes printing on paper, while printing on other materials would be classified on that basis - i.e., plastics, textiles, etc.

Machine vision applications can be found throughout these industries, but mostly in converters (those that add value to paper), packaging and label printers. It is applied to both sheet-fed and continuous operations. Machine-vision-based techniques, more so than other sensor-based controls, are recognized as having the ability to build quality into the printing process. Statistical process control and ISO 9000 accreditation are factors that have increased the demand for machine-vision-based systems because they can capture and tabulate data automatically.

For years press controls have been available to control the registration of the paper during printing for both cross-web wandering and down-web changes. Registration controls for print registration have also been available for some time. These use a variety of electro-optical techniques including linear-array and area-array based machine vision processing and interpretation. While earlier versions provided cross web controls or edge to print registration control, the newer versions using machine vision techniques also offer print-to-print registration control.

Often these controls are an integral part of the printing press. Based on fiducial patterns, these type systems provide both color-to-color and print-to-print process registration. In general they provide closed loop control involving modifying movement of the web to correct for misregistration.

The printing industry has also adopted machine vision for defect inspection (Figure 10.7). These systems automatically make comparisons against a referenced image created at the beginning of a run alerting the operator when defects are detected. The systems incorporate image magnification capability that enables monitoring for: scumming, streaking, hickeys, fill-ins, traps, print-to-die registration, bar code quality, etc. In some cases, closed loop control has been achieved to initiate corrective action on the basis of the type defects detected.

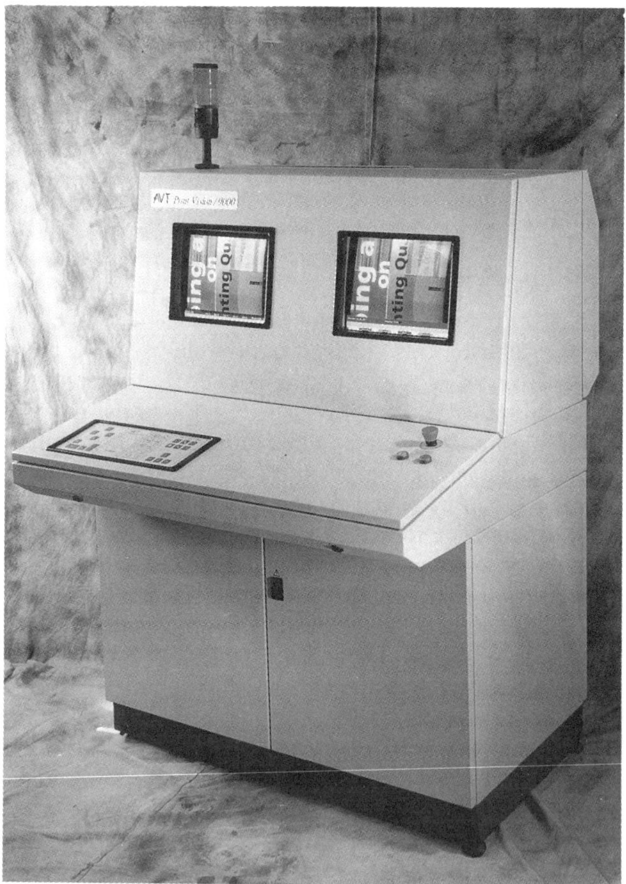

Figure 10.7 - AVT system inspecting printed materials.

These latter systems are generally color-based and use either single-chip or three-chip area color cameras. Systems are available with cameras that have nominal resolutions of 500 X 500 up to 2000 X 2000. In some cases, image acquisition is based on a linear array camera. Also in some cases the cameras are generally mounted on a motorized traversing mechanism and scan only a small piece of the image being printed at any one time using a strobe lighting arrangement to capture the picture. Usually these come with a split screen display to enable the operator to compare roll-to-roll or job-to-job color to a predetermined standard in addition to performing this comparison automatically.

These systems typically digitize the colors into RGB space rather than the CMYK colors used on the press. Most systems have built in conversion programs in their software with the ostensible result of the ability to monitor color trends based on the perceived press color. Some systems are only suitable for inspecting sheet-fed stock and others are designed to operate on the running web.

Some systems have emerged that view the entire image across the web for defects and, when necessary, view both sides of the printed stock. In some cases this capability is limited to narrow webs. In some cases these systems are offered by the printing press manufacturer as a value-adder to their press and not as a stand-alone system.

10.8 Wood

10.8.1 Why Automate?

The driving force for mill modernization is to increase productivity and yield through automation. Plant modernization is being driven by the continued decline in the supply of large diameter, old growth trees in North America. To maintain production volumes, mills have to process smaller diameter logs more quickly. This means more care because the margin for error is even smaller. There is a major concern over available timber supply.

10.8.2 Scanning/Optimization Systems

The objective of scanning/optimization systems used in the wood products/forest products/sawmill industries is to maximize yield out of the sawing operation. Mother Nature does not make square trees and the users of sawmill products require square- or rectangular-shaped pieces of wood most of the time.

There are six basic operations where scanner/optimizer systems are employed in the wood products industry: bucking, primary breakdown, cant, edger, trimmer and planer. Volume-measurement-based optimizers are used in the first five operations and grade-based optimizers are used in the last two operations. The first three operations are often considered the primary breakdown area and the last three the secondary breakdown area.

All six operations may or may not exist within the same sawmill or complex. Significantly, additional secondary operations may be conducted at a second mill on pieces that have already been through secondary operations at another mill. For the most part, the hardwood dimension and flooring segments of the industry log only get involved secondary operations. These are referred to as remanufacturing operations.

10.8.2.1 Volume-Measurement-Based Optimization. The term "optimization" takes on several meanings in the wood products industry. As far back as the early 1970's the term was associated with yield optimization based on volume measurements. The impetus for installing this type of equipment did not occur until the mid-1980's, however.

Adoption required advances in: computer/PLC price/performance, a transition to cubic measure of input scale, log modeling such as BOF (best opening face), and simulation software. The BOF programs test all possible ways to break down a log and typically generate the optimum sawing solution via a lookup table by: small end diameter, length, taper class and machine set.

Yield/volume optimization was dictated by data on the size and shape of the piece being cut. In the case of the bucking and primary log breakdown operation, essentially the actual log was being cut (Figure 10.8). In the case of secondary operations, the final product was being cut. In both cases it reflects the requirement to obtain as many pieces of final product as possible out of a single log. The benefits from an optimizer control system can include: increased production, improved machine utilization by reducing set-ups, recovery improvements, less economy lumber, less planermill trim loss, or a higher chip factor.

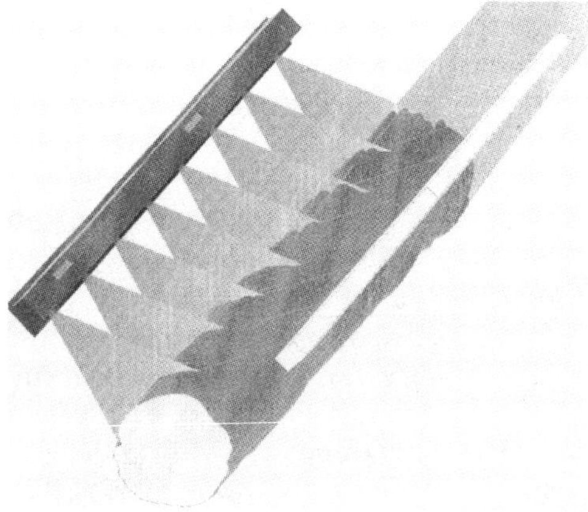

Figure 10.8 - Representation of 3-D system applied to log optimization application (courtesy of LMI/DynaVision).

In addition, one can place a value on data and reporting capabilities. This data can be used to identify the ideal product mix and raw material based on: selling prices of finished products, raw material cost and supply, processing rates and equipment capabilities, lumber recovery and manufacturing cost. The data also lends itself to being used in simulations that can predict impacts on the mill.

10.8.2.2Value-Based Optimization. The ultimate in optimization may be that based on value. On any given day the value of the pieces of wood that could be cut from a log (based on product possibilities and price) are introduced into the equation. In addition to absolute number of pieces, the number of pieces of a specific cut may be dictated by the value of the specific cuts on that day.

Value-based decisions are more useful than those based on volume only. However, performance improvement based on value optimization is dependent on the product mix and the limitations of each machine center (range of cant sizes that can be cut or handled) as well as understanding the marketability of any particular product or its relationship to current inventory needs. Maximizing value is often based on length decisions that must eliminate crook, minimize sweep and be at diameter points that produce the most valuable lumber sizes.

10.8.2.3 Grade-Based Optimization. Grade optimization is a term that reflects a refinement of optimization based on quality of wood being cut and ultimate use of the piece being cut. It reflects the requirement to obtain not only the most pieces out of a log but also the most pieces of the highest grade. It is observed that companies have more graders on their payroll today than they did five years ago. Many are grading to customer specifications. Smaller batch sizes and frequent specification changes are drivers to automate this process.

In planermill operations, human graders visually inspect each board before final trimming, stamping and sorting. As a board passes in front of the grader, he evaluates all four sides by rolling the board over. He then writes on the board, the grade and trim symbols or marks. Later as these boards are ready to enter the trimmer/sorter, an operator reads the instructions and enters them into his panel.

Today there are grade mark readers that can automatically read the symbols and enter them automatically into the trimmer/sorter system. The specific location of the symbols on the panel is important. Different colors are used to distinguish different graders.

The ideal would be a system that automatically does the grading. There has been some work in this area and some products have been offered with compromised performance. Grade optimization takes into consideration the detection of not only functional concerns associated with the ultimate use of the piece of wood but also appearance which may be perceived as quality by the user. Conditions detected in grade optimization include: rot, splits, knots, stain, pitch pockets, etc.

The number of defects, type, sizes and location determines the grade of a piece in relation to the highest grade. Defects can be detected based on photometric data, geometric data or some combination of the two (Figure 10.9).

Figure 10.9 – Ventek/AMVC vision system identifying "defect" conditions in green veneer to generate optimum clipping strategy.

Wood surface defects are of various kinds. Some stand out due to differences in contrast; dark knots, scars and bark pockets, for example. Some stand out because of a color difference. Other defects show a more intricate deviation from the background. Decay of various kinds (compression wood, pitch wood, and sound knots represent this more difficult group.

In some cases conditions such as checks and shakes have to be detected. A check is essentially a crack that goes across the growth rings and is caused by seasoning. A shake is a lengthwise separation of the wood that occurs between the rings of annual growth. A third group of defects contains defects that are not possible to detect visually. These include damage resulting from bacteria attacks whereby cell walls are destroyed.

Any automatic grader must be able to handle appearance variables within a species and from species to species. There are appearance variables that stem from roughness and how dry the sample is. Lumber could be graded right after it is first sawn to a time after it is kiln-dried. There is also an appearance variable that stems from length of time exposed to ultraviolet light and, consequently, to where in a stack a panel might be. Handling operations can also introduce dirt, etc.

The drying process can also introduce differences in how a specific reject condition appears. Grading after planing is a much easier set of conditions as it removes most of the appearance variables cited. A grading system should have a capacity to analyze at least two linear feet per second, ideally, 120 linear feet per minute.

Systems that combine volume and grade optimization are expected to be adopted in trimmer and planermill and/or remanufacturing operations.

10.8.3 Bucking Optimizers

Logs cut in the forest are typically delivered to the sawmill in 60' lengths. The first cutting operation is to cut the 60' lengths into smaller lengths. Because of the nature of the material handling associated with this first cutting procedure, it is frequently referred to as "bucking". The output of the bucking operation is a log/stem/bolt (pretty much all synonymous) for a sawmill.

Operations whose output could either be a log for a sawmill, plywood or wood chips are called merchandisers. For the most part today these bucking operations are at the sawmills. There has been, however, some bucking activity recently taking place in the fields owned by the larger sawmills. Bucking involves positioning the stem/log at the saw for cutting. Often more than one log can be cut at a time.

Where there are two logs, the pair of logs move transversally to a centrally positioned bucking saw that cross cuts each of them into two individual logs. The pair can be of two different lengths made possible by different positioning mechanisms. Logs are then sorted into appropriate bins by length and diameter. While most mills would only have one bucking operation, some of the larger mills could have as many as three. In the bigger mills, a 60' length stem would be cut into three pieces in about one third of a minute or less.

Today programmable logic controllers and a variety of photoelectric switches control much of this process. Saw infeed and positioning conveyor is controlled by the PLC. Chain speed is ramped down automatically as logs approach the bucking length target. Shaft encoders and optical lit switches mounted along the conveyor typically correct for log slippage and enable accurate log positioning.

Optimizers are being used in this first cutting operation since the shape of the log can dictate what the best lengths should be to get the most yield. For the most part these have been based on simple techniques to measure diameter, photocell beam breaking. A few more comprehensive real-shape scanners are understood to have been installed in a couple of mills but their cost has, until now, generally precluded their serious consideration.

10.8.4 Primary Breakdown Optimizers

In the case of yield/volume optimization at the primary breakdown operation, the following takes place in a typical operation. Logs or "stems" (cut logs) are loaded into an in feed line where crooked, oversized and otherwise defective logs are removed. Stems are logs with butts that may flare out exceeding the maximum diameter that can be handled by a saw line. In this case the butt must be removed first.

All logs are typically then run through a debarker. In some cases, the remaining logs are sized and sorted by size range (both diameter and length) into one of two or three lines. Throughput is apparently improved if the mill can process in batches based on diameter, especially when handling small diameter logs. While being debarked, in some scenarios the logs are scanned for length, diameter and taper. In the primary breakdown area there can be different mechanisms used to position the log for sawing: carriage, sharp chain, dogging, chip and saw, etc. Each utilizes a different approach to positioning the log/stem and so different approaches have emerged to optimization measurements.

Optimizers use the diameter, length and taper data obtained from the scanner and production planning information, such as the orders to be processed and what logs are on inventory, to decide on the optimized cutting operation. In these cases a computer can display a log-cutting solution. The operator, however, can override the solution based on gross defect observations he makes, such as scars or crooks that any scanner does not recognize. Most systems can then automatically derive an alternative solution accordingly. These systems typically provide reports such as diameter distribution by length.

While many optimizers have already been deployed in the primary breakdown areas, most are of the simpler beam breaking or light screen design and do not provide real shape data. The value of real shape data is recognized as well as the value of optimization. Consequently, there appears to be an opportunity to replace earlier optimizers with new ones that provide more accurate real shape data.

In the primary breakdown area, some of the larger mills could have as many as three lines. The number of primary breakdown lines is generally the same as the number of bucking lines. Each primary breakdown line operates handling 10 to 15 logs per minute or so.

10.8.5 Cant Optimizers

Optimizer features typically include the ability to relate log orientation, log center line to sharp chain center line and offset the bands to compensate for log offset caused by sweep, etc. These optimizers can generally base yield on number of cants (preliminary cuts with angled sides) or final cuts with squared-off sides at least in one axis (called live sawing - simply sawing the log through with parallel saw lines) or combinations of the two. In live sawing there will still be cants that result if decisions are based on only one axis.

10.8.6 Edger and Trimmer Optimizers

After the cant operation, optimizers are also used in edger and trimmer operations. In these secondary operations - final cutting operations - board optimizers scan to detect wane (local size changes due to board coming from piece of the outer circumference of the log), thickness, and, in some cases, defects. The result is optimization of edging and trimming operations. These optimizers define the geometry of each board by top and bottom profile scanning at 1" to 2" intervals along the length of the board.

The edger operation is typically a combination machine that profiles the edge of the sideboards out of the cant and removes the sideboards. The cant is then rotated 90 degrees and moved into another profiling machine which produces the edges of the sideboards from the cant where the first faces had been produced. The cant is then run through another sawing operation that produces the second set of sideboards and lumber from the center of the cant. The objective is to get as many complete 8" wide boards as possible (4" x 8", 2" x 8"). Each two- or four- inch slice is called a "flitch" - a board before it reaches its final size.

The next operation is the trimming operation, and finally sorting. Significantly, virtually all edger and trimmer saws today are offered with scanner/optimizer options.

Again, there is typically one edger line for each cant line and one trimmer line for each edger line. The throughput at the edger is typically 30 per minute and 60 per minute at the trimmer.

10.8.7 Log Scanners

Log scanners have been in sawmills since the early 1960s. The earliest optimizers were single-axis light screens or light-calipering scanners. They used a single circular cross-sectional representation to model the log as a truncated cone. This is reasonably effective on logs with good form, little sweep and crook and essentially round in shape. Significantly, even these simple optimizers yielded recovery gains of 20% over operator-eyeballed judgments.

Today, for the most part, the simplest optimizers are those which use an X-Y photocell beam-breaking arrangement or a light screen arrangement in combination with encoder data. This was made possible by improvements in infeed mechanisms that can orient and hold logs in a stable condition as they are conveyed through the scanner and cutting heads. The photocells are spaced at 6"-12" intervals.

These are effective if the log can be moved longitudinally down a conveyor for its full length without any lateral motion. The dimensional measurements are dependent on encoders as well as log stability. Sweep and crook may not be dealt with effectively with this arrangement. Two axis scanning and log modeling techniques are based on the elliptical cross-sectional representations permitting the independent positioning of saws in the X- and Y-axes to account for ovality and sweep in logs.

Today "snapshot" or camera-based scanners are popular. Two-dimensional camera scanners or edge-detection scanners measure the outside edges for diameter and offset data. These do not provide profile data. Camera-based systems that operate on structured light images are also available which do provide profile data in addition to edge data.

The advantage of camera-based systems is that they can strobe to take a picture. They have an advantage in close-coupled situations where there are space constraints, where bucking and corresponding log motion destroy data quality and in log straightening applications, where logs must be straightened before they are dogged.

The ultimate scanner for volume/yield optimization could be one that generates a complete three-dimensional map of the log. These typically deliver more data that can be important for wane measurements. They typically create a plane of laser light which forms a line of light all around the log. A single camera module can scan 120 degrees around the circumference of the log. A four module unit delivers overlap in measurements.

However, it is noted that although there may be an infinite number of log shapes, there are only a finite number of decisions that are reasonable to consider. The ability of the equipment at a given mill to handle single axis or multiple axis log information is another factor. To use multiple axes data the sawing plane should be modifiable based on the more extensive log shape information. Where the sawing plane is fixed, then single axis information is adequate.

To take advantage of multiple plane information requires an ability to rotate the log into a plane that maximizes recovery. This permits fully using multiple axes data to determine which cutting plane produces the best results. Three approaches today to capturing this data are: multiple snapshots based on structured light and laser scanner, where the log is passed longitudinally through the scan zone; or an arrangement of triangulation based linear arrays surrounding the log.

These approaches can determine conditions such as whether the logs have crook or sweep in more than one plane.

10.8.8 Current Change-Impacting Optimizers

One major change, as noted above, is the increased use of small diameter logs. Coinciding with this is the increased use of sizing the logs before primary breakdown to perform this operation in batches based on specific size.

Today there also is a growing interest in being able to saw on the curve in edging operations to further improve yield. This enables recovery of longer length material and a higher grade of material. In general, existing optimizer scanners can provide the required data. What has to be modified are the optimizing algorithms themselves and the saw controls and saws.

10.8.9 Non-Optical Scanning

Today there has been some work to demonstrate internal scanning of logs based on X-ray CAT and MRI to detect and locate knots and other defects so that logs can be oriented to avoid them to raise the value yield. This takes on increased importance in large, old growth forests where the lumber is being exported to high priced Japanese and European markets.

Using X-rays, however, does present a challenge given the inhomogeneity of wood. Wood typically experiences local discrete changes in density in the form of earlywood and latewood growth bands which can selectively attenuate the penetrating radiation and cause corresponding variation in density in the detection scheme used. When delaminations, cracks, or other latent defects transcend the inspection areas, their images superimposed on the already confusing radiographic images make interpretation of the images difficult.

X-ray based systems are being used in the lumber industry today to measure strength based on wood density for those products aimed at the structure business where strength is important. This may be more of a non-destructive test based on a sampling basis than on 100% of the wood produced.

10.8.10 Miscellaneous Applications

Truck Stack Volume

In Portugal, for stock management purposes, the pulp mills need to measure incoming timber by volume, whether they own the forest sources or are buying by volume. The Optics Division of the National Institute for Engineering and Industrial Technology has developed a machine vision system which looks at both sides and the rear of the truck to establish the length of the logs. From the measurements made, the apparent volume of the wood stack can be calculated. It is even possible to estimate the proportion of wood and free space in the stack based on image analysis, thus giving a "solid volume" figure. From this, the production efficiency of the mill can be calculated.

Fiber Size Measuring

Advanced Fiber Imaging offers an online automatic fiber size measurement system for oriented board manufacturers. The PC-based system captures a quantity of pulp, dilutes it and presents the fibers within a transparent block which is backlit. The vision system then measures the length, area and perimeter of every fiber within the field of view. These measurements are then combined to form a single value which is sensitive to changes in raw materials and refiner settings, and is an indicator of the finished panel strength.

Color-Based Vision for Color Matching

MacMillan Bloedel, a sawmill operator, developed in-house a color scanner to match short pieces of siding, mainly for grain, so they can be fingerjointed into standard lengths of siding. The scanner is an adaptation of a scanner used in the baking industry. Matching similar color and stains permits more uniform treatment when paint or stain is applied to wood. The scanner recognizes three distinct variables: color, grain and length - or combinations, and sorts accordingly.

The system is designed to operate on a line that runs 300 fpm. Five times a second the camera freezes an image of 8 inches of board. The software bases its decision on: color, grain, defects, lengths, and waste and then clears and sends the right cutting prescription to the chop saws to maximize values. Defect detection is based on detecting markings applied by graders who use fluorescent chalk to mark the defects above and below each defect. While the defect detector is camera-based, the color and grain scanner may not be.

The color scanner can be taught a range of hues associated with a specific color and sort accordingly. Grain judgment incorporates guidelines for judging early and late wood. The system can also distinguish between vertical and flat grain.

11

Common Generic Applications Found in Manufacturing

11.1 ALIGNMENT

Virtually all GPMV companies offer some kind of alignment or "find" routine. Table 8 summarizes the offerings by the respective companies. All suggest their approach yields sub-pixel accuracy and repeatability, typically 1/3 - 1/4 a pixel. This is generally accepted as the 'state-of-the-art' for industrial applications although much better performance can be shown in a laboratory.

Most companies include both blob analysis or correlation tools that they use as the basis of their alignment routines. The blob analysis is more likely to be used when aligning a part based on the geometric features of the part itself. Correlation is generally used when aligning off a fiducial mark. The more robust approaches base their alignment on gray scale processing versus binary. So for example, the blobs would be determined based on edge segmentation rather than thresholding; correlation would be based on normalized gray scale versus binary correlation.

While all GPMV companies claim a "find" routine, for most it is an integral part of the their inspection routine: first find a part before performing the inspection. Not all GPMV companies have actually interfaced the data to a motion controller.

The company with the most experience in alignment is Cognex. They have the most OEM agreements for alignment applications, mostly with production equipment suppliers of equipment used in the manufacture of semiconductors and in populating printed circuit boards.

11.2 METROLOGY AND MACHINE VISION - OVERVIEW

No matter what instrument is used to measure a parameter there are two critical factors: accuracy and repeatability. A basic rule-of-thumb is that the measuring instrument should be at least ten times better than the process specification it is to measure. In other words, it should be at least ten times as repeatable and accurate as the process.

All measuring instruments have a scale made up of a number of "ticks" or markings along the scale. In the case of machine vision, the distance between "ticks" is the size of the pixel (subpixel) or alternatively the distance between pixels (subpixels). In machine vision a "tick" corresponds to resolution and may, but not necessarily, correspond to the sensitivity of the machine vision system -the smallest change in the measured quantity that the system is capable of detecting. In machine vision this corresponds to the pixel (subpixel) increment or pixel (subpixel) resolution.

When using machine vision to gauge a part, one is faced with the dilemma that the edge of the part feature generally does not fall precisely on a pixel or precisely between two pixels. The effect of an edge is generally felt over several neighboring pixels. Different machine vision algorithms take advantage of this property of an edge to calculate to within a pixel (subpixel) the position of an edge. Significantly, different algorithms do yield different results in terms of the size of the subpixel increment.

Accuracy is dictated by the calibration procedure. In machine vision, as in most digital systems, the "calibration" knob can be changed one "tick" (one pixel or subpixel distance) at a time. Each "tick" represents a discrete value change in the system's output, the discrete value being a physical dimensional increment.

The procedure to determine system accuracy requires the operator to place a "standard" in the correct location - a referenced position established during the initial calibration procedure. With a machine vision system, the operator adjusts the calibration until the measured value is as close as possible to the "standard's" value. This then determines the pixel (subpixel) dimension.

Most metrologists prefer to have at least ten "ticks" (pixel or subpixel units) across the tolerance range. With ten "ticks", the most the system can be miscalibrated from the standard's value is one-half a "tick" at each measurement point on the standard. Therefore, the most the norm of the readings (accuracy) can differ from the true (standard) value is one twentieth of the total span of the specification.

For example, given the nominal dimension of 0.1" with a tolerance of +/- .005". (Total tolerance range of .01"). Therefore, each "tick" (pixel or subpixel distance) of the calibration knob should have a value of .1 of .01" or .001". One half of each step is thus equal to 0.0005". In other words, the accuracy of the machine vision system should be equal to or better than 0.0005".

Since the rule-of-thumb for repeatability is the same as that of accuracy, the system requirements for repeatability are the same, i.e. the repeatability should be equal to the dimension of a "tick" - 0.001" in this application where the tolerance on the part is +/- 0.005".

While accuracy may not be as critical in a given application because it can be derived by calibration, repeatability is more critical since it can not be corrected by calibration or otherwise. It is observed that the above analysis is considered conservative by many. Consequently, some suggest relaxing repeatability from 10/1 to 5/1. This suggests in the application example given an acceptable system repeatability performance would be 0.002". This should be at least the one sigma repeatability of the machine vision system in the application.

In some cases the rule-of-thumb that is used is that the sum of accuracy and repeatability should be less than one-third the tolerance band. No matter what "rules" are observed, the accuracy or repeatability of the measuring instrument should not equal the tolerance of the dimension being measured and, in fact, must be appreciably smaller!

While machine vision with subpixel capability can often be used in many metrology applications satisfying such "rules", in some cases such performance approaches the practical limit of machine vision in an industrial setting regardless of the resolution of a system or the theoretical pixel size (field- of-view divided by number of pixels in horizontal/vertical direction).

In the above example application where the part dimension to be measured is 0.1", given the full field-of-view of the camera/machine vision system is applied across this dimension, the theoretical subpixel resolution could be 0.1"/1500 (based on a 500 x 500 area camera based machine vision system and a subpixel capability of 1/3 the pixel resolution) or .000066" -well within the required 0.00050".

However, it is observed that it is not necessarily the case that the more pixels there are the better. There are other potential physical limits. For example, resolutions below 0.0001" may be limited by the diffraction limit of optics.

Diffraction is fundamental to the propagation of electromagnetic radiation and results in a characteristic spreading of energy beyond the bounds predicted by a simple geometric model. It is most apparent at the microscopic level when the size of the optical distribution is on the order of the wavelength of light.

For imaging, this means the light collected from a single point in the object is focussed to a finite, rather than infinitesimal spot in the image. If the radiation from the object point fills the lens aperture uniformly, and the lens aperture is circular and unobstructed, the resultant spot distribution will appear as a central disc surrounded by concentric rings. According to the wave theory of light the central disk contains 83.8% of the total energy in the spot distribution.

The most immediate image effect is that adjacent points blur together and therefore, are unresolved, one from the other. In a machine vision application, the points referred to are the subpixel "ticks". The diffraction limit is defined by the Rayleigh criteria as:

$R = 1.22\lambda N$

Where

N = Numerical aperture of the lens

λ = Wavelength of light

For example, based on

N = f/2

λ = 500 Nm (approx. avg. of white light)

R = 1.22 microns or .000048"

What this suggests is that while games can be played by using blue light and f/.8 lens, for example, the theoretical limit is on the order of 0.00002".

This limit, however, is exacerbated by application conditions and variables such as: light level and spectral changes, optical distortions and aberrations, camera sensitivity non uniformity, vision algorithm interpretation variations, temperature, vibrations, etc. Not to say anything of part appearance and presentation variables. The result is that in any given machine vision application the practical limit is on the order of 0.00008 - 0.0001".

This is analogous to having a ruler with a scale in 0.0001" increments. The measurement sensitivity is half this (0.00005") - i.e., the scale is read to one or the other of two neighboring hash marks or "ticks". Another observation made is that measurement sensitivity in conjunction with making a dimensional check between two edges on a part relates to the determination of the position of one of the two measurement points. There is a similar sensitivity in conjunction with the determination of the other position. This, too, contributes to repeatability errors.

11.2.1 METROLOGY AND MACHINE VISION - COMPONENT ANALYSIS

In metrology applications of machine vision it is not unusual to deal in dimensions that have tolerances that require .0001" repeatability and accuracy. These turn out to be very demanding and, therefore, require attention to detail.

Lighting:

The ideal lighting is a back light arrangement using collimated lighting to provide the highest contrast with the sharpest defined edges. This will only work if the features to be measured can be observed in the silhouette of the object. Another consideration may be that the ideal light should be one with blue spectral output. A xenon strobe has a blue output given the visible IR are filtered out. A strobe offers the additional advantage of reducing the effects of motion and vibration on image smear.

Using a strobe has the advantage of high efficiency and an ability to accurately control tuning of the light pulse either to the camera or the moving object. Strobes reduce the effect of smear to the duration of the strobe cycle. However, ambient light must be controlled to avoid "washing out" the image produced by the strobe. A camera with an electronic shutter will minimize this effect.

The alternative, a top lighted arrangement may not result in measuring the absolute dimensions because of radii artifacts, fuzzy edges, etc. Measuring using a top lighted arrangement, again ideally a collimated light arrangement, should use blue light to optimize the measurement.

Another issue in top lighting is that the lighting should be as uniform as possible. A circle of light is one possibility. This can be accomplished with an arrangement of fiber optic light pipes. These arrangements are commercially available. The light pipe can be connected to a strobe light.

Optics:

The collecting optics should be a telecentric design to obtain the least edge distortions and be tolerant of potential magnification changes due to positional and vibration variations.

In some applications microscopic optics can be employed; that is, optics that magnify the image since the imager size is on the order of 8 millimeters and the part is smaller. Not withstanding the magnification, the issue associated with the optics is the resolution of the optics. In the case of optics, resolution refers to the ability to distinguish the distance between two objects. In measurements this is analogous to the ability to distinguish between two "ticks".

Under idealized laboratory conditions, a microscope can be designed which has a resolution to 0.000020". In a practical industrial environment, however, one is unlikely to get better than .00005 - .0001" resolution.

The presentation of the part and fixturing of the cameras should be such that the part is always presented squared or parallel to the image plane of the camera. This will avoid pixel stretching due to perspective and keystoning effects. It is noted that other properties of optics can also affect resolution, especially off-axis resolution: distortion, astigmatism, field curvature, vignetting, etc. In some cases these will become a substantial percentage of error compared to pixel size, and, consequently, compensated for accordingly by the machine vision system. Alternatively, better optics may be required.

Camera:

The imager that is used in the camera should have a resolution, in terms of the number of photosites that it incorporates, on the order of at least 500 x 500. In the case of the imager/camera, resolution in machine vision is often equated to the number of photosites in the imager. The camera should have as high a response to the blue spectrum of the illumination as possible to yield an image with as high a contrast as possible.

In a back lighted arrangement this is less critical but may be an issue. Different image sensor designs have different spectral responses. A CID sensor generally has a higher sensitivity than interline transfer CCD sensors, although a frame transfer CCD sensor may be comparable because it's higher fill factor results in a high effective quantum efficiency.

A preferred camera would be one that has a square pixel so horizontal and vertical values are the same. While this can be corrected in software, it requires additional compute power and in turn, more time.

The camera itself should have an asynchronous scanning capability especially if a strobe operation is anticipated. An exposure control feature would also be useful to further reduce the effects of background illumination.

A camera with an asynchronous scanning ability also has advantages to assist in synchronizing the event to the camera. By synchronizing the camera to the event ensures that the object will always be physically in the same location in the image. This may minimize the need for translation correction, an algorithm that adds processing time to execute and will, therefore, slow down the machine vision system performance.

Issues that affect resolution in cameras include lighting, electronic noise, mechanical noise, optics, and aliasing.

The A-D/frame buffer should be compatible with the number of photosites in the imager. There should also be a capability to synchronize the camera from the frame buffer in order to eliminate the effects of pixel jitter - a practice common in commercially available machine vision systems that make such issues transparent to the user.

Pixel jitter can result in an error in the mapping of a pixel to object space repeatedly. This is more of a factor in higher resolution cameras, such as those that should be used in metrology applications, where jitter can be a one pixel error.

Vision Computer:

The vision computer should have the capacity to do a subpixel algorithmic analysis. Significantly there are many different approaches to subpixel processing and these yield results which differ in robustness. While many companies purport to have an ability to in effect subpixel to $1/10^{th}$ of a pixel, it is commonly agreed that the best one would generally achieve in an industrial application is on the order of one part in three or one part in four of a pixel.

Subpixelling approaches take advantage of the fact that the edge of an object will actually fall across several pixels and result in effectively a gray scale curve. Any number of different mathematical approaches operating on that curve will yield an ability to interpret the position of the edge to within a pixel. Essentially a subpixel represents the finest distance between two "ticks" and relates to the vision system's ability to make measurements and not detect attributes that are smaller than a pixel.

The vision computer should have the capability to compensate for modest translation (.005") and rotation (5 degree) errors. It should also have the ability to operate on edges based on subpixel processing techniques (to at least 1:3). Ideally it should be able to trace an edge based on best fit of the pixel data and make measurements between two such lines. All this processing and a measurement decision must be done at production rates.

Significantly, a vision computer with higher processing speeds (3000 per minute or greater) may have an advantage in that it could average a number of pictures taken on a single part. Such averaging may improve the signal-to-noise and, therefore, effectively the resolution of the system typically by the square root of the number of samples.

11.2.3 Summary:

Given this attention to detail in a gauging application, the best repeatability and accuracy that can be achieved with a machine vision system is on the order of 0.000050- 0.0001".

11.3 OPTICAL CHARACTER RECOGNITION (OCR)

There are at least four fundamental approaches to OCR: correlation-based, essentially geometric shape scoring and matching; nearest neighbor classifiers/decision theoretic, essentially using specific geometric features/feature sets and matching based on proximity to specific multidimensional model; syntactic or structural, essentially using specific features and relationships between the features; and neural network/ fuzzy logic classification based on train-by-showing and reinforcing based on correct decision.

Some systems come with factory installed fonts which are pre-trained: e.g. semi, OCR-A, OCR-B. Some have the ability to be trained at the factory on the specific font style being used. Others require that the vendor train new font styles.

Different executions can yield different performance so it is difficult to suggest that one approach is the most superior. All but the syntactic approach is font specific. Requiring a system to be trained to read more than one font style at the same time with these exacerbates the ability to provide correct reads as more characters can become the victim of a confusion pair. The syntactic approach is generally conceded to be the most robust when it comes to multi-font applications. With this approach specific character features and specific relationships are used which are generally font-style independent.

For example, an "E" might be characterized: start at (0,0) if a vector is generally easterly and another vector southerly, and if the vector in the southerly direction meets and intersection from which there is a vector also in the easterly direction and a vector in the southerly direction and the vector in the southerly direction intersects with a vector in the easterly direction, then the character is an "E".

The approach requires that each character be uniquely defined based on vectors and arcs and their directions. These conditions would generally exist regardless of the font style, although there could be font styles that may cause confusion by their very nature. These should be avoided in any application.

Read accuracies are related to many factors: quality of print, application conditions, such as lighting uniformity/consistency, font style and potential for confusion pairs, how well the system is trained, consistency of spacing between characters,

Often systems have more than one approach to reading characters. If the characters can be reliably read with high confidence (as determined by the system) only one set of algorithms are enabled. If the system determines a degree of uncertainty for a given character, the better systems can then enable another set of algorithms to read that specific character. If still a concern, some systems can enable even more recognition algorithms. The final decision may be then based on "voting" - the number of times the different approaches suggest that it is the same character results in the decision regarding the character. Even where there is only one suite of character recognition algorithms, the system may have the ability to "vote" by reading the character 5 - 10 times. For example, if ten times, the threshold might be set such that at least six times the system must agree on the specific character and when that is the case that is the character read.

False reads are usually controllable by establishing conditions during training that err in favor of a "no read". In the case of a "no read" the system will generally display the unidentified character in a highlighted fashion so an operator can decide and input the correct character via a keyboard.

False reads can also be reduced by the addition of check sum number in a character string. This number must agree with the other number read by some rule: e.g. the last digit of the sum of all the numbers read must be the same as the check sum number. It is understood that this is routinely done in the wafer side of the semiconductor business but not in the packaging side.

Comparing read rates is also not straightforward. Most companies are reluctant to expand too much on their read rates or throughputs. For the most part it can be assumed that the read rates claimed are based on optimal conditions. In most cases, one must also add times associated with taking the picture and finding the string/finding the character before reading takes place. Again, these are dependent on quality of print and font types as well as whether rotation as well as translation must be handled.

Lighting:

In the case of characters and bar codes laser scribed on wafers, lighting has been determined to be very critical. This is especially the case when wafer alphanumeric is being read to track product during work-in-process. It seems that as one adds value to the wafer, the various steps may have a tendency to deteriorate the quality of the characters. Some lighting arrangements have a tendency to wash out the characters after some of these steps. Companies that offer OCR systems for this application have come up with optimized lighting arrangements often based on some degree of programmability: levels, angles of incidence, etc. For newly formed characters this should be less of a concern. Nevertheless, lighting may be as critical as the algorithms in achieving rigorous OCR.

In general in OCR applications the lighting yields a binary image of the character strings. Either the characters appear white on a dark background or vice versa. Ideally the lighting yields a consistent binary image. As a consequence, the algorithms used typically operate on binary images rather than gray scale images, which accelerates processing and reading.

11.4 OPTICAL CHARACTER VERIFICATION (OCV)/ PRINT QUALITY INSPECTION (PQI)

11.4.1 PRINCIPLE REVIEW

Besides engineering details, there are two basic issues related to the optical character verification (OCV) application. They are: is the application satisfied by verifying that the character is correct; or is it also the requirement that the system must be able to assure the quality of the character PQI). In the way companies have approached this application these are not mutually exclusive.

As observed in our comments about OCR, it is also true of OCV applications that lighting is critical. The objective is to yield a binary image. Most of the approaches to OCV exploit the fact that the image is binary. Virtually all who offer GPMV systems offer a binary correlation approach to verifying characters. As long as a binary state can be insured, this approach generally produces adequate results for verifying correct characters.

Some systems use a gray scale template as the basis for establishing a character's pattern and subsequent comparison. The correlation routine uses degree of match or match score as the basis of determining if the character is correct. Being based on a gray scale template it is the most tolerant of conditions that affect gray scale, in effect normalizing them. Hence, such conditions are less likely to cause false rejects.

Some such conditions include: print contrast itself, print contrast differences that stem from different background colors, variations in contrast across a stroke width, variations in contrast stemming from lighting non-uniformity across the scene, from scene to scene, etc.

On the other hand, shape scoring may not be the best at checking the quality of the characters. Where conditions vary that affect the shape of the character stroke but are still acceptable in terms of legibility (stroke thickening or thinning, minor obliteration), establishing a shape match score that tolerates these conditions may effect the character verification reliability. By loosening the shape score match criteria to tolerate conditions that reflect modest character deterioration but still legible characters, it may even be possible that the shape of a different character may become acceptable or conditions that may yield a character that could be misinterpreted could become acceptable to the system.

One special condition of concern is handling stroke width variation that is apparently a potential concern in laser coding. One approach to handling this is to 'fool' the system by training it to accept a character provided it is a match with any similar character in one of three font styles. The font styles would correspond to thick, thin and nominal stroke width. Two drawbacks with this approach are: most likely to make the system even more tolerant of quality concerns and speed; each match scoring comparison takes a finite amount of time; correlating to three matches would take three times as long as doing a single correlation.

Incidentally, while some approaches actually do an optical character recognition (OCR) during operator set-up, in the run mode they do an optical character verification (OCV). During set-up such systems read each character based on correlating the gray scale template to each and every character in the font library. The character in the font library with the best match score or correlation to the character being read is deemed to be the character read. This is displayed on the screen for operator verification of correctness. During the run mode, the system knows which character is in each position.

The gray scale template derived from the current label is matched to the gray scale template stored for the specific location in the current trained set. As long as the match number (or correlation number) is greater than the previously established value (on a scale of 1-1000 with 1000 being a perfect match), the system acknowledges the character as correct.

Verifying characters that could constitute "confusion pairs" may require additional logic than just match scores to be reliable. One approach is to establish regions of interest (ROIs) where there could be confusion and to look for contrast changes. Another approach uses logic that includes different tools that are automatically enabled based on the character sets involved. So, for example, while one system would apply an ROI at three locations along the left hand stroke of a "B" which are areas that distinguish it from an "8" and look for contrast changes, another might use a direction vector property - the direction in which the gray scale is changing at each pixel along the boundary. This is perceived to be more robust. Other rules they would use for other character pairs include number of holes, character breaks and pixel weighting.

While several companies basically use a binary template as the basis of their character comparisons, their respective executions may differ dramatically. For example, one approach may be based on using a gray scale region around a nominal value as the basis of the binary image. That is, all pixels that have a gray shade value between 60 - 120 (for example) are assigned to the black or foreground region. All pixels outside that range are assigned the background or white region. The nominal value itself is adaptive; that is, it uses the gray scale distribution in an ROI within the current scene to correct the value.

Another approach might establish a single threshold above which all pixels are white or background and below which all pixels are black or foreground. Their approach might use an adaptive threshold to compensate for light level variations, and be based on performing a correlation routine to establish best character match for a nominal threshold as originally determined and for each shade of gray +/-10 shades around nominal. This is all performed automatically during operator set up. The threshold is not adapted on a scene to scene basis.

Another fundamental difference between approaches might be that one bases the window search routine on blob analysis while another uses a normalized gray scale correlation. The blob analysis is based on bounding box distribution of the character pixel string and locating its centroid and correcting to a referenced position accordingly. This approach will be sensitive to any text, graphics or extraneous "noise" that may get too close to the character strings being examined.

After the region is found, one system might look for specific features on specific characters to perform a fine align. Another system might establish via a correlation match usually based on a piece of a character - "a gray scale edge template". This is done automatically during operator set up and the system includes certain rules to identify what piece of what character it should use as the basis of the correlation template.

After binarizing the image, a system might do an erosion and then base its decision on a foreground and background comparison to the learned template on a sub-region by sub-region basis. Another might do a character by character align, in addition to the greater string align, before doing the image subtraction. This is followed by a single pixel erosion to eliminate extraneous pixels and then another erosion whose pixel width is based on thick/thin setting established during engineering set up. This is designed to compensate for stroke width variations.

One approach might base its decision on the results of a single sub region, while another base its decision on the pixel residue results associated with the template subtraction for a specific character. In both cases "sensitivity" for rejection is based on a pre-established percent of the pixels that vary. Some systems might have the ability to also automatically reject characters whose contrast is less than say 50% of the contrast established during operator training. This per cent can be adjusted during engineering set up.

Some systems also include a built in a set of rules to handle confusion pairs. Different systems may also require somewhat more operator intervention than others during set up. In some cases properties such as: threshold, string locate, character align, contrast property for rejection, character scale and aspect ratio and rules to enable for confusion pairs, are all performed automatically totally transparent to the operator.

Another approach found in some products is based on binary images and a syntactic analysis based on the extraction of localized geometric features and their relationship to each other. This approach is better suited to OCR applications and less suitable to applications that involve character quality as well. While such an approach should be less sensitive to scale or stroke width thinning or thickening it is probably more sensitive to localized stroke shape variations.

Several of the executions are actually based on OCR as the means to verify characters. These tend to be somewhat slower and be less amenable to print quality inspection (PQI). In addition to specific algorithms executed, throughput is a function of: positional repeatability of the character string, number of characters in a string, number of strings, expected variables in appearance and whether PQI is also required. In general, systems that have executed morphological operations are better at PQI.

One difference between executions is that some perform their template matching on the basis of sub regions within a larger region that includes several characters (possibly even the entire string) at one time versus matching on a character by character basis. Some suggest that this may be adequate for laser marking since the failure is generally a "no mark" condition for the entire set of characters (all strings and all characters in the string) rather than a character specific "no mark".

Where several characters are verified as a pattern confusion pairs are more of a problem since not performing analysis on a character by character basis. It is also more susceptible to characters whose character to character spacing is varying. In general the systems do offer an ability to establish their regions of interest on a character basis but that this may be more awkward to do during operator set-up and may require more time.

What does all this mean? Basically while there are many ways to perform reasonably reliable OCV, systems that apparently perform more image processing are better suited to perform PQI as well. The more robust systems generally operate on gray scale data and do more than binary pixel counting. The simpler type systems tend to have a higher incidence of false rejects when set up to avoid false accepts. Some approaches are somewhat less suited to character quality evaluation than others. Some are considered better suited to handling confusion pairs.

11.5 REVIEW OF DEFECT DETECTION ISSUES

11.5.1 OVERVIEW

Since the late 60's companies have applied a variety of techniques to inspect products for cosmetic concerns, either geometric flaws (scratches, bubbles, etc. or reflectance (stains, etc.). Some arrangements are only able to detect high contrast flaws such as holes in the material with back lighting, or very dark blemishes in a white or clear base material. Some arrangements are transmissive, others reflective, others a combination. Some arrangements are only able to detect flaws in a reflective mode that are geometric (scratches, porosity, bubbles, blisters, etc.) in nature; others only those based on reflectance changes.

Some systems only have detection capability, others have ability to operate on flaw image, and develop descriptors and, therefore, classify the flaw. These latter systems lend themselves to process variable understanding and interpretation and ultimately automatic feedback and control. As machine vision techniques improve in image processing and analysis speeds, there will be opportunity to substitute these more intelligent systems for those earlier installed with only flaw detection capability.

Advances in sensors are also emerging which will improve the signal-to-noise out of the cameras making it possible to detect even more subtle reflectance changes which in some processes are considered flaws. These advances alternatively make it possible to use less light and/or operate at faster speeds. Among these advances are time delay and integration (TDI) linear array based cameras which are more suitable in situations where there is motion - either the product moves under the camera or the camera moves over the product.

Adoption of these systems is being driven by increased customer pressures to improve product quality as well as competitive pressures to keep costs down with improved process controls. In some industries systems with improved performance, especially with the addition of defect classification may command a higher price and be purchased as a substitute to an already existing system.

The problems associated with the inspection for flaws in products have two primary considerations: data acquisition and data processing. The data acquired by the sensor must be reliable (capable of repeatedly detecting a flaw without excess false rejects) and quantifiable.

Ideally it should also provide sufficient detail so one can classify the defect. This should ultimately make it possible to interpret the condition causing the flaw to render corrective action to maintain the process under control. In order to quantify the defects the data must be in a separable form-separating depth, size and location information. It is anticipated that this type data can be quantitatively related to numbers with engineering significance and analyzed in a manner that can be correlated to human perception parameters.

What follows is an attempt to characterize the different approaches to flaw detection. One observation is that the flaws to be detected could require 3D information. However, video images are two-dimensional projections in which the third dimension is lost and must be recovered in some manner. The most developed techniques are those that operate on gray scale photometric data in a scene. Significantly, there are many different implementations based on different data acquisition schemes as well as different data processing and analysis schemes.

11.5.2 GRAY SCALE/PHOTOMETRIC

Techniques to detect "high frequency" or local geometric deviations based on capitalizing on their specular properties have been studied more than any other approaches. Complementing these techniques are those that operate on the diffusely reflective properties of an object. These have been successful in extracting texture data - color, shadows, etc. Using this approach "shape from shading" (low frequency or global geometric deviations) can detect bulges. The diffuse and specular components can be distinguished because as one scans light across a surface the specular component has a cosine to the nth power signature while the diffuse component a cosine signature.

In other words, scanning a light beam across the surface of an object creates specularly reflected and diffusely scattered signatures. When light passes over a flaw having a deformation or a protrusion, the reflected light at the sensor position "dissolves" in amplitude as a result of changing the reflection angle. The energy density within the light bundle collapses. The flaw geometry redirects much or most of the reflected light away from the specular sensor. The sensor detects a lack or reduction of light.

All this is compounded by the fact that in addition to a change in angle of reflection these flaws also cause a change in the absorption of the material at the defect. For example, in metals this is particularly so for scratches, though less so for dents. In other words, in theory one should be able to separate out the two components (absorption and reflection) to arrive at a proper understanding of the defect. So far this has only been reduced to practice in a limited way.

Surface defects that essentially have no depth at all, such as discolorations, are detected as differences in reflectivity of the surface material. Stains, for example, do not change the light path but will change the amount of absorption and reflection in relation to the nominally good surrounding area.

11.5.3 DATA ANALYSIS

Techniques depend on the relationship of the signal from one pixel to its neighbors and to the whole image. The analysis to process gray scale data is computationally intensive. This factor in combination with the processing rates required to handle the typical throughput requirements and the size defects one wants to detect for a given span puts severe requirements on any data processing scheme.

Early approaches relied on fixed thresholds. That is, defect detection was a function of the value of the analog signal from the detector. If a flaw caused a change in the value of the signal greater than a certain pre set value than it was characterized as a flaw condition. Using registers and other techniques the number of times the flaw condition was detected in a given area, a threshold could also be set for flaw size.

Today various adaptive thresholding techniques are in widespread use, which reduce the incidence of false rejects experienced in fixed threshold systems. They are designed to compensate for variations in illumination, sensor sensitivity, surface variables, etc. Significantly, different wavelengths may be influenced differently by films, oils, etc. Similarly, today techniques exist to compensate for variation in illumination that might be experienced. Similarly, in systems using detectors with an array of photosites, photosite to photosite sensitivity compensation is performed. These type corrections have also made these type systems more reliable, experiencing fewer false alarms.

When it comes to analysis - simple techniques look for light/dark changes, run length encode to establish duration of change in the direction of scan, and essentially segment with connectivity to assess length of flaw perpendicular to the direction of scan - are frequently used.

Data based on arbitrary thresholds or the particulars of a specific inspection device may essentially be subjective in nature and could prove to be difficult to calibrate. Consequently, data processing must be on the gray scale content. More sophisticated processing involves operations on the signal to remove noise. Where two-dimensional images are used, the opportunity exists to perform neighborhood operations (either convolution or morphological) to enhance images and segment regions based on their boundaries or edges. Gradient vectors (slope of the curve or first derivative of the curve describing the gray scale change across a boundary can characterize edges). These vectors can include magnitude and direction or angle.

Image processing with these techniques often uses two-dimensional filtering operations. Typical filters may take a frequency dependent derivative of the image in one or two dimensions to act as low pass filters. In others, one might take a vertical derivative and integrate horizontally. In these cases one capitalizes on the features of the defect that might exhibit distinctive intensity gradients. Still other filters might correlate to the one or two dimensional intensity profile of the defects.

In terms of detecting and classifying surface defects one would like to have as much resolution as possible. However, for economic reasons, one establishes the resolution of the system to provide just enough data to classify a defect. In general a number of sensor photosites must be projected on the defect to adequately determine its dimensions.

The actual resolution required for defect detection is dependent on the characteristics of the defect and background. It is generally agreed that sub pixel resolution is of no consequence to detection of flaws but related to ability to measure flaw size. The size flaw one can detect is a function of contrast developed and "quietness" of background. Under special conditions it may be possible to detect a flaw smaller than a pixel but one can not measure its size. Nyquist sampling theorem, however, suggests reliable detection requires a flaw be greater than two photosites across in each direction. Using this "rule of thumb" alone usually results in unsatisfactory performance. The problem is that flaws do not conveniently fall across two contiguous pixels but may partially occlude several neighboring pixels, and in fact only completely cover one pixel.

Resolution may be a function of optics. For most surface inspection the requirement for resolution, coverage and speed are in conflict with each other. Defect characterization may be possible using convolution techniques. It may be possible to develop specific convolution filters for individual defects. It may also be possible to characterize some defects based on their gradient vectors: magnitude, direction and angle.

Other specific features may become the basis - signal change in specific channel (specular or diffuse) vs. width of pulse. Parameters to be used could include the shape of the analog signal waveform in the region of the defect, the position of the defect on the object, and its light scattering characteristics.

Good, repeatable signals are required, however, for reliable classification of defects. Hence, the need for an optimized data acquisition system. Experience with data acquisition approaches based on characterizing the specular/diffuse properties of a scene have been found sensitive to lighting variations, reflectivity variations of the product, oils, dirt, and markings. In other words, these approaches have produced data that are not reliable. These approaches also suffer as they are not equally sensitive to flaws regardless of direction of the flaw.

11.5.4 DETECTING DEFECTS IN PRODUCTS

In terms of the requirement to detect flaws in discrete products, there are five factors that strongly interact: size of anomalous condition to be detected, the field of view, contrast associated with the anomalous condition, consistency of the normal background condition and speed or throughput.

Where contrast is high, (e.g. black spot on white "quiet" background naturally or appears so due to lighting tricks), anomalous condition is relatively large, (greater than 1/8"), field of view on the order of 15" and throughputs of 400 FPM or less, techniques are available that in general are very successful. While in principal the same techniques should be extendible to conditions with lower contrast, smaller anomalies, larger fields of view and higher throughputs, extension is not trivial.

In general these techniques are not generic but rather application specific. For example, today machine vision systems routinely detect flaws on high-speed paper making line where throughputs are up to 5000 FPM. Lighting, data acquisition and data processing are dedicated specifically to the paper product being produced, however, and may not easily transported into other web scanning applications.

There are several issues that strongly interact: ability to uniformly illuminate wide areas, sensor/data acquisition, compute power and requirements in terms of being able to measure and classify in addition to detect. To see smaller detail in the same sized product requires sensors with more pixels. To handle higher speeds requires faster data acquisition. This in turn effectively reduces the photosite signal which, especially when low contrast changes have to be detected, results in poorer signal-to-noise. In other words, the quality of the input going into the computer is less reliable.

Such less reliable data requires that more processing be done on the signal. Where contrast is high, say black and white exists, simple thresholding techniques can be employed to ignore all data except that with a value below the threshold or the data associated with the flaw. Significant data compression is the natural effect. Similarly, the analysis can be simple - a count of the number of pixels can be correlated to size.

Where contrast is low or background "busy", on the other hand, simple thresholding is virtually ineffective. Some form of gradient processing is required. That is, tactics which detect small gray shade changes corresponding to the edges of the anomaly. Since edges typically fall across several pixels, this means that the anomaly must cover an area of at least 25 pixels to have a reasonable confidence it will be detected and classified reliably. In other words, cameras with a larger number of pixels are typically required to detect lower contrast anomalies.

Again, this can exacerbate the data acquisition rate. Furthermore, gradient processing is far more compute intensive and the higher data rate means even faster compute power is required. Significantly, all the above discussion has revolved around detection.

Another issue is the requirement to recognize the anomaly for purposes of reliable classification. Such classification by the system would differentiate between: holes, dirt, bubbles, etc. The ability to so classify generally requires more pixels than simple detection as well as more image analysis or compute power.

The ability to uniformly illuminate a product may also be a factor. Another issue associated with lighting is that it be sufficient to provide good signal to noise property out of the camera. This is especially critical in applications where wide variation in opaqueness is experienced which manifests itself as noise in the signal. This will also be exacerbated in this if high-speed data acquisition is required.

Lighting "tricks" that may be able to mitigate some of these issues include the use of directional lighting to exploit the fact that flaws like bubbles and dents are geometric. Such a lighting arrangement takes advantage of light scattering off geometric surface permutations. Higher contrast conditions, such as holes can be easily detected based on photometric changes -light level changes.

In a transmissive mode, for example, a hole will result in more light a contrast change going from dark to lighter- and dirt will result in less light-a contrast change going from light to darker. In both cases a delta threshold can be set. Pixels above this threshold can be counted and if enough contiguous pixels exceed the threshold that corresponds to the setting of a hole or dirt, the condition flagged.

11.6 TWO DIMENSIONAL SYMBOLOGY

A standardized format for marking components too small for conventional bar codes is critical for telecommunications, automotive, semiconductor, and other electronic manufacturers. These industries have been working together to establish space-constrained marking standards based on two-dimensional (2D) symbologies for three years. Small product marking is also critical to the medical device and pharmaceutical industries. In this case, the Uniform Code Council (UCC) has been working on a multi-industry standard for three years.

In 1996, strides were made in adopting standards that should result in accelerated adoption in the next few years. The Electronic Industry Association (EIA) introduced a standard (ANSI MH 10.8.3) that specifies the use of DataMatrix ECC200 protocol for use in all inter-industry component marking. This standard also specifies the use of 2D symbols for shipping labels. The telecommunications industry has been favoring PDF 417, while the automotive and semiconductor groups have been supporting DataMatrix. The Automotive Industry Action Group has issued a specification requiring DataMatrix for all automotive parts marking.

One challenge overcome in 1997 for the DataMatrix code was the absence of a verifier. Several companies have introduced machine vision products that can be used to verify the integrity of the DataMatrix code in a QC environment. Verification typically includes calibration to NIST standards. These verifiers check the reference decode, symbol contrast, print growth (ink spread), axial nonuniformity, and unused error correction.

Instead of acknowledging the DataMatrix code and the work of the other standards committees, the UCC has added to the confusion by sanctioning a new 2D code dubbed IPC –2D. The pharmaceutical industry is also using the DataMatrix code for packaging reconciliation. AIM International announced the final approval of the International Symbology Specification for Aztec code and QR Code 2D bar code symbologies. These are the third and fourth symbology standards published by AIM International. QR Code was developed in Japan and is the first such symbology developed outside the U.S. to receive an International Symbology Specification.

In addition to the competition from other 2D codes, there continues to be competition from stacked linear codes. A new PDF417 code (MicroPDF417) has been introduced, and the UCC devised a new NK linear code called IPC-13/14. These codes are perceived to have the advantage that they can be read with conventional laser bar code techniques and do not require machine vision based techniques.

The need for 2D is being driven by the desire to include more data in a machine-readable code, while not taking up any more "real estate." In some cases, the interest stems from understanding that a 2D code takes up much less real estate than a conventional code. The fact that more data can be available is inconsequential. Industries that are embracing 2D codes include electronics, semiconductors, automotive, pharmaceuticals, and medical devices. In addition, there is interest in these codes for warehousing operations.

With the issuance of the standards in 1996, the electronic and semiconductor industries clearly shifted to the adoption of 2D symbology. These industries include manufacturers of electronic components such as silicon wafers, semiconductors, printed circuit boards, and assemblies as well as major users of those components (including the consumer electronics, telecommunications, computer, and automotive industries). These industries, as well as others, are interdependent because of electronic parts. Standards for marking and tracking components and assemblies result in efficiencies such as just-in-time manufacturing, work-in-process monitoring and logistics processing that keep these industries competitive in a global marketplace.

The semiconductor industry has standardized on DataMatrix EC200 for wafer identification. The EIA's SP3132 standard specifies DataMatrix ECC200 for all inter-industry component marking: bare-printed circuit boards, integrated circuits, capacitors, etc. For product marking, on the other hand, the EIA Bar Code Committee is looking at MicroPDF 417. In this case, users want a scanning range of 1 to 24 inches, as well as the ability to read the shipping container symbol which favors the stacked linear codes.

Major applications in these industries include inventory and production tracking, enhancing security and traceability, work-in-process monitoring, and product identification marking. The EIA, Semiconductor Equipment and Materials International, Telecommunications Industry Forum, and the Automobile Industry Action Group have been working on an inter-industry series of compatible standards for wafer, component, and product marking. Several of the codes (PDF 417, Maxicode, and DataMatrix) are in the public domain, lending themselves as standards.

Some of these codes have the ability to include up to 5000 characters. Most are scaleable from sizes as small as 0.0002" on up, and are highly redundant which makes them readable even if they're partially obliterated. Hand-held readers are available even for the DataMatrix code.

There are at least two major approaches to these higher density bar codes. One is a stacked conventional bar code (e.g. PDF417) compatible with hand-held laser scanners or linear array-based or area camera-based scanners. The matrix code is based on markings arranged in a 2D array. Two-dimensional bar code readers or matrix code readers typically incorporate features associated with machine vision applications: find routines to locate the code, enhancement routines to sharpen markings, and analysis routines to decode and read the code. While hand-held readers exist for reading matrix codes, the matrix code reader often is designed for overhead reading at a fixed location.

The matrix code readers are essentially machine vision systems. These scanners integrate the CCD area imager, image processing hardware/software for image enhancement, "find" and de-coding, as well as the necessary lighting and optics into a single package. They are becoming similar in packaging and use, at only a slightly higher cost to their linear/laser equivalents.

Several 2D codes that exist include PDF417, Codablock, Code 16K, Code 49, DataCode, Vericode, MaxiCode, Aztec, QR Code, and Softstrip. Several of these are stacked codes that generally require more real estate or sacrifice the number of characters that can be read. They generally have less redundancy than the matrix codes. The matrix codes are generally scale and rotationally invariant. Those so far favored in the standards are DataCode, PDF 417, and Maxicode.

In addition to several of the traditional bar code reader suppliers who have adapted their readers to reading the stacked codes, a number of suppliers of general purpose machine vision systems are now offering matrix code readers.

11.7 COLOR BASED MACHINE VISION

Color based machine vision is benefiting from the advances that have been made in the price/performance of computers and color TV cameras. Leading the adoption of these advances has been the graphics display sector of the electronic imaging industry. In addition, the widespread adoption of camcorders by consumers and competition in that market has resulted in bringing color TV cameras down to a price that is quickly approaching monochrome cameras. While these cameras lack the quality required for industrial use and modified versions are emerging adapted for industrial uses, the fundamental camera technology and sensors used in industrial cameras benefit from the competition in the home video market.

In a similar way the lure of the consumer market and potential for interactive graphic displays with television is fostering major advances in image board technology. That is, boards that are aimed at acquiring and displaying images. The purposes envisioned include: interactive video games, desktop publishing, motion analysis, animation, simulation and visualization.

While these advances in image board products are mostly aimed at the part of the electronic imaging industry that operates on computer generated images, there are a growing number of suppliers that are offering video boards capable of acquiring and storing images as well as operating on the stored images for purposes of computer enhancement, etc.

With all this product technology emerging, it is only a matter of time before machine vision becomes a major adopter. Significantly, as machine vision is frequently applied as a substitute for human vision and given that humans do operate on images of scenes based on color hues, when color based image processing essentially becomes "free" then the machine vision industry will see widespread substitution of monochromatic processing with color processing. By "free" is meant that the costs, in terms of dollars, of processing color approach the costs of processing monochrome machine vision. At present rates of improvement in compute power, in less than five years color machine vision processing should be as technically and economically feasible as monochrome machine vision is today.

11.7.1 LIGHT, COLOR AND HUMAN VISION

It was observed years ago that from three "primary" colors all other colors can be derived. Typically, the primary colors employed are red, green and blue. Color as we see it is a single wavelength or a combination of the RGB wavelengths that reflect off an object. For example, a yellow object absorbs the entire blue wavelength and reflects equal amounts of red and green. Yellow is the complement of blue, while magenta (equal amounts of red and blue) is the complement to green and cyan (blue and green) complements red.

Reflected color has been observed to have three properties: hue, saturation and intensity (HSI). Hue corresponds to what we commonly call the color, blue, violet, for example. Saturation refers to the percent reflection of one primary color in the hue relative to the other two primary colors (or the degree of absence of white light), while brightness corresponds to the quantity of photons being reflected. The color pure red is a hue that reflects only the red wavelength (about 700 nanometers) and is, therefore, highly saturated. The brightness value is a function of the quantity of "red" photons being reflected by the object that is in turn dependent on the photons generated by the source of illumination. Red becomes lighter when the object reflects equal amounts of green and blue which when combined with red are seen as white.

Issues associated with human vision and color:

1) Human perception of lightness or darkness does not vary uniformly with the amount of color or light applied.

2) Human perception of color is inconsistent (by an individual himself, or from observer to observer).

3) Standard objective RGB models have been found better for color matching than subjective hue, saturation and intensity models.

4) The proximity of other colors to the color(s) being observed influences perception of color, i.e., backgrounds influence perception of color.

5) The surface texture underlying the color affects perception of color.

6) Humans enjoy "color constancy", that is, regardless of intensity and spectrum of color, perception of color is the same.

7) Color perceptions such as HSI do not have fixed relationships to the physical world and the perception of a given color has no unique relation to a particular color stimulus.

8) Human vision is inherently trichromatic versus spectral in nature - all color stimuli are analyzed by the retina into only three types of responses.

9) People accept slight mismatches in lightness or chroma more readily than an equivalent mismatch in hue.

10) The color of a surface depends on the direction or directions of illumination and the direction from which the surface is observed.

11) Sensitivity of the eye is almost constant over a wide range of light intensity.

12) While there is much knowledge about color and spatial vision, no comprehensive model of color appearance exists that takes viewing conditions into account.

Issues associated with HSI:

1) Do not completely resolve problems of nonuniformity of human color vision.

2) Do not take into account the effect of viewing conditions on color appearance.

Issues associated with CIE

1) Colors match only under identical conditions.

2) Lighting constancy is essential.

3) Tristimulus values have no direct perceptual correlations themselves.

4) Tristimulus values quantify characteristics of a color stimulus but do not specify the appearance of that stimulus. These coordinate systems re frequently misused and misunderstood as color appearance spaces.

11.7.2 MACHINE VISION TECHNOLOGY AND COLOR

Color cameras are now available which focus light through a lens onto a system of optics that break the light into the three primary colors - red, green and blue. These are then focused each onto its own sensor array arrangement. These signals are amplified, encoded and transmitted. The encoded color video signals contain separate signals for each color's hue, saturation and intensity as well as synchronization and "color burst" signals that contain data required by the decoder to reassemble the image.

The stability of solid state cameras over vidicon based cameras is what makes it appropriate to consider industrial machine vision applications. Significantly, there are both one chip and three chip cameras. In the latter case the camera handles the primaries individually, delivering power signals to a destination which: may retain them as components or encode them in composite. In any application, the camera details should be worked out as part of a comprehensive systems analysis.

In machine vision the camera signals are typically digitized (sampled and quantized) and stored in frame buffers, one for each color. Typical color video frame grabber cards capture an image in real time rates (30 hz) and store three frames 500 x 500 or so each, 8 bits or 256 levels deep. Typically available color image processing systems use both RGB and HSI models to address image analysis/machine vision applications. In these cases the execution of the conversion of RGB space to HSI space is done in hardware. No company is known with product that uses hardware to convert color image date to CIE space. These conversions are done in software.

Because of the amount of data handled in color, most machine vision systems go into a level of data compression, the exact nature of which is application dependent. Typically, vendors of color based machine vision products speak of ability to handle 16,000,000 colors (a figure arrived at as a multiple of a 24 bit system -8 bits per R, G, & B - and 8 bits or 256 different levels of intensity, and process properties into a hypercube arrangement from which a "fingerprint" can be extracted representative of a "golden" model. The fingerprint corresponds to a statistically based evaluation of the color image. Based on an analysis of samples a tolerance can be developed around the golden model and as long as the statistics stay within the tolerance band the system will record the product as acceptable.

Most color based machine vision applications have involved segmentation based on color that in a color scene is far more robust than segmentation based on gray shades. Conventional machine vision operations are then performed on the segmented image - thresholding, geometry extractions, etc. In other cases, histograms associated with specific hues serve as the basis of product sorting - separating green bottles from amber, etc. Issues associated with machine vision and color:

1) Fact that light is inversely proportional to distance can be a factor.

2) Color cameras are typically linearized to improve their color display properties, i.e., compensate for CRT phosphor non-linearities. They are also compensated for viewing in dim areas where the average luminance level of the surrounding area is somewhat lower than the average luminance level of the displayed image.

3) YIQ space processing may be an alternative - Y - intensity of luminance characteristic of a color image; I the "in phase" and Q the "quadrature phase" of the chrominance data. YIQ composite system is the broadcast standard and lends itself to RGB conversion.

4) Both the spectrum of illumination and quantity (intensity) of light will influence resultant interpretation of color.

11.7.3 APPLICATION ISSUES

As with all machine vision systems, the technical issues revolve around application issues. In each application there are variables associated with both appearance and position that often can make the difference between success and failure in a machine vision application. Appearance issues that stem from normal variations in production processes, as well as environmental effects, such as, lighting, both intensity and spectral.

The net result is that basically there is an inherent background noise that the machine vision system must be tolerant of and ignore before processing the data and normalize somehow before processing the data of consequence to the application. So, for example, an apparently straightforward application such as can end inspection has many variables. On a given line, one manufacturer may in fact be making can ends for different companies. As a consequence, they may have different coatings that are in fact different colors or specularity. There may also be certain embossings on the can ends, etc.

Consequently, different lines may wind up with different application engineering content even though one is selling a turnkey system. The system itself may also have to have a capability to be fine-tuned to handle different products on the same line. The ability to cope with application idiosyncrasies in a user-friendly manner is essential for success.

11.7.4 TECHNICAL CONCERNS

One major concern is that of packaging the system into a "factory hardened" system and the effort that may entail. This not only requires production engineering to assure reliable performance over time in relatively harsh industrial environments, but also requires the preparation of test procedures, system documentation, software documentation, engineering change procedures, etc. as well as policies with respect to: acceptance testing, warranty, service/maintenance, spare parts, software upgrades, training, etc.

Another concern is the ability to take out the idiosyncrasies associated with all of the components used in a machine vision system based color interpretation system. In the case of lighting, in addition to spatial and spectral uniformity, another concern with the lighting has to be lamp life and the implications this has on machine downtime to replace the lamp and maintenance cost stemming from frequency of bulb replacements as well as the cost of bulb replacement. Many companies consider the life-cycle cost of a system and not just the basic cost of the system when making purchase decisions, especially if there is competition.

Most of these lamps are only warranted for 2000 hours. In many applications in process industries where such systems are in continuous use, this would mean that a bulb would have to be replaced on a line every 15 weeks. Depending on the expense of the bulb, this might be objectionable.

Another concern is the quality of the optics required. Effects such as chromatic aberrations, flaring, distortions, etc. must be eliminated. In the cases of the camera/imagers, concerns include: interpixel differences in spectral responsivity and linearity of response.

Another concern has to be with the throughput limits in any implementation. While designed for on-line applications, what are the constraints and trade-offs associated with throughput. In any given application will the sampling frequency be adequate? Will it be that greater throughputs can only be achieved if only a smaller piece of the object can be monitored to observe color changes. In other words, by processing less information (fewer pixels). While this may be a concession that is acceptable in certain applications, it would not be acceptable in others where 100% inspection is required of 100% of the product.

Resolution of the system is limited to that associated with one camera. This is consistent with the resolution limits of other machine vision systems as well. Nominally a camera resolves a picture into 500 X 500 pixels. Given a certain field of view, one can get a measure of the size of a pixel in object space by dividing the maximum dimension of the object by 500. This may or may not be sufficient discrimination for a given application.

It is recognized that resolution can be enhanced just by adding cameras and hardware to provide the ability to process those cameras. Similarly, throughput can be addressed by adding compute power to perform parallel processing. However, such projects would entail engineering content and greater cost to implement.

Another concern is system calibration. Ideally this should be automatic and transparent to the line operator. Undoubtedly calibration may become an application specific implementation. Still another concern is integration with a display and the implications for colors observed on the display vis-a-vis those of the real world.

12

Evaluating Machine Vision Applications

So you think you have a machine vision application. Do you want to somehow determine if the project is at least remotely feasible and you don't want to use a company's salesman to do the evaluation? How does one go about doing that? Well there are several "rules of thumb" that can be used to at least get some measure of feasibility. These all start with having at least a fundamental understanding of how a computer operates on a television image to sample and quantize the data. Understanding what happens is relatively straightforward if one understands that the TV image is very analogous to a photograph.

The computer operating on the television image in effect samples the data in object space into a finite number of spatial (2-D) data points that are called pixels. Each pixel is assigned an address in the computer and a quantized value that can vary from 0 to 63 in some machine vision systems or 0 to 255 in others. The actual number of sampled data points is going to be dictated by the camera properties, the analog to digital converter sampling rate, and the memory format of the picture buffer or frame buffer as it is called.

Today more often than not the limiting factor is the television camera that is being used. Since most machine vision vendors today are using cameras that have solid state photo sensor arrays on the order of 500 or so by 500 or so, one can make certain judgements about an application just knowing this figure and assuming each pixel is approximately square. For example, given that the object you are viewing is going to take up a one-inch field of view, the size of the smallest piece of spatial data in object space will be on the order of 2 mils, or one inch divided by 500. In other words, the data associated with a pixel in the computer will reflect a geographic region on the object on the order of 2 mils by 2 mils.

One can so establish what the smallest spatial data point in object space will be very quickly for any application: X (mils) = largest dimension/500. Significantly this may not be the size of the smallest detail a machine vision system can observe in conjunction with the application. The nature of the application, contrast associated with the detail that you want to detect, and positional repeatability are the principal factors that will also contribute to the size of the smallest detail that can be seen by the machine vision system.

The nature of the application refers to exactly what you want to do with the vision system: verify that an assembly is correct, make a dimensional measurement on the object, locate the object in space, detect flaws or cosmetic defects on the object, read characters or recognize the object. Contrast has to do with the difference in shade of gray between what you want to discriminate and the background, for example.

The organizational repeatability is in effect just that - how repeatable will the object be positioned in front of the camera. If it can not be positioned precisely the same way each time then it means that the field of view will have to be opened up to include the entire area in which one can expect to find the object. This will in turn mean that there will be fewer pixels covering the object itself. Vibration is another issue which can impact the size of a pixel as does typically motion in the direction of the camera itself since then optical magnification may become a factor - increasing or decreasing the size of the spatial data point in object space.

Let's take the generic applications one at a time.

12.1 ASSEMBLY VERIFICATION

You want to verify that all of the features on an assembly are in place. If you can perceive that there is a high contrast between each of the features and the background or when a feature is in place or not in place, then the smallest feature one can expect to be able to detect would have to cover a two pixel by two pixel or so area. If, on the other hand, the contrast is relatively low then a good rule of thumb is that the feature should cover at least 1% of the field of view, or in the case of 500 by 500 pixels a total of some 2500 pixels. So knowing the size of a pixel in object space one can multiply that value 2 or 2500, depending on contrast, to determine the area of the smallest detectable feature.

12.2 DIMENSIONAL MEASUREMENTS

In the case of making dimensional measurements with a machine vision system one can consider the 500 pixels in each direction as if they were 500 marks as on a ruler. Significantly, just as in making measurements with a ruler a person can interpolate where the edge of a feature falls within lines on a ruler, so, too, can a machine vision system. This ability to interpolate, however, is very application dependent. Today the claims of vision companies vary all the way from one-third of a pixel to one-tenth or one-fifteenth of a pixel. For purposes of a rule of thumb, you can use one-tenth of a pixel.

What will this mean in conjunction with a dimensional measuring application? Metrologists have used a number of rules of thumb themselves in conjunction with measuring instruments. For example, the accuracy and repeatability of the measurement instrument itself should be ten times better than the tolerance associated with the dimension being checked. Today this figure is frequently modified to one fourth of the tolerance. The other rule of thumb that is often used by metrologists is that the sum of repeatability and accuracy should be a factor of three or one-third the tolerance.

So how does one establish what the repeatability of a vision system should be? Given the sub-pixel capability of one tenth of a pixel mentioned above and as in the example an object that is one inch on a side, the discrimination (the smallest change in dimension detectable with the measuring instrument) associated with the machine vision system as a measuring tool would be one tenth of the smallest spatial data point or two mils or .0002". Repeatability will be typically +/- the discrimination value or .0002".

Accuracy, which is determined by calibration against a standard, can be expected to run about the same. Hence, the sum of accuracy and repeatability in this example would be 0.0004". Using the three to one rule, the part tolerance should be no tighter than 0.0012" for machine vision to be a reliable metrology tool. In other words, if your part tolerance for this size part is on the order of +/-.001" or greater, the vision system would be suitable for making the dimensional check.

As you can see, as the parts become larger and with the same type tolerances, machine vision might not be an appropriate means for making the dimensional check, that is, based on the use of area cameras that only have 500 x 500 discrete photosites. Conversely, if the tolerances were tighter the same would be true.

12.3 PART LOCATION

Using machine vision to perform a part location function one can expect to achieve basically the same results as making dimensional checks. That is, most vendors whose systems are suitable for performing part location claim an ability to perform that function to a repeatability and accuracy of +/- one-tenth of a pixel. Using our example again, namely a one-inch part, one would be able to use a vision system to find the position of that part to within +/-.0002". Some companies that offer products based on geometric modeling claim to be able to locate a part to $1/40^{th}$ of a pixel. This capability applies only to part location and not dimensional analysis.

12. 4 FLAW DETECTION

For applications involving flaw detection, contrast is especially critical in determining what can be detected. Where contrast is extremely high, virtually white on black, it is possible to detect flaws that are on the order of one-third of a pixel. Significantly, one can detect these flaws but not actually measure them or classify them. When detecting flaws that are characterized as geometric in nature, for example, scratches or porosity, it is noted that the presence of such flaws can frequently be exaggerated by creative lighting and staging techniques. So if those were the only flaws one wanted to detect and detection was all that was necessary, a rule of thumb would be that the flaw has to be greater than one-third of a pixel in size.

Where contrast is moderate, the rule of thumb associated with assembly verification, namely that the flaw cover an area of two by two pixels would be appropriate. Classifying a flaw with moderate contrast would require that it cover a larger area, on the order of 25 pixels or so. Again, where contrast associated with a flaw is relatively low as is the case with many stains, the 1% of the field of view rule would hold or it should cover 2500 or so pixels. Significantly, if it is a question that one is trying to detect flaws in a background that is itself a varying pattern (stains on a printed fabric, for example), the chances are that one would only be able to detect very high- contrast flaws.

12.5 OCR/OCV

For applications involving optical character recognition (OCR) or optical character verification, the rule of thumb is that the stroke width of the smallest character should be at least three pixels wide. A typical character should cover an area on the order of 20-25 pixels by 20-25 pixels. The critical issue here then is the length of the string of characters that one wants to read. At 20 pixels across a character and two pixels spacing between characters, the maximum length of the character string would be on the order of 22 characters in order to fit into a camera with a 500-photosite arrangement. In optical character recognition/verification applications, a bold font style is desirable. In general it is also true that only one font style can be handled at a given time.

Another rule-of-thumb is that the best OCR systems have a correct read rate on the order of 99.9%. In other words, one out of every thousand characters will be either misread or a "no-read". The impact of this should be evaluated. For example, if 300 objects per minute are to be read, and 0.1% are sorted as "no reads", in one hour you would have approximately 20 products to be read manually. Is this acceptable? This is the best-case scenario. The worst case would be if they were misread.

When it comes to pattern recognition applications, a reasonable rule of thumb is that the differences between the patterns should be characterized by something on the object that is greater than 1% of the field of view or again on the order of 2500 pixels. Significantly, the gray shade pattern can be a major factor in making it possible to see pattern differences or to recognize patterns that have differences of far less than 2500 pixels. This would be the case, for example, where both geometry and color are factors.

Significantly, where more than one generic application is involved in the actual application, the worst case scenario should be determined and used as the criteria to establish feasibility. Throughout this rule-of-thumb analysis, the dictating factor has been the number of photosites in the camera. Significantly, today solid state cameras do exist that have up to 1,000 X 1,000 photo sites. These cameras, however, are not cheap. It is even possible that they would be more expensive than the vision system itself. Furthermore, few commercialized machine vision systems have the capacity to process so many pixels and make vision/decisions at anywhere near real time rates.

12.6 LINE SCAN CAPTURE IMPLICATIONS

An alternative, however, to capturing images of an object with an area camera would be to use a linear array camera. There are several vision companies who offer linear-array-based vision systems where the linear arrays have up to 2,000 photosites. Using a linear array one would have to move an object under the camera or move the camera over the object in order to capture a two-dimensional image. Significantly, if the object is going to be moved under the camera, the speed with which it passes must be well regulated and the operating speed of the camera in combination with the speed of the object as it passes underneath the camera will in effect dictate the size of the pixel in the direction of travel.

Typically, vision systems that use these principles will operate at up to 2 megahertz rates. For a 2,000 element array that means that you will be scanning 1,000 lines per second (2,000,000/2000) in the direction of travel. For example, given an object speed of 10 inches per second (10,000 mils per second), at a sample rate of 1,000 lines per second, the effective pixel size in the direction of travel will be 10 mils (10,000/1000). So when evaluating machine vision applications, you may want to consider the possibility that the application can be addressed with a linear-array-based technique. In these instances all of the size details one can discriminate in object space would be proportionally better. For example, with a 2,000 element linear array, everything would be four times better than using an area camera with 500 X 500 photosites.

12.7 SUMMARY

Significantly, these are meant to be rules of thumb and should be only used as such in the evaluation of an application. Having performed this type of evaluation, however, it would be more reasonable for you to decide whether or not to pursue an application. It will avoid your wasting time with salesman trying to convince you that your application is "a piece of cake".

13

Application Analysis and Implementation

13.1 SYSTEMATIC PLANNING

There are many opportunities for machine vision within a plant. To date there has been little or no experience with machine vision in most companies. To get meaningful experience and ensure success, the introduction of this technology must be done systematically (Table 13.1). As indicated in Chapter 4, the successful installation of machine vision can yield significant gains in productivity and quality. In addition to sorting reject conditions, the application of machine vision offers the opportunity for process control.

Today's machine vision technology is sold on the basis of its "configurability" and, consequently, in theory, its ability to be reapplied as plant conditions change. In fact, most often the performance envelope of specific systems limit the applications they can address, and those with more capacity require nontrivial application engineering content to address different requirements. So, for example, techniques well suited to gaging sheet metal assemblies can be reconfigured with moderate engineering content for differently shaped assemblies. They cannot, however, be engineered to address surface inspection applications.

What this suggests is that the treatment of the purchase of machine vision systems as if one is purchasing a commodity item can be a serious mistake. At

TABLE 13.1 Application Analysis and Implementation

Systematic planning
Know your company
Develop a team
Develop a project profile
Develop a specification
Get information on machine vision
Determine project responsibility
Write a project plan
Issue request for proposal
Conduct vendor conference
Evaluate proposals
Conduct vendor site visits
Issue purchase order
Monitor project progress
Conduct systematic buy-off

tention must be paid to evaluating an application for both near- and long-term requirements to develop a comprehensive specification thoroughly reflecting the requirements. Vendors with appropriate technology and experience must be identified, solicited, and properly evaluated as suppliers. Since the technology is advancing rapidly and similarly the applications it can address are increasing rapidly, a conscious effort must be made to stay "on top of it." Vendors selected must have demonstrated an ability to keep up with advances.

Complicating the decision about machine vision is the fact that it is unlikely that one supplier will have the capacity to satisfy all applications identified in a facility. Recognizing that in the long term products from different vendors will be operating side by side, a consideration in the vendor selection process is how many different vendors' products can a facility with limited technical resources support, especially since machine vision is only one of many advanced manufacturing technologies now being introduced.

One approach to addressing these concerns was taken by General Motors in their evaluation of machine vision requirements. Internal plant surveys designed to evaluate the need for machine vision were conducted. These requirements were distilled into generic requirements, such as precision measurement, sheet metal gauging, two- and three-dimensional robot guidance, character reading, and so on. Machine vision companies were then evaluated. It was established that there were roughly 30 types of machine vision techniques. The applications were then assessed to determine which machine vision techniques could satisfy them. Companies offering a range of suitable capabilities were thus identified.

13.1.1 Know Your Company

There are many factors that should be taken into consideration before proceeding with a machine vision installation (Table 13.2). Fundamental factors include recognizing the short- and long-term manufacturing philosophy of the company. For example, is there already in place or under consideration the wherewithal to tie together the manufacturing process via a hierarchy of controllers and computers? Is there an overall advanced manufacturing technology plan? Should this be anticipated? This therefore dictates the use of a machine vision system with compatibility, the ability to be interfaced to and communicate with an arrangement of computers. Essentially, then, the machine vision system becomes a computer peripheral.

This is an important consideration when one must accommodate the inspection of products manufactured in small batches. If no means is available to download inspection programs, the system will have to be retrained at the beginning of each run. The result could be a significant setup time that would interfere with efficient batch production. Even if provision is made for local program storage on a cassette or floppy disk, are there so many models to be concerned with that a second system will be required for "training" - the development of the "golden" files? Interfaceability back to a CAD database that dictates the inspection criteria on the basis of design rules or its own golden file will be far more efficient.

TABLE 13.2 Know Your Company

Manufacturing philosophy
Productivity improvement versus capital expansion
In-house skills
Age of capital equipment
Innovations
Materials
Processes
Industry
Technology "leap"
Build or buy
Risk-taking organization
Design for inspection

Other considerations include the following: Is the contemplated installation intended for productivity improvement or to expand production capacity? In either case, is the machine vision system to be delivered a "stand-alone" system or is the installation to be "turnkey"? The former case implies that the ultimate responsibility for making the system work rests with the buyer. This in turn implies that the buyer must be prepared to train staff to become reasonably familiar with image-processing theory as well as system properties so he or she can optimize the performance of the system.

As a turnkey, the machine vision supplier or a systems house assumes total responsibility for making it work. The end user never has to understand why it works, simply that it does work. In this case the supplier must become familiar with the buyer's manufacturing process. In other words, a successful turnkey installation requires the buyer develop a specification that correctly establishes the criteria characterizing a reject condition. The development of the specification may also provide focus to the type of machine vision technology that will be required for the installation. For example, if color variation must be tolerated and only shape monitored, a system that does not operate on shades of gray may be more appropriate. Similarly, a backlighted arrangement might be more appropriate so only silhouetted properties are captured and operated on.

Other "systemic" considerations might include the following: How old is the capital equipment and the manufacturing process itself? For example, does it make sense to augment the capabilities of the equipment if it has already been fully depreciated and may be replaced in a year or two because of technological changes in materials. Are there manufacturing technology breakthroughs that may be taking place that will result in wholesale replacement of capital equipment. Keyboards represent a good example. For years the characters on the key caps were developed by injection molding techniques. Now the keyboard industry has largely adopted a transfer printing process that in one shot transfers all the legends onto the key caps. The inspection problem is completely different. Whereas before, key transposition was a major cause for rejection and legend quality only of secondary concern, with the printing technique, transposition problems are all but eliminated, but legend quality is more difficult to control. On the other hand, will it be possible to get payback from the system by using it over in another application?

Another issue revolves around the "technology leap." This is a two-sided issue. On the one hand is the concern that the company does not have adequate resources to support the technology when it is introduced. This, of course, can be overcome by training or hiring personnel with appropriate skills. The flip side of the issue is that for a given application, the technology is not quite ready.

Other managerial philosophies must be examined. Is there a "build-or-buy" decision contemplated? That is, is there a possibility the company may seek outside vendors in the future rather than produce it internally? Is the emphasis of management being placed on making it right in the first place and, therefore, monitoring the production process to avoid rejects as opposed to culling rejects in final inspection?

Are there risk-takers in the organization willing to stick their necks out to change a situation or is it strictly a laissez-faire organizational philosophy that prevails? (We beat the Germans in World War II without robots in the battlefields so why do we need them now?) Is there a management concern for the employee that is the motivation for considering machine vision automation - one's health - avoiding hazardous or hot environments?

Another consideration involves product design changes or possibly even manufacturing changes that could make it more viable to inspect or perform optical operations automatically on the object. "Design for inspection" should be a philosophy wherever possible. For example, where color does not impact function, light-dark colors can enhance the contrast associated with the task. The addition of machine-vision-readable codes can be useful for identification purposes. The addition of fluorescent dyes may also be a means of providing contrast to an otherwise difficult scene.

Experience has shown that by paying attention to the production process, a new and better understanding often results in modest redesigns to make the product easier. Similarly, points for process control with simpler sensors (e.g., proximity switches) are often identified that serve to prevent the reject conditions one is buying a machine vision to detect. A machine vision system then becomes an insurance policy.

All the preceding factors play a role in specifying a system. Similarly, understanding these factors beforehand can make the difference between a white elephant" and a successful installation. The message should be clear - know the company before proceeding with the identification and feasibility assessment of a machine-vision installation.

13.1.2 Developing a Team

Given that a decision has been made to deploy machine vision, what procedure does one follow? The first thing to do is to identify all those who can provide input. People involvement is critical and "Jack of " is frequently associated with project failure. Both "bottom-up" enthusiasm and "top-down" directives are doomed to failure.

TABLE 13.3 Machine Vision Team

Management
Manufacturing
Manufacturing engineers
Quality control engineers
Line supervision
Support engineering
Plant
Industrial

Involvement should include all those that will be affected by the installation (Table 13.3): line supervisors, foremen, and operators. Others that should be included are plant engineering, maintenance, quality control, manufacturing engineering, industrial engineering, and so on. While meeting as a "team" may not be necessary, one should plan to involve each as appropriate in the project. Involvement results in a sense of ownership associated with the change. Communications and education can accomplish this. An orientation seminar attended by all those to be involved in a machine vision project can significantly reduce resistance to change stemming from an apprehension about the unknown. Such seminars will not only spark enthusiasm on the part of some participants but will also result in the less enthusiastic not standing in the way.

13.1.3 Develop a Project Profile

The team should be used to develop a "profile" for a machine vision project. This can be based on input about experiences with manufacturing technology of complexity and costs comparable to machine vision and one example for a first time installation is summarized in what follows (see also Table 13.4):

1. Perceived Value. Corporate staff and plant operations should share in the perception that a successful machine vision installation will have a value. Similarly, at the installation facility quality control, manufacturing engineering and plant engineering should share in a similar perception of value.

2. Cost-Justifiable. The benefits of an installation should be tangible. A post installation audit should be possible or a set of measures developed to evaluate the application.

TABLE 13.4 Project Profile Example

1.	Perceived value
2.	Cost-justifiable
3.	Recurring concern
4.	Straightforward
5.	Corrective action possible
6.	Technical feasibility
7.	User-friendly potential
8.	Dedicated line
9.	Long line life
10.	Operator champion
11.	Management commitment

3. Recurring Concern. The application of the machine vision system should be associated with the detection of a condition that is experienced with some frequency (ideally, several times a shift). In other words, there should be an opportunity for "instant gratification" stemming from improved quality in goods shipped.

4. Straightforward. The installation of the machine vision system should not require extensive line rearrangements or line modifications associated with delivering the product to the vision station. Ideally, an idle station should exist in a line that holds the parts well organized and in a repeatable position. Indexed motion may be preferable and touching and overlapping parts should be avoided. Room should be available to install the system.

5. Corrective Action. It should be possible to do something about the condition being detected. Detection should not be an end unto itself.

6. Technical Feasibility. The first installation of machine vision systems should employ "off-the-shelf" technology that has been applied in similar applications.

7. User-Friendly Potential. The machine vision technology deployed should not be intimidating to the operator or to plant engineering who must maintain the equipment. Ideally, the technology will be virtually transparent, and a computer language should not have to be learned.

8. Dedicated Line. Ideally, the first installation, although of a system with potential for reconfigurability, should not be required to be reconfigurable. Essentially, a fixed automation scenario would be preferred.

9. Long Line Life. Installations of the machine vision system should be associated with a new model or one that has been introduced recently. This should guarantee the payback from the system. Ideally, the system should be incorporated with new tooling.

10. Operation Champion. The plant selected for the installation should have someone, preferably in manufacturing engineering, who wants to see it work and will ensure it does.

11. Management Commitment. Management must agree to the value of the application and be committed to doing something about it.

13.2 SPECIFICATION DEVELOPMENT

Having established the objectives along the lines of those depicted in Chapter 4 for the project, the next step is to develop a set of detailed specifications for the application. Specifications are technical data that describe all the necessary functional characteristics that the system must have to perform the required job. They should not state how the vendor is to do the job. Specifications must state the required productivity and capability of the system as well as all significant operational requirements. They should also explain the requirements of the related fixtures, material handling, and so on. Similarly, requirements for machine control and line interfaces should be detailed as well as plant requirements and limitations.

The importance of specifications cannot be overstressed for the time and effort put into proper specifications will be more than repaid in terms of reduced start-up time and a reduction in maintenance and quality problems.

By analyzing the present methods, observing the present operation, and getting input from all those involved, a detailed description of the operation and a review of all anticipated variables can be developed. For example, it may appear that an operator is simply performing a sorting function, separating containers by their size or shape. On closer scrutiny, one will also observe that once in a while the operator throws a container into a scrap bin. While probably not classified as an inspector, by virtue of innate intelligence, the operator knows enough to separate incomplete or misshapened containers.

By virtue of increased sensitivity to exceptions, an operator can become more sensitive to specific conditions. For example, because training has emphasized the importance of separating all green containers, the person has a heightened awareness when such objects pass. Similarly, a machine vision system might have to somehow include a weighting factor (e.g., a complementary color detector) that will increase sensitivity to a specific factor - color in this example.

The converse is also true. An operator may be trained to be tolerant of color shade variations: For example, all yellows, regardless if pale gold or virtually orange, might be acceptable since it is the basic color itself that provides the distinction. In this case, therefore, the machine vision system must be equally tolerant of these normal variations while also maintaining sensitivity to the fundamental defects the system is supposed to capture.

Developing the specifications includes a quantification of existing procedures in terms of productivity and quality: number of shifts, scrap, warranty repairs, machine downtime, and so on.

When examining the operations, the following should be considered:

How are goods routed for the vision inspection task presently?

Does the operation require automating loading/unloading?

Is the operation inventoried?

Are products stored in bins? Magazines? And so on?

What are the actual inspection functions that must be performed by the system?

Is it gauged by "eyeball" or with instruments (micrometer)?

Is cosmetic inspection done by detail or is it performed by a cursory look for more gross appearance differences?

Does the inspection first require identification? If so, is it by shape? By reading characters? And so on?

Is the inspection itself one of verifying shape conformance? Again, with or without instruments?

Is the operation one of just verifying that an assembly is complete before another operation is performed or that objects are oriented properly?

If, in fact, it is a complete assembly operation that is performed, does the operator locate parts and guide them into place?

Is this operation a combination of several of the preceding tasks: location guidance, cosmetic examination, and gauging, for example? Is the present task 100% inspection or sample inspection?

Does the operator perform tasks other than vision (e.g., assembly and machine loading)?

Is there contrast in what must be observed, that is, can you visually see the condition without picking up the part to manipulate it in the light? How small is the smallest detail you want to see? How big is the object (field of view)? As in photography one can see both large and small objects with television cameras, but the detail that can be detected reliably is proportional to the field of view. Where necessary, machine vision systems can employ more than one camera so detail versus field of view need not be a factor that would preclude considering machine vision.

Are there normal variations in the appearance of the object that are ignored? A vision system will somehow have to normalize those conditions. Are there variations in the appearance of the background? Is the part repeatably and consistently located? If not, image capture will dictate a requirement for an even larger field of view, reducing the detectable detail. It will also require location analysis to reconcile the image captured with respect to position.

Are parts overlapping? Touching? Jumbled? Can they be presented registered and not touching?

Are parts moving or indexed where they can be held in place in front of the scanner? If in motion, at what conveyor speed? How well regulated is the speed?

Other considerations are as follows:

Does the operator perform three-dimensional analyses? Is the decision based on color interpretation?

How much time is there to make a decision?

What of the performance of any system substituted for a person? What percentage of the reject objects will one tolerate to pass as good? What false reject rate (number of good units that are rejected) will be allowed? No system is perfect!

How much start-up time will be allowed? If the system is not dedicated forever to a specific task, how much time will be permitted to get the system ready between product changeovers? How much floor space is available? Overhead space that may be needed to mount cameras and/or lighting arrangements? Will much equipment rearrangement be required? Are power, air, and so on available? How much downtime can be tolerated for other routine maintenance?

In what kind of environment will the equipment operate? In the presence of dirt and dust? Grease and lubricants? Water? Shock and vibration? Temperature and humidity? Electrical noise?

To support the specification of a system can the following be made available to the prospective vendors for bidding: job descriptions, present specifications, drawings, samples, photographs and/or videotapes of the facility and inspection area? What kind of personnel will be available to operate and service the system following installation?

If a system is in place today to perform the function, answer the following:

(a) What technology is used?
(b) What are the system's capabilities and limitations?
(c) What actual performance is being experienced?
(d) What problems are being experienced that hamper productivity and effectiveness?
(e) What is the utilization factor?
(f) If starting today, what would be done differently?
(g) What problems were experienced and how were they overcome?
(h) What impact did the project have on the user organization? Was it measurable? If not, why not?
(i) Is the impact more or less than target levels? Why?
(j) Does upper management perceive improvements attributable to the project?
(k) What is currently perceived as the system's greatest contribution or benefit? Is it the intended one? If not, why not?
(1) Was the budget maintained? The time schedule? If not, why not?

The following can be used as a guide in the "system" analysis:

1. Straightforwardness of installation
    a. Parent equipment modifications required
    b. Rearrangements
    c. Floor space restrictions
    d. Material handling
2. In-house skills required and available
    a. Ability to do material handling
    b. Ability to do installation
    c. Personnel available to operate
    d. Personnel available to service
    e. Environment
3. Availability of
    a. Job descriptions
    b. Specifications as now performed
    c. Samples
    d. Photographs of floor space
    e. Management support
    f. Labor/union support

What should you use to obtain the comprehensive insight reviewed in the preceding? This can come from the following:

(a) Written description of the operation that might be available. These should be reviewed to determine if observed activities agree with the written descriptions.

(b) Review any drawings: equipment layout drawings (are there critical dimensions?), those that affect functional capabilities, and part drawings. What are the tolerances?

(c) Job descriptions - do they agree with observations?

(d) Part specifications.

(e) Samples - do they experience corrosion or discoloration or other change with time?

(f) Photos of the prospective installation site as well as parts.

The following checklist can serve as a guideline in systematically examining the requirement as observations are made:

1. Scene complexity
   - (a) Number of stable states
   - (b) Number of parts in view at one time
     - (1) Touching
     - (2) Evenly spaced
     - (3) Registered
     - (4) Overlapping
     - (5) In a bin
   - (c) Number of features and description
   - (d) Contrasts
   - (e) Field of view vs. detail for gaging, part tolerances
   - (f) Part positioning
   - (g) Variations in acceptable appearances
     - (1) Color, saturation
     - (2) Specularity, texture (markings, lubricants, corrosion, dirt, perishability)
   - (h) Sizes
   - (i) Temperature
   - (j) Other
   - (k) Positional variations
     - (1) Registration (*x*, *y*, z, *theta*)
     - (2) Object distance
   - (l) Part sensitivity to heat
2. Cycle time
   - (a) $<$ 100 min
   - (b) 100-300 min
   - (c) $>$ 300 min
   - (d) Hand loaded
   - (e) Indexing (triggering possibilities) still time
   - (f) Continuous (triggering possibilities)

3. Line speed
    (a) Regulated speed
    (b) Synchronous operations
4. Perspectives (camera, subject)
    (a) Multiple perspectives
    (b) Variable perspectives
    (c) Vision access
5. Background contrast good
    (a) Backlighting possible
    (b) Gray scale, front light
    (c) Structured lighting possible
    (d) Color interpretation required
    (e) Three-dimensional and/or depth contrast
    (f) IR and/or temperature contrast
    (g) X-ray contrast
6. Image precision
    (a) Subject observation area
    (b) Subject size
    (c) Minimum feature, flaw size
    (d) Height variations (depth of field)
    (e) Other
7. For character-reading applications
    (a) Reading or verification
    (b) Inked, painted; molded, cast; stamped; raised; laser-etched; dot matrix; acid-etched; engraved; recessed; other
    (c) Describe single-font style and type and multiple-font style and types for each font style:
        (1) Stroke width
        (2) Aspect ratio
        (3) Character height
        (4) Character width
        (5) Character depth
        (6) Center-to-center spacing
        (7) Space between characters
        (8) Characters per string and per line
        (9) Numbers of strings and lines
    (d) Read rate (characters per second)
    (e) Character positioning, repeatability
        (1) Vertically
        (2) Horizontally
        (3) Skew
    (f) If "no read," what to do?
    (g) Percentage of misreads allowed per character string

8. System performance
    - (a) Percentage of subjects flawed
    - (b) False rejects allowable
    - (c) Escape rate allowable (how many bad ones can be allowed to pass?)
    - (d) Warm-up time allowable
    - (e) Setup time permissible per batch (changeover time)
    - (f) Skill level of person performing changeover
    - (g) Training time allowed
        - (1) For new subjects
        - (2) Modified subjects
        - (3) Frequency of new and modified designs
    - (h) System reliability
    - (i) Response if triggered with no part in view
9. Physical and interface requirements
    - (a) Floor space
    - (b) Camera location relative to the subject, feet per inches
    - (c) Light source
    - (d) Processor distance
    - (e) Mechanical
        - (1) Parent equipment modifications
        - (2) Rearrangements
        - (3) Air available for cooling (filtered)?
        - (4) Vacuum available
        - (5) Floor vibrations
10. Electrical
    - (a) Power available, power preferred
    - (b) Line conditioned and/or regulated
    - (c) EMI/RFI in environment
    - (d) Computers and program controllers
11. Interfaces
    - (a) Interconnected equipment
    - (b) Communications protocols required
    - (c) Trigger signals, time delays
    - (d) Reject signal delays
    - (e) Data log; specify details required (CRT display, hardcopy/printer, indicator lights)
12. Operator interface
    - (a) Preferred CRT/keyboard, menu, and icons
    - (b) Report generation, describe
    - (c) On-line programming skills available
    - (d) Setup and calibration
13. System reliability and availability

(a) Number of hours used per week (up time)
(b) Number of hours available for maintenance per week
(c) Maximum hours to wait for repair
(d) Redundancy, back-up required

14. Environment hostile
    (a) Washdown
    (b) Corrosive
    (c) Dirt
    (d) Oil mist
    (e) Shock and vibration
    (f) Ambient temperature
    (g) "Storage" temperature (during shutdown)
    (h) Humidity
    (i) Electrical noise: EMI, RFI
    (j) Ambient light
15. Miscellaneous
    (a) Layout drawings available?
    (b) Material handling to be whose responsibility?
    (c) Special enclosures to be specified
    (d) Personnel available
        (1) To operate
        (2) To service
    (e) Material handling
    (f) Training: operator, service
    (g) Service spares
    (h) Future modifications and enhancements desired

13.2.1 Specification Review from Machine Vision Perspective

Having distilled the preceding and written a specification, give some thought to the application from the machine vision perspective and review and revise the specification accordingly. The following can serve as a guideline.

13.2.1.1 Lighting. Is special lighting needed to provide even illumination? Will reflections be a problem? Are strobes required? Can lighting exaggerate contrast of attribute to be observed?

Gauging. Exaggerate edges, structured edges, backlighting, directional light and/or shadow.

Flaws. Dark- and light-field illumination, UV fluorescence.

Recognition. Back illumination, structured light, directional light.

Identification. Light trapped, light scattered.

Verification. Structure, specular, directional.

Guidance/location analysis. Backlighting, structured light.

13.2.1.2 Optics. Adequate field of view is needed to compensate for positional variations. Distortion and magnification influences accuracy. Polarizers reduce unwanted specular reflections. These are considerations regardless of application.

13.2.1.3 Sensors and Cameras. The significance of delectability versus resolution should be recognized, that is, the ability to observe a single object versus the ability to separate objects. The relationship of pixel arrangement to minimum detail observable should be understood.

Gauging may require use of line scanners to obtain required accuracy. It may be necessary to use a multiple-camera arrangement to make differential measurements, and stereo techniques may be needed to compensate for magnification changes due to part motion in the field of view.

Flaw detection is required for field-of-view analysis. The smallest reliably detectable flaw will typically be on the order of 0.3% of the field of view, not the object size.

For reliable verification or recognition the minimum size of the attribute upon which the decision should be based is 1% of the field of view assuming relatively high contrast associated with that attribute.

13.2.1.4 Preprocessing and Processing. Preprocessing and enhancement needs can be eliminated or reduced by the following conditions: registration, contrast (real or artificial), and windowing unwanted areas.

The application dictates the amount of preprocessing required: noise removal, smoothing, thresholding, segmentation, and so on. If segmentation is required, it is for the gray level, edges, color, texture, or changes between images. For example, an analysis outlined in a 1984 Carnegie-Mellon University dissertation, entitled "The Potential Societal Benefits from Developing Flexible Assembly Technologies," by Jeffrey L. Funk, for location analysis is as follows (Bi, binary; ED, edge; GFA, global feature analysis; LFA, local feature analysis):

A. Rigid parts (same stable state), repeatable position
   1. Separated
      (a) High contrast Bi/GFA
      (b) Low contrast ED/GFA
   2. Touching
      (a) High contrast Bi/LFA
      (b) Low contrast ED/LFA
   3. Overlapping
      (a) High contrast Bi/LFA
      (b) Low contrast ED/LFA

B. Rigid parts (several stable states), nonrepeatable positions
   1. Separated
      (a) High contrast Bi/GFA
      (b) Low contrast ED/LFA
   2. Touching
      (a) High contrast Bi/LFA
      (b) Low contrast ED/LFA

3. Overlapping
    (a) High contrast Bi/LFA
    (b) Low contrast ED/LFA
4. Tilting
    (a) High contrast Bi/3D
    (b) Low contrast ED/3D

13.2.1.5 Image Analysis. Again dictated by application, image analysis refers to the specific features that contain the data upon which decision is based. Is the number of features constant or do they vary by samples, batches, and so on? What will the program do if a feature is missing or an unwanted feature is present? Can statistical decision-processing techniques work or does it require syntactic analysis?

13.2.1.6 Mounting and Interface. Where will camera(s) and lighting be mounted and how? Do custom mounts need to be designed and built? What will produce the electrical trigger that initiates the vision system to capture and process the images? How will reject products be removed from the line? When will they be removed, immediately or downstream?

Does the system need to communicate with printers? Computers? Controllers? Other equipment?

13.2.1.7 Reports. Are there reports that the machine vision system should generate? Tally of production statistics? Trend analysis? Histogram analysis? Or will data be uploaded to a cell controller or host computer?

13.2.1.8 Access Controls and Visual Displays. Which plant personnel should have access to the system during setup and operation? Is there need for password and security system? Key lockout?

Should the system display inspection results for a run on demand? Should the display be a graph with icon-like indicators of trends? Should the system monitor trends from nominal conditions to indicate shift in process that could lead to rejects, an indication that corrective action is needed before rejects are produced. Should this be a bell?

13.2.2 Writing Final Specification

Having evaluated the process from the different perspectives, one can finalize the specification. Above all, the specification should not tell the vendor how to do the job; it should not specify the design or image processing techniques, and so on. Rather it should specify the application. The specification should include the following.

1. Scope. This section should include an overview of the existing process and the overall system requirements as well as what the vision system will be required to do. Material handling and interface requirements to existing machines and control systems as well as future anticipated interfaces should be explained.

2. Part Description and Specification. This should include the distillation of all the considerations reviewed in the preceding about the process, accept-reject criteria, escape rate-false error tolerances, environmental considerations, utilities, data storage and transmission, operator interfaces, cycle times, changeover times, and so on.

3. Acceptance. Acceptance criteria and the method to be used to prove out the system should be defined. The equipment to be used for the acceptance test procedure should be listed. Different acceptance test plans could be defined: one for use before shipment, one for preliminary off-line trials at the site, and one for final on-line acceptance.

The acceptance test might include number of parts, total hours of operation, accuracy or repeatability, acceptable standard deviation for measurements or other statistical details about the test, and so on.

13.3 GETTING INFORMATION ON MACHINE VISION

13.3.1 Company Personnel

Staff
Line personnel with experience

13.3.2 Vendor Representatives

Data to aid in selection can be secured from vendors. Besides technical advice and assistance, vendor representatives can furnish written materials and brochures. Recommendations generally have to be taken with a grain of salt since they may lack objectivity.

13.3.3 Consultants

Consultants who do not design and build machine vision systems or act as sales agents for machine vision vendors or system integrators can objectively examine plants to identify technically feasible and cost-effective applications. In addition, they can assist in developing specifications and identifying and evaluating the most appropriate vendors to be solicited.

13.3.4 Technical Society Meetings and Papers

Meetings and proceedings can provide insight into experiences with machine vision as well as provide information concerning new developments:

Automated Imaging Association, 900 Victors Way, Ann Arbor/Ste 132, MI 48108

SME/MVA, the Society of Manufacturing Engineers and Machine Vision Association, One SME Drive, P.O. Box 930, Dearborn, Michigan 48121

IEEE, the Institute of Electrical and Electronics Engineers, 345 East 47th Street, New York, New York 10017

SPIE, the International Society for Optical Engineering, P.O. Box 10, Bellingham, Washington 98225

13.3.5 Trade Journals

Virtually every trade periodical with editorial content on any aspect of manufacturing is publishing papers on machine vision regardless of industry focus. The MVA periodical *Vision* as well as those with robotics flavors often include papers with an application orientation, e.g., *Robotics World.*

General publications covering factory automation will sometimes run articles on machine vision. These include *Manufacturing Engineering, Industrial Engineering, Managing Automation, to* name a few.

13.4 PROJECT MANAGEMENT

13.4.1 Determine Project Responsibility

The best manner to handle a project should be found. For example, if the division or company does not have experienced staff in machine vision or related automated equipment, it may be necessary to look to the outside for assistance. There are a variety of alternatives:

1. Rely on the machine vision equipment supplier for overall direction of the project.

2. If line automation and/or material-handling equipment is involved that must be integrated with the machine vision system, a system integrator may be the one to take on overall project direction. The system integrator should be one with experience integrating machine vision with line equipment and, ideally, experienced with the machine vision vendor whose equipment is best suited for the application.

3. Consider using a consultant in machine vision to act as a project leader and carry the program through to completion. Such consultants can conduct feasibility assessments, assist in the development of the specifications, identify the vendors with the most relevant experience, qualify the vendors, and recommend the most appropriate vendors. In addition, they can manage the project through training, installation, and acceptance testing. They can also provide training on machine vision that is appropriate for all involved in the project.

In general, it is not a good idea to have the machine vision vendor serve as the project director if there is any material handling or other specialized equipment needed. The small machine vision companies generally do not have the personnel trained in product handling. They are for the most part computer hardware- and software-trained.

13.4.2 Writing a Project Plan

Having analyzed the application and written the specification, a project plan should be written. A project timetable should be established. The schedule should be realistic, not too fast that projecting dependence on the system can cause delays in overall production start-up and not so extended that it will become difficult to maintain a project commitment.

Three to five prospective vendors should be identified with demonstrated product capabilities consistent with the requirement. If gauging, they should have systems installed performing gauging. If flaw detection, applications of their techniques should have been installed performing, for example, cosmetic inspection.

Keeping the number of prospective vendors down reduces the number of proposals that have to be evaluated as well as avoids too many vendors making an investment in a proposal.

During this prospective vendor identification phase a telephone survey should be the mechanism used. After reviewing the project, the vendor's interest and related experience should be ascertained. In addition, the price for similarly installed systems should be determined. The average of these figures can serve as a "budgetary estimate" for the cost of the project. This figure, in combination with estimates of material handling, training, line modifications, and so on, can be used to perform a rough return-on-investment analysis. If the results are reasonable, a more serious bid cycle should begin.

13.4.3 Request for Proposal

A formal request for proposal (RFP) should be written that includes a description of the project in detail along with a brief discussion of the business of the operation. The description of the system should be as follows:

1. Review in detail the specific functions the machine vision system is to perform. Describe part(s) in detail indicating variables in parts, process, suppliers, and so on. Make sure tolerances are realistic from a production point of view and are in fact currently being met. Review all defects anticipated and outline their properties.

2. Describe how the machine vision system will be integrated into the line, especially reviewing cycle times.

3. Review environmental considerations.

4. Benchmark criteria and acceptance test procedures (if not part of the specification already), both for vendor facility buy-off and post installation buy-off, should be spelled out clearly.

5. If necessary, because of technical uncertainty, outline a feasibility demonstration requirement at the vendor facility. This should be treated as a separate line item in the cost proposal.

The rationale for the project solicitation should be reviewed as well as the schedule. The functional requirements (what, why, when, where) should be spelled out in detail in the specification. Where possible, separate the "needs" from the "wants" - those requirements that are essential (otherwise the system is worthless) from those requirements that are less critical.

In addition, the RFP should define the response wanted with respect to the following:

1. Cost. The cost of a system to satisfy the application should be separated from the costs of training, warranty, and installation. In the case of specifications with needs and wants, the costs should be separated, so the incremental costs associated with enhancements can be assessed. Engineering charges should be separated from system charges. If several units are anticipated in the future, the cost for the quantity should be requested at this time. Feasibility studies should be cost out separately.

2. Schedule. A project schedule should include important milestones for anything that is not "off the shelf." These might include scheduled preliminary and final design reviews and specific benchmarks associated with building and proving out. Potential intermediate approval points during the project could be identified in such a schedule.

3. Description Solicited. This should include a review of lighting, optics, fixture or staging designs, and image-processing techniques. The detail should be sufficient to assess whether resolutions and image-processing throughput are adequate for the application. Both operation interface and engineering interface should be reviewed as well as issues such as memory, enclosures, environmental considerations, diagnostics, electrical interfaces, and calibration procedures.

The description requested should do the following:

(a) Ask to review the specific functions the machine vision system is to perform.

(b) Ask for a description of how the machine vision system will be integrated into the line, especially reviewing cycle times of the computer and processing as well as mechanical cycle times.

(c) Ask for a review of the environmental conditions.

4. Policies to be reviewed in the RFP should include the following:

Service
Spares
Warranty
Training
Installation
Software upgrades
Documentation

5. Relevant Experience. The proposal should be asked to include a review of both specific and generic installations: number of systems, how long operational, types of installations (beta sites, laboratory, shop floors, etc.).

Along with the RFP photographs of the application site, part prints and representative samples should be forwarded to each of the solicited vendors. If setup costs are involved to evaluate samples, a fee should be anticipated to cover those costs. The samples should be representative of all production and should include good and bad samples as well as samples that are marginally good and marginally bad.

13.4.4 Vendor Conference

A vendor conference should be conducted at the prospective installation site 10-14 days after the RFP is received. Ideally, the vendor's technical personnel will be present as well as the local salesman. The vendor's conference performs a number of functions:

It is an opportunity to clarify objectives and requirements.

It demonstrates the vendor's interest.

It permits a facility review and first-hand appraisal of the operation on the part of the vendor.

Being a one-time session, it avoids the need to tie up staff for individual visits by each of the vendors.

It assures the same information is given to all in attendance.

The agenda for the vendor conference should include the following:

a. Briefing on company/operation's business
b. Review of project
   1. Why it is under consideration
   2. How it fits into overall CIM plans
   3. Management support
c. Review of specifications, line by line
d. Review of response required
e. Review of proposal and vendor evaluation criteria
f. Facility tour with emphasis on installation site
g. Time for one-on-one question-and-answer period

Participants at the vendor conference should include representatives from each of the departments that will be involved in the installation. A project manager should conduct the session, though individuals from, for example, management and purchasing should be used to provide briefing on the specific issues.

13.4.5 Proposal Evaluation

Upon receipt and review of proposals, where necessary, clarification of details, especially disparities, should be obtained from the vendors. A technical evaluation should be conducted to narrow the field to two, possibly three, vendors. The evaluation should include an assessment of the feasibility of the proposed approach. This should include an assessment that the application and the environment are understood.

From the perspective of machine vision equipment, the following are considerations:

1. Is the system's resolution adequate?
2. Is the hardware and software suitable to conduct the analysis required?
3. Is processing speed adequate for throughput?
4. How is application developed?
   a. Train by showing
   b. Panel buttons
   c. Pendant (hand-held terminal)
   d. Light pen, touch screen, or cursor menu
   e. ICONS
   f. CRT terminal, off-line programming
   g. Language knowledge (BASIC, FORTRAN, Forth, etc.)
5. How does operator interface?
   a. Panel buttons
   b. ICONS
   c. Pendant
   d. Light pen or touch screen
   e. CRT terminal or cursor commands
6. Application program storage, provision for data storage in event of power outage
7. Adequacy of interfaces, 1/0, etc.
8. System diagnostics
9. Model changeover times
10. etc.

In addition to cost and delivery, other considerations in the vendor evaluation should revolve around acceptance testing and policies: training, maintenance, spare parts, field service, warranty, and documentation. What terms and conditions for acceptance should a vendor be prepared to adhere to? Demonstration and prove out testing at the vendor facility? At plant site? Off-line? On-line? The actual acceptance test plan should be mutually developed and should include as comprehensive a test as possible, including a way of addressing variables such as part positioning, lighting, and environmental variations.

TABLE 13.5 Decision Matrix

		Vendor A	Vendor A	Vendor B	Vendor B	Vendor C	Vendor C
Criteria	Relative Weight	Rating	Cumulative	Rating	Cumulative	Rating	Cumulative
Price	5	9	45	8	40	6	30
Delivery	8	10	80	10	80	8	64
Schedule disclaimers cited	5	10	50	10	50	10	50

Installation policy	8	10	80	7	56	5	40
Warranty (30 days + 1, 2 yrs = 10	5	10	50	5	25	5	25
Postwarranty service cost	5	8	40	8	40	8	40
Documentation	5	10	50	8	40	10	50
Spares policy	5	10	50	10	50	10	50
Maintenance /operation training	8	5	40	10	80	10	80
Future system upgrade policy	5	3	15	10	50	10	50
Internal technical expertise	9	9	81	8	72	10	90
Previous application	8	7	56	8	64	10	80
Experience:							
Company history	5	8	40	8	40	10	50
Market synergism	8	7	56	7	56	10	80
User friendliness	5	7	35	7	35	10	50
Project management	5	5	25	6	30	10	50
Sub total			793		808		879
Fixed threshold Vs. adaptive							
Fixed 5, adaptive 10							
Number of windows	10	10	100	9	90	8	80
Speed versus windows	5	10	50	5	25	5	25
Number of geometric features	10	8	80	6	60	6	60
Speed versus features	5	10	50	10	50	10	50
Lighting design	10	10	100	10	100	10	100
Software compensation for light variations	5	10	50	3	15	3	15
Subpixel processing	8	10	80	10	80	10	80
Product changeover difficulty	8	5	40	8	64	8	64
"Wants" operator friendliness	10	8	80	10	100	8	80
Engineer friendliness	5	10	50	10	50	10	50
Self-diagnostics	10	7	70	8	80	8	80
Program storage	5	10	50	10	50	10	50
Data backup	5	0	0	10	50	10	50
Subtotal			1618		1672		1713
Needs							
Throughput	0/1	1	1618	1	1672	1	1713
Resolution	0/1	1	1618	1	1672	1	1713
Enclosure rating	0/1	1	1618	1	1672	1	1713
Interfaces	0/1	1	1618	1	1672	1	1713
Calibration procedure	0/1	1	1618	1	1672	1	1713
Escape rate	0/1	1	1618	0	0	1	1713
False reject rate	0/1	1	1618	1	0	1	1713
Accuracy	0/1	1	1618	1	0	1	1713
Repeatability	0/1	1	1618	1	0	1	1713
SPC package	0/1	1	1618	1	0	1	1713
Confidence factor		80%	1294	-	0	60%	1029

Who is responsible for installation? Are there specific maintenance requirements - hardware, software? What level of sophistication is required to replace lights, adjust cameras, calibrate, reconfigure 1/0, or do basic troubleshooting? How much training is included with the purchase order and at what level? What material is covered in the training? How much is hands-on? How long is the training? Where will it be conducted? How often and what type of training classes are available? At what cost for additional training?

What spares should be inventoried, even delivered, with the system? What are the costs? Availability? Is there a basis for meantime between failures? Meantime to repair? How "local" is field service? Spare? Is 24-hr service provided? What are the normal service response times? Service costs? Maintenance contracts available? What is the standard warranty? (Parts and labor?) Is an extended warranty available? At what cost and terms?

What documentation is provided? What are the application program source codes? Schematics? What is the quality of the documentation and how complete is it? How many copies are provided with the system? How application-specific is the documentation? How well is the software documented within the vendor's facility?

The company itself should be evaluated. What is the company's overall experience? How many systems have they installed? In your industry? Related to the specific application? What is the experience of staff in terms of installations and application? Company data should include the following: time in business, time in machine vision business, number of people by skills, and financial situation.

A decision matrix (Table 13.5) is a useful tool in evaluating proposals by reducing subjectivity from the decision-making process. A typical matrix lists the vendors across the top, and the criteria are in the left column. The criteria include pricing, system properties, and vendor characteristics as determined from the proposals. On a scale of 0-10, each criterion is rated as to its relative importance, 0 being the least important and 10 the most important. These relative weights for each of the factors are independent of the vendors' performance and should be established even before the proposals are received.

In the columns under each vendor, on a scale of 1-10, evaluate how the performance to the criteria relates to the need or to each other, whichever is appropriate. Multiply this value by the criteria weight and place this value in the appropriate column.

Sum the scores of the criteria for each vendor and subtotal the results.

These subtotals give an indication of the relative capabilities of the vendors evaluated. It does not tell whether the vendor can do the specific application.

The decision matrix should include specification criteria. In the case of wants, the rating of 1-10 can be assigned in accordance with how closely a vendor satisfies the want. A subtotal associated with nonspecification criteria and wants should be determined as the addition for each vendor of all of the factored ratings in the column.

Below the subtotal, all specification criteria that are needs should be listed. In these cases the issue is not degree of responsiveness but rather yes-no or can-cannot perform factors. The ratings, therefore, should be either 0 or I (for can-yes answers). Each of these factors would then be multiplied by the subtotal. If a vendor cannot satisfy a need specification and receives a 0, this total score will be 0, and the vendor should be disqualified from the application.

Technology criteria may also be included in the decision matrix. A value might be assigned to binary, thresholded binary, intensity-contrast, or gray scale-edge systems. There are many technologies that can be quoted. Some may just do the job; some may be an overkill but may be more flexible (a consideration possibly for future requirements), and some may be just perfect for the requirement at hand. Most often the technology should be rated relative to the specific requirements.

Having concluded this procedure, the next step is to assign a confidence factor associated with the confidence that the vendor can meet each of the specification needs and perform as specified. Confidence may be dictated by the extent of experience the vendor indicates in the proposal, especially relevant experience. This confidence is then multiplied by the previous subtotal to achieve a figure of merit for each vendor for the application being quoted. This figure should lead one to two or three vendors that should be visited.

Table 13.5 is an example of a decision matrix for the evaluation of three vendor offerings. The first part evaluates the company, policies, and so on, generally. The second part is an example of the technical aspects that could be important. These will vary depending on the application. Needs and wants are evaluated differently. Needs must be satisfied so they either have a 0 or a 1 weighting. This is simply multiplied by the cumulative score so far. The confidence factor is based on the cumulative "gut feeling" for a vendor based on dealings with representatives, thoroughness of response, and so on. In our example, the result of this analysis would lead us to favor vendor A with a contract.

Before this is cast in concrete, however, it would be in order to review with vendor B why it was not responsive to the need to satisfy our escape rate criteria (in this example). Similarly, where vendors A and B do not compare favorably, it is appropriate to review those issues with the respective vendors to make certain there were no inadvertent oversights in the preparation of the proposals. In any event, using such a tool will reduce the subjectivity associated with evaluating bids and improve the probability of selecting the most qualified vendor.

13.4.6 Vendor Site Visits

These visits provide an opportunity to view any concept-proving demonstrations possible on existing equipment. In addition, these visits allow one to assess the following about a vendor:

1. Technical resources (optics, television, computer, mechanical)
2. Technology, design, philosophy, product line breadth
3. Understanding of application
4. Capital and human resources to support and service installation

5. Physical facilities
6. Business philosophy with respect to
    a. Warranty
    b. Training
    c. Installation support
    d. Service
    e. Spares
    f. Documentation
7. Project schedules and possible conflicts with other projects
8. Financial stability and staying power
9. Review experience and obtain references associated with similar installations
10. Quality of work
11. Knowledge of business
12. Review typical documentation
13. Quality control practices, product bum-in procedures, etc.

Another decision matrix might be in order to evaluate systematically the vendors based on the site visit. From this analysis a vendor of choice should be determined. At that time the references should be contacted. If possible, a visit should be scheduled to the installation site. In any event it is important to assess the reference's opinion of the vendor:

Quality of work
Ability to meet schedules
Policies
Support

The leading question should be: If they had to do it all over, what would they do differently?

The final consideration with respect to a vendor revolves around the number of different machine vision systems that may ultimately be installed at a facility. Will they come from so many different vendors that the facility will experience difficulty servicing and maintaining spares?

Having made a decision on a vendor, it is now possible to fully assess the project's cost and to conduct a return-on-investment (ROI) analysis. An example of such an analysis follows.

13.4.7 ROI Analysis

Machine vision system costs are based on each requirement. Number of systems required:

1. Estimated basic system cost
2. Estimated cost for fixturing
3. Estimated cost for material handling

4. Application engineering cost
5. Other costs (e.g., utilities)
6. Estimated shipping charges
7. Estimated installation charges
8. Estimated annual maintenance costs
9. Estimated costs for training (operators, maintenance staff, etc.)
10. Estimated annual spare-parts costs

The estimated annual savings are as follows:

1 Direct labor savings
2. Indirect labor savings
3. Estimated savings due to reduced scrap
4. Estimated savings due to reduced in-warranty repairs
5. Estimated savings due to reduced tooling and fixturing costs
6. Estimated savings due to increased utilization of equipment
7. Estimated savings due to reduced setup times
8. Estimated savings due to reduced paper and affiliated activities
9. Estimated savings due to reduced labor-training costs
10. Estimated savings due to reduced liability claims
11. Other estimated savings

ROI Summary

Payback period required

Labor cost worksheet

Number of operators performing task per line

Labor category

Number of shifts

Number of lines

Total dollar direct labor savings

Total dollar indirect labor savings

Dollar value of goods inspected

In-warranty cost analysis

Cost of units, c

Number of units per month, N

Cost per month, $C = cN$

Warranty period, t

Number of units requiring in-warranty repair per month (monthly failure rate), X

Average unit repair cost (including shipping, labor to diagnose, repair, new parts, documentation, loaners, etc.), R Delta production cost for improvement per year, $XR\ (12)$

Project cost analysis: total costs associated with installation of system

Total annual savings anticipated

Annual operating costs: miscellaneous labor, maintenance and repair costs, taxes and insurance (2.5% times capital expenditure), depreciation (calculated over 5 years), total annual operating costs

Pretax cost reduction (annual operating savings, total annual operating costs)

After-tax cost reduction (based on 34% tax rate)

Investment base (50% of total expenditure, average capital cost over 5 years)

After-tax ROI (after-tax cost reduction divided by investment base)

Net cash savings (after-tax cost reduction plus depreciation)

Payback period (total capital expenditure divided by total annual savings)

Summary Sheet

Department

Operation description

Supervisor

Criteria for selection of project

Payback

Total annual dollar potential savings

13.4.8 Issuing Purchase Order

A contract should be issued that includes all the details associated with the project reviewed in the RFP and should include a "buy-off " procedure.

A payment schedule should be considered with a reasonable deposit up front (up to 40%) to cover application engineering costs that are unique to the project. Payment based on well-defined milestones could be a reasonable alternative. However, enough should be retained to make sure the vendor's interest is sustained so the project is completed in a timely manner. For example, a comprehensive concept-proving demonstration could be the first milestone.

Where an application solution is not "off the shelf," a phased study plan or development contract should be considered. In this case, one should avoid developing a new technique or at least be aware of the risks. It should be understood that there is a distinction between "knowledge" and "application." Moving knowledge into applications involves risk that both the buyer and seller must understand. In other words, phased procurements are appropriate in these cases.

If dealing with proven techniques, however, a more conventional contract is appropriate. In this case the responsibilities of both the seller and buyer should be spelled out, especially with respect to installation and support: training, service, etc.

13.5 PROJECT-PLANNING ADVICE

In conjunction with hands-on machine vision clinics offered by the Society of Manufacturing Engineers, panel sessions were conducted. One panel consisted of known users and the other of vendor-applications people. They were asked to share their experiences in dealing with machine vision and to offer recommendations to prospective users that might guarantee a successful installation. The following represents a distillation of the comments.

Good opportunities for machine vision exist at the following points:

1. Where the product has the least value added,
2. Where process control is involved to prevent reject conditions,
3. Where a robot with complementary vision avoids expensive fixturing and adds to system flexibility, and
4. Where an installation on a new line is involved versus line retrofit.

One vision system should not be expected to cover all applications. Systems with a degree of flexibility or reconfigurability should be sought so that if the immediate application goes away, with new fixturing, the system has a chance of being reconfigured for a new application, at least one within the same class of problems.

Project planning should be a reasonable and realistic process, and one should avoid trying to inspect a part that is not inspectable. The system purchased should be a cost-effective and technically viable solution to the specific and real problem. For a first-time application, the esoteric should be avoided, even if it may have the best payback possibilities. The project should definitely be worth doing, however, and have a payback.

One should learn how to specify a system and develop as comprehensive a set of specifications as possible, quantifying parameters and avoiding subjectivity. It is necessary to distinguish between wants and needs. The application should be understood. What specifically is the task for which the vision system is to be substituted? What are the process variables? Vendors expect the user to bring process knowledge about the application. They will bring vision technology knowledge. Design changes could be effected to make it easier to perform the vision task.

Simplification may make the application technically feasible with less risk or uncertainty as well as may make possible a more cost-effective solution. In addition to product design changes, process procedure changes may be required to make the installation successful. Then the full burden is not on the machine vision system. The vendor should not be told how to do the job, nor should the technology be specified.

One should be prepared to provide a realistic range of samples - good, marginally good, marginally bad, bad - and those exhibiting as many of the variables as possible - different vendors, different machine outputs, and so on. An attempt should be made to become objectively educated about machine vision. The technology should be understood so a decision can be made about which vendors have the technical approach in their product required to address the application. The differences between vendors should be known as well as the limitations of their respective "engines." The number of vendors solicited should be limited to 4-6.

Involvement as early as possible in the project cycle of all that will be affected by the system or who can make a contribution or who will have something to do with the system after installation is crucial. It is important to provide all with a sense of "ownership" associated with project. Communication is essential, especially to all levels of management. Communication throughout the project with all on the team, management, and vendor will avoid surprises. A task team could be formed, including line supervisors and controls, process production, industrial, plant, electrical, quality, and engineering disciplines, as well as a "vision research engineer." The objective is to get all the variables out into the open as early as possible. Orientation training for all the aforementioned may also aid in making the project go smoother, well before installation is anticipated. Management should have a level of understanding of the technology.

Do not just anticipate issuing a purchase order with specifications. One should be prepared for sustained guidance, recognizing that the adoption of new technology requires a "partnership," a cooperative arrangement between vendor and user. Assure that project benchmarks are being met by vendor site visits and project reports. Establish a realistic schedule. See that the vendor assigns someone with project responsibility. Avoid scheduling invention. In a vendor, bigger is not necessarily better. Responsiveness is a better criterion for vendor selection. Check vendor references.

Make certain both user and vendors understand the differences associated with the requirement - if development is required or application can be satisfied by a standard product. Similarly, the extent of the custom content or applications engineering should be understood.

Where risk is involved, consider a phased procurement cycle with the first-phase concept proving demonstrations and, subsequently, breadboard and first-piece demonstrations.

Avoid "creeping" expectations as the project evolves, and when they occur, recognize the impact on the vendor in terms of design changes and schedule.

If possible, avoid becoming a systems integrator. Look to the machine vision vendor or a designated systems house to provide a turnkey solution. Define fully the areas of responsibility.

Train all service and operating personnel on details before the system is delivered.

Have spare parts delivered with the system.

Make certain the system works off-line to avoid interfering with production as well as influencing production's opinion of the system if exposed to repeated difficulties in start-up. Document the acceptance test procedure and criteria.

Anticipate that some post-installation changes will be involved. For a first-time application and/or user, be prepared for a "mutual mystification" period, which may take several months.

Do not just turn the system over to production without some period of supervised on-line operation.

Specific technical advice includes the following:

1. A system with built-in climate control may avoid maintenance problems in certain applications.
2. Avoid requiring unnecessary peripheral equipment to be included in the system; this will just complicate the application.
3. Define system interface requirements fully.
4. Avoid applications that require extended lengths of cable.
5. If possible, incorporate a manual mode to exercise the system for one full cycle to allow an easy test mode for servicing.

Expect that the vendor knows the process involved so he or she can make independent assessments of variables and reflect an awareness of the environment. Expect that the vendor will provide training, documentation, and technical support after as well as before installation.

Vendors should recognize that the application of machine vision technology is a learning experience for the user; this could lead to new expectations for the equipment, especially where new knowledge about the production process itself comes about as a consequence of being able to make observations only for the first time with such machine vision equipment.

Recognize that software is not a "Band-Aid" for otherwise poor staging designs. As a last piece of advice, one user panelist suggested, "Never trust a machine vision vendor that uses the phrase 'piece of cake.' "

REFERENCES

APPLICATIONS

Abbott, E. H., "Specifying a Machine Vision System," Vision 85 Conference Proceedings, Machine Vision Association of the Society of Manufacturing Engineers, March 25-28, 1985. Revised for SME Workshop on Machine Vision, November 1985.

Abbott, E., and Bolhouse, V., "Steps in Ordering a Machine Vision System," *SMEIMVA Vision 85,* March 1985.

Funk, J. L., "The Potential Societal Benefits from Developing Flexible Assembly Technologies," Ph.D. Dissertation, Engineering and Public Policy, Carnegie Mellon University, December 1984.

LaCoe, D., "Working Together on Design for Vision," *Vision,* September 1984. Quinlan, J. C., "Getting Into Machine Vision," *Tooling and Production,* July 1985. Robotics Industries Association, "Economic Justification of Industrial Robots," pamphlet.

Rolland, W. C., "Strategic Justification of Flexible Automation," *Medical Devices (MD &* DI), November 1985.

Sephri, M., "Cost Justification Before Factory Automation," *P&IM Review and APICS News,* April 1984.

Zuech, N., "Machine Vision: Part I-Leverage for CIM," *CIM Strategies,* August 1984; "Machine Vision: Part 2-Getting Started," *CIM Strategies,* September 1984.

Zuech, N., "Machine Vision Update," *CIM Strategies,* December 1984.

14

Alternatives to Machine Vision

14.1 LASER-BASED TRIANGULATION TECHNIQUES

These sensors (Figure 14.1) project a finely focused laser spot of light to the part surface. As the light strikes the surface, a lens in the sensor images the point of intersection onto a solid-state array camera. Any deviations from the initial referenced point can be measured based on the number of sensor elements deviated from the referenced point. Accuracy is a function of standoff distance and range. Figure 14.2 depicts an integrated system performing both 2-D and 3-D measurements using sensor data based on laser triangulation principles.

These techniques can be extended to making contour measurements (Figure 14.3). In this case, light sections or structured light sheets are projected onto the object. The behavior of the light pattern is a function of the contour of the object. When viewed, the image of the line takes on the shape of the surface, and a measurement of that contour is made. Again, a referenced position is measured, and deviations from the referenced position are calculated based on triangulation techniques. Determination of the normal-to-surface vectors, the radius of curvature,

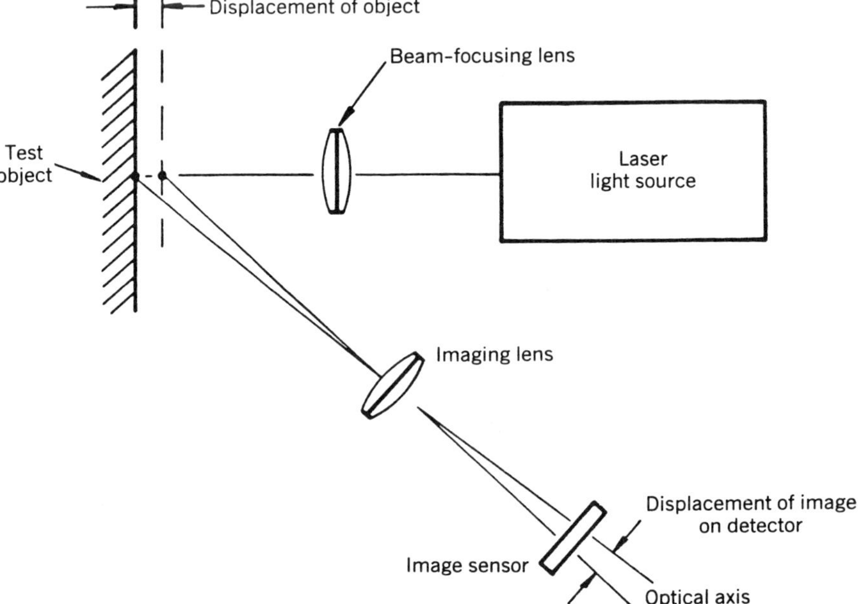

Figure 14.1 – Laser-based triangulation technique.

and the distance from the apex to the sensor (range) can be made in a single measurement.

Arrangements of multiples of such units can be configured to accommodate virtually any combination of shapes and sizes.

14.2 SIMPLE PHOTOELECTRIC VISION

Optical methods can be used to provide edge guidance, typically associated with opaque web products (paper, rubber, etc.). Two photoelectric "scanners" are used, one above and one below the web. Each scanner includes an emitter and receiver arranged so that when the two units are in operation, each receiver sees light from the other's emitter. By phase-locking techniques, the two beams developed can provide edge-guidance feedback.

14.3 LINEAR DIODE ARRAYS

An alternate approach is to use two linear diode arrays positioned at the edges (Figure 14.4). Differences in edge locations are simultaneously detected and used to determine edge positional offset.

Linear array cameras are well suited to making measurements on objects in motion both perpendicular to and along the line of travel. Perpendicular measurements are derived by pixel-counting techniques. The resolution of measure-

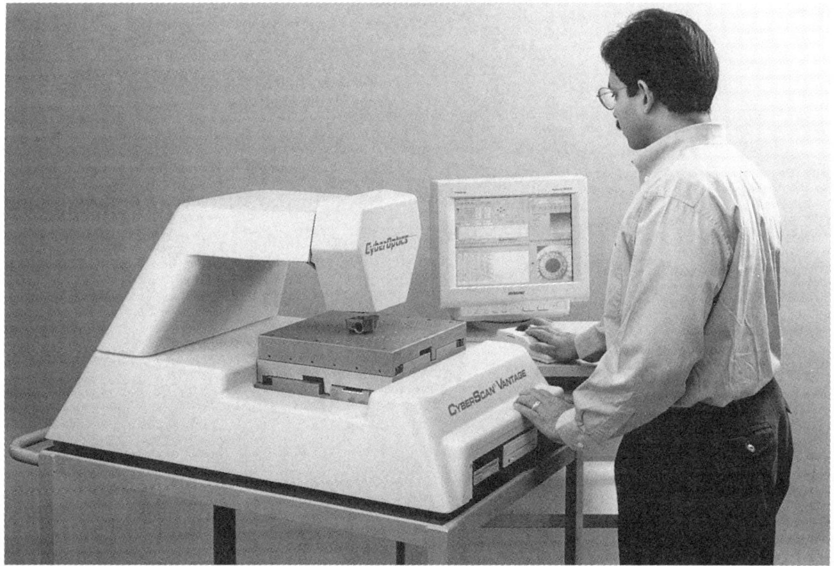

Figure 14.2 - System offered by CyberOptics that employs laser triangulation principles to make dimensional measurements.

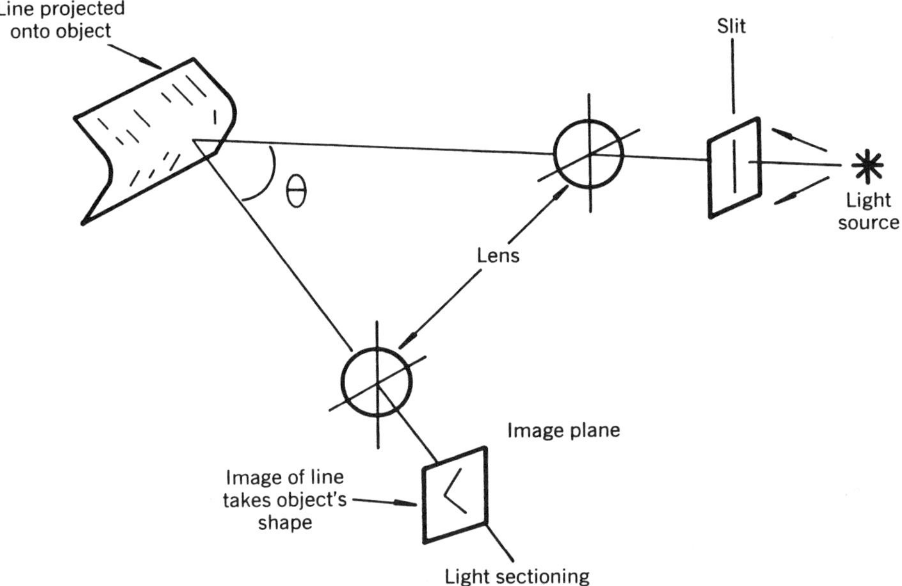

Figure 14.3 - Depiction of light-sectioning principles.

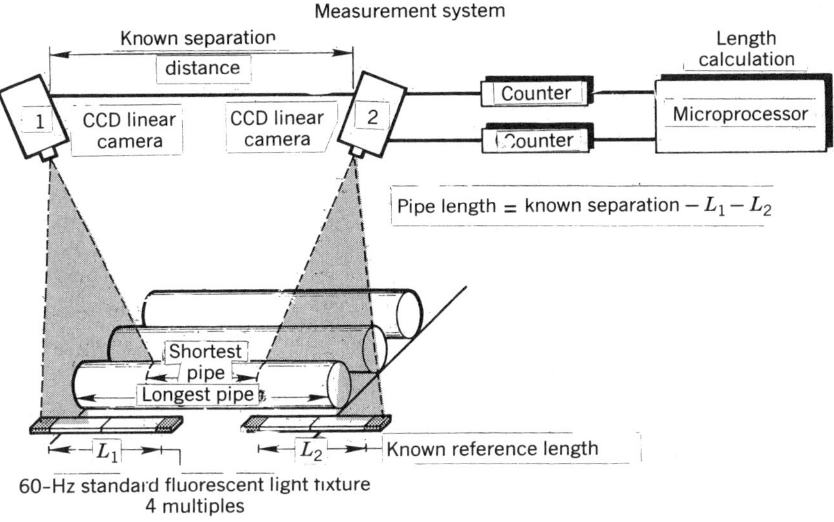

Figure 14.4 - Gauging with linear array cameras.

ment along the axis of travel is determined by the scan rate of the system. A higher resolution can be achieved by increasing the frequency of data gathering with increasing number of pixels in the array.

Another application for which linear arrays are well suited is pattern recognition to control the amount of spray material released. In these systems the array scans the product as it passes on the conveyor. The image with data associated with the object's extremities are stored and fed back to the spray mechanism to control the spray pattern. This is especially useful where different sizes and shapes are comingled on the conveyor.

14.4 FIBER-OPTIC ARRANGEMENTS

Fibers within a bundle can be custom arranged for specific applications (Figure 14.5). For example, to detect the presence of the edge of a moving web and to control its position, versions with three bundles can be used. Using this arrangement and a special photoelectric switch with one emitter and two receptors, two relay outputs can be obtained capable of controlling web width and position.

14.5 LASER SCANNERS

In laser scanner systems a laser beam is deflected along a line across the object under investigation. A detector will measure the irradiance

1. transmitted through a translucent or transparent object or not intercepted by an opaque or strongly reflecting and/or absorbing object,

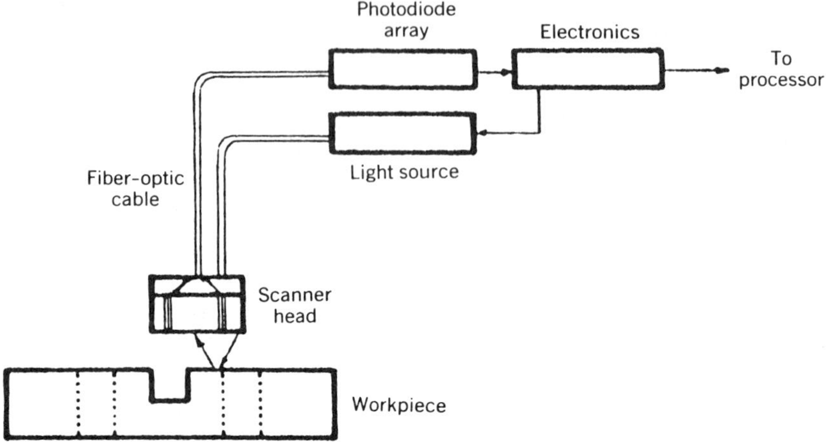

Figure 14.5 - Simple fiber-optic photoelectric sensor.

2. intercepted by the object and scattered or specularly reflected by it, and
3. evident at the surface of the object at the incidence point on the line.

These techniques can be used to make measurements, check for presence and absence, and assess surface quality (e.g., pits, scratches, pinholes, dents, distortions, and striations).

In the case of making measurements, a typical laser gaging system (Figure 14.6) uses a rotating mirror to scan the laser across the part. The beam is converged by a lens into a series of highly parallel rays arranged to intercept the part being measured. A receiver unit focuses the scanning rays onto a sensor. Because the speed of the scanning mirror is controlled, the time the photodetector "sees" the part shadow can be accurately related to the dimensional characteristic of the part presented by the shadow.

In the case of surface characterization, a similar laser scanner arrangement projects light across the object. By positioning the photodetector properly, only light scattered by a blemish will be detected. Analysis of the amplitude and shape of the signal can, in some cases, provide characterization of the flaw as well as size discrimination.

14.6 LASER INTERFEROMETER

Interferometers function by dividing a light beam into two or more parts that travel different paths and then recombine to form interference fringes. The shape of the interference fringes is determined by the difference in optical path traveled by the recombined beams. Interferometers measure the difference in optical paths in units of wavelength of light.

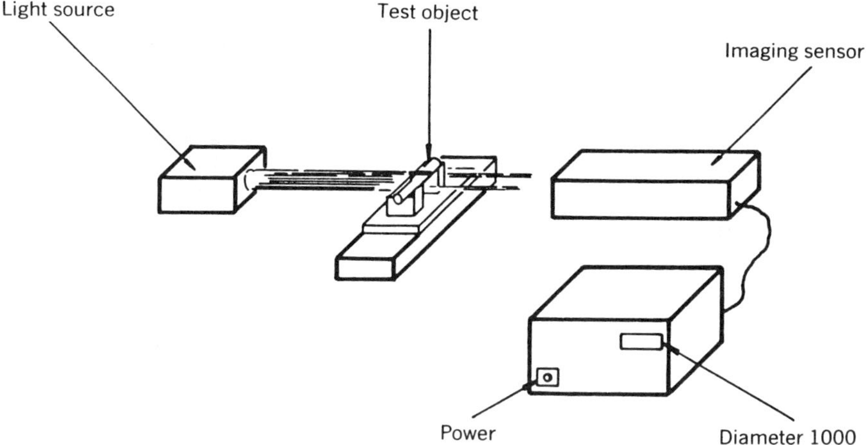

Figure 14.6 - Principles of laser-gauging approach to dimensional measurements.

Since the optical path is the product of the geometric path and the refractive index, an interferometer measures the difference in geometric path when the beams traverse the same medium, or the difference of the refractive index when the geometric paths are equal. An interferometer can measure three quantities:

1. Difference in optical path,
2. Difference in geometric path, and
3. Difference in refractive index.

Laser interferometers are used to perform in-progress gaging on machine tools. The laser is directed parallel to the Z-axis of the machine toward a combination 90' beam bender and remote interferometer cube. The beam bender-interferometer is rigidly attached to the Z-axis slide and redirects the optical beam path parallel to the X-axis and toward the cutting position at the tool turret.

The beam is thus directed at a retroreflector attached to the moving element of a turret-mounted mechanical gage head. The actual measured distance is between the retroreflector on the gage head and the interferometer.

14.7 ELECTRO-OPTICAL SPEED MEASUREMENTS

Laser velocimeters exist for the noncontact measurement of the speed of objects, webs, and so on. Some of these are based on the Doppler effect. In these cases, a beam splitter breaks the laser beam into two identical beams that are directed onto the surface of the object at slightly different angles with regard to the direction of motion.

Both beams are aligned to meet at the same point on the object's surface. The frequency of the reflected light beam is shifted, compared to the frequency of the original light, by the movement of the object. The shifted frequencies are superimposed so that a low-frequency beat (interference fringe pattern) is produced that is proportional to the speed of the moving object.

14.8 ULTRASONICS

Ultrasonic testing equipment beams high-frequency sounds (1-10 MHz) into material to locate surface and subsurface flaws. The sound waves are reflected at such discontinuities, and these reflected signals can be observed on a CRT to disclose internal flaws in the material.

Cracks, laminations, shrinkage cavities, bursts, flakes, pores, bonding faults, and other breaks can be detected even when deep in the material. Ultrasound techniques can also be used to measure thickness or changes in thickness of materials.

14.9 EDDY CURRENT

Eddy current displacement measuring systems rely on inductive principles. When an electrically conductive material is subjected to an alternating magnetic field by an existing magnetic coil, small circulating electrical currents are generated in the material. These "eddy currents" generate their own magnetic field, which then interacts with the magnetic field of the existing coil, thereby influencing the impedance.

Changes in impedance of the existing coil can be analyzed to determine something about the target: to evaluate and sort material; to measure conductivity of electrical hardware; to test metals for surface discontinuities; and to measure coating thickness, thermal conductivity, as well as the aging and tensile strength of aluminum that its alloys. Measurements are useful in finding defects in rod, wire, and tubing.

14.10 ACOUSTICS

Acoustic approaches based on pulse echo techniques (Figure 14.7), where emitted sound waves reflected from objects are detected, can be used for part presence detection, distance ranging, and shape recognition. In the case of part presence, if the part is present, there is a return signal to a detector. In the case of ranging, the sensors detect the time of flight between an emitted pulse of acoustic energy and the received pulse reflected from an object.

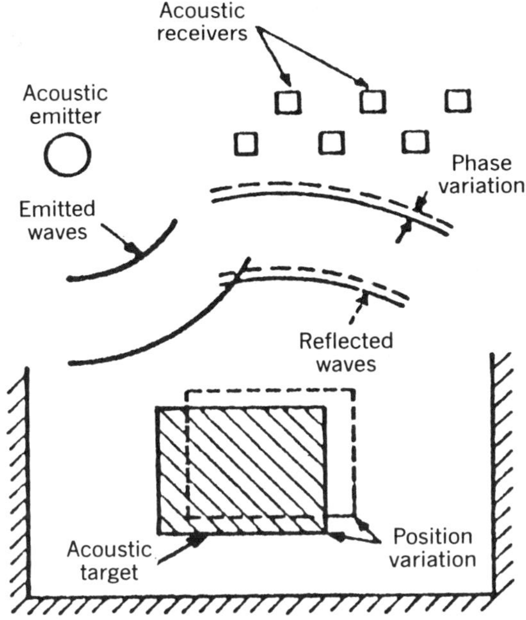

Figure 14.7 – Acoustic-based pattern recognition.

In shape recognition, the system uses sound waves of a fixed frequency usually at 20 or 40 KHz. The fixed-frequency sound wave reflects off objects and sets up an interference pattern as the waves interfere constructively and destructively. An array of ultrasonic transducers senses the acoustic field set up by the emitter at a number of distinct locations, typically eight. Pattern recognition algorithms deduce whether the shape is the same as a previously taught shape by comparing interference patterns.

14.11 TOUCH-SENSITIVE PROBES

Touch-sensitive probes employ some type of sensitive electrical contact that can detect deflection of the probe tip from a home position and provide a voltage signal proportional to the deflection.

When such probes are mounted on a machine, the electrical signal corresponding to probe deflection can be transmitted to a control system. In this manner they can serve as a means to determine where and when the workpiece has been contacted. By comparing the actual touch location with the programmed location in the part program, dimensional differences can be determined.

Appendix A

Glossary

Aberration. Failure of an optical lens to produce exact point-to-point correspondence between an object and its image.

Accuracy. Extent to which a machine vision system can correctly interpret an image, generally expressed as a percentage to reflect the likelihood of a correct interpretation; the degree to which the arithmetic average of a group of measurements conforms to the actual value or dimension.

Acronym. Model-based vision technique developed at Stanford University that uses invariant and pseudoinvariant features predicted from the given object modes; the object is modeled by its subparts and their spatial relationships.

Active Illumination. Illumination that can be varied automatically to extract more visual information from the scene; for example, by turning lamps on and off, by adjusting brightness, by projecting a pattern on objects in the scene, or by changing the color of the illumination.

A/D. Acronym for analog to digital; A/D converter converts data from analog form to digital form.

Algorithm. Exact sequence of instructions, with a finite number of steps, that tell how to solve a problem.

Aliasing. Effect caused by too low a sampling frequency for the spatial frequencies in an image. The effect is that the apparent spatial frequency in the sampled image is much lower than the original frequency. It makes repetitive small features look large.

Ambient Light. Light present in the environment around a machine vision system and generated from sources outside of the system itself. This light must be treated as background noise by the vision system.

Analog. Representation of data as a smooth, continuous function.

Analog-to-Digital Converter. Device that converts an analog voltage signal to a digital signal for computer processing.

Angle of Incidence. Angle between the axis of an impinging light beam and perpendicular to the specimen surface.

Angle of View. (1) Angle formed between two lines drawn from the most widely separated points in the object plane to the center of the lens. (2) Angle between the axis of observation and perpendicular to the specimen surface.

Aperture. Opening that will pass light. The effective diameter of the lens that controls the amount of light passing through a lens and reaching the image plane.

Area Analysis. Process of determining the area of a given view that falls within a specified gray level.

Area Diode Array. Solid-state video detector that consists of rows and columns of light-sensitive semiconductors. Sometimes referred to as a matrix array.

Array Processor. Programmable computer peripheral based on specialized circuit designs relieves the host computer of high-speed numbercrunching types of calculations by simultaneously performing operations on a portion of the items in large arrays.

Artificial Intelligence. Approach in computers that has its emphasis on symbolic processes for representing and manipulating knowledge in solving problems. This gives a computer the ability to perform certain complex functions normally associated with human intelligence, such as judgment, pattern recognition, understanding, learning, planning, classifying, reasoning, self-correction, and problem-solving.

Aspect Ratio. Ratio of width to height for the frame of a televised picture.

The U.S. standard is 4 : 3. Also, the value obtained when the larger scene dimension is divided by the smaller scene dimension; e.g., a part measures 9 X 5 in.; the aspect ratio is 9 divided by 5, or 1.8.

Astigmatism. Lens aberration associated with the failure of primary and secondary images to coincide.

Autofocus. Computer-controlled function that automatically adjusts the optical system to obtain the sharpest image at the image plane of the detector.

Automatic Gain Control. Camera circuit by which gain is automatically adjusted as a function of input or other specified parameter.

Automatic Light Control. Television camera circuit by which the illumination incident upon the face of a pickup device is automatically adjusted as a function of scene brightness.

Automatic Light Range. Television camera circuit that ensures maximum camera sensitivity at the lowest possible light level as well as provides an extended dynamic operating range from bright sun to low light.

Automatic Vision Inspection. Technology that couples video cameras and computers to inspect various items or parts for a variety of reasons. The part to be inspected is positioned in a camera's field of view. The part's image is first digitized by the computer and then stored in the computer's memory. Significant features of the stored image are than "compared" with the same features of a known good part that has been previously placed in the computer's memory. Any difference between the corresponding characteristics of the two parts will be either within a tolerance and hence good or out of tolerance and therefore bad. Also see Computer Vision and Machine Vision.

Back Focal Distance. Distance from the rearmost element in a lens to the focal plane.

Backlighting. Condition where the light reaching the image sensor is not reflected from the surface of the object. Often backlighting produces a silhouette of an object being imaged.

Back Porch. That portion of a composite picture signal that lies between the trailing edge of a horizontal sync pulse and the trailing edge of the corresponding blanking pulse.

Barrel Distortion. Effect that makes an image appear to bulge outward on all sides like a barrel. Caused by a decrease in effective magnification as points in the image move away from the image center.

Bayes Decision Rule. One that treats the units assigned by a decision rule independently and assigns a unit u having pattern measurements or features d to the category c whose conditional probability P(c) given measurement d is highest.

Beam Splitter. Device for dividing a light beam into two or more separate beams.

Bimodal. Histogram distribution of values with two peaks.

Binary Image. Black-and-white image represented in memory as 0's and 1's. Images appear as silhouettes on video display monitor.

Binary System. Vision system that creates a digitized image of an object in which each pixel can have one of only two values, such as black or white on or 1 or 0.

Bit (Binary Digit). Smallest unit of information that can be stored and processed by a computer. In image processing the quantized image brightness at a specific pixel site is represented by a sequence of bits.

Bit Mapped. Method of storing where one data item (pixel in the case of an image) is stored in 1 bit of memory.

Bit Slice. Rudimentary building-block-type processor where one defines the instruction set.

Blanking. Suppression of the video signal for a portion of the scanning raster, usually during the retrace time.

Blob. Connected region in a binary image.

Blob Addressing. Mechanism used to select a blob, such as sequential addressing, XY addressing, family addressing, and addressing by blob identification number.

Blob Analysis. Vision algorithm developed by SRI International that identifies segmented objects according to geometric properties such as area, perimeter, etc.

Blob Labeling. Method of highlighting an addressed blob on the displayed image by, e.g., shading, cursor marking, or alphanumeric labeling.

Blooming. Defocusing experienced by a camera sensor in regions of the image where the brightness is at an excessive level.

Blur Circle. Image of a point source formed by an optical system at its focal point. The size of the blur circle is affected by the quality of the optical system and its focus.

Boolean Algebra. Process of reasoning or a deductive system of theorems using a symbolic logic and dealing with classes, propositions, or on-off circuit elements such as AND, OR, NOT, EXCEPT, IF, THEN, etc., to permit mathematical calculations.

Bottom-up Processing. Image analysis approach based on sequential processing and control starting with the input image and terminating in an interpretation.

Boundary. Line formed by the adjacency of two image regions, each having a different light intensity.

Boundary Tracking (Tracing). Process that follows the edges of blobs to determine their complete outlines.

Brightness. Total amount of light per unit area. The same as luminance.

Brightness Sliding. Image enhancement operation that involves the addition or subtraction of a constant brightness to all pixels in an image.

Burned-In Image (Burn). Image that persists in a fixed position in the output signal of a camera tube after the camera has been turned to a different scene.

Calibration. Reconciliation to a standard measurement.

CCD. Acronym for charge-coupled device, a solid-state camera.

CCTV. Acronym for closed-circuit television.

Cellular Logic. Same as neighborhood processing.

Centroid. Center; in the case of a two-dimensional object the average X and Y coordinates.

Chain Code (Chain Encoding). Method of specifying a curve by a sequence of 3-bit (or more) direction numbers; e.g., starting point *(X, Y)* and the sequence of 3-bit integers (values 0-7) specifying the direction to the next point on the curve.

Change Detection. Process by which two images may be compared, resolution cell by resolution cell, and an output generated whenever corresponding resolution cells have different enough gray shades or gray shade n-tuples.

Character Recognition. Identification of characters by automatic means.

Charged-Coupled Device. Technology for making semiconductor devices including image sensors. The device consists of an array of photosensors connected to an analog shift register. In the analog shift register, information is represented by electric charge (quantity of electrons). The charge packets are shifted (coupled) from one stage of the shift register to another each clock cycle of the shift register.

Charge Injection Device (CID). Conductor-insulator-semiconductor structure that employs intracell charge transfer and charge injection to achieve an image-sensing function using a matrix address technique for address.

Child. Term sometimes used in the SRI algorithm to denote the relationship between one object (the child) wholly contained within another object (the parent). For a washer the hole is the child and the entire object, including the hole, is the parent.

Chromatic Aberration. Optical defect of a lens that causes different colors (different wavelengths of light) to be focused at different distances from the lens.

CID. Acronym for charge injection device.

Classification. See Identification.

Closed-Circuit Television. Television system that transmits signals over a closed circuit rather than broadcasts the signals.

C Mount. Threaded lens mount developed for 16-mm movie work; used extensively for closed-circuit television. The threads have a major diameter of 1.000 in. and a pitch of 32 threads per inch. The flange focal distance is 0.69 in.

Code Reading. Actual recognition of alphanumerics or other set of symbols, e.g., bar codes, UPC codes.

Code Verification. Validation of alphanumeric data to assure conformance to qualitative standard subjective.

Coherent Radiation. Radiation in which the difference in phase between any two points in the field is constant while the radiation lasts.

Collimated. Rays of light made parallel.

Collimator. Optical device that produces collimated light.

Color. Process that stems from the selective absorption of certain wavelengths by an object.

Color Saturation. Degree to which a color is free of white light.

Coma. Abberation in imaging systems that makes a very small circle appear comet-shaped at the edges of the image.

Compactness. Measurement that describes the distribution of pixels within a blob with respect to the blob's center. A circle is the most compact blob; a line is the least compact. Circular objects have a maximum value of 1, and very elongated objects have a compactness approaching zero.

Compass Gradient Mask. Linear filter based on specific weighting factors of nearest neighbor pixels.

Complementation. Logical operation that interchanges the black and white regions in an image.

Composite Sync. Combination of horizontal and vertical sync into one pulse.

Composite Video. Television, the signal created by combining the picture signal (video), the vertical and horizontal synchronization signals, and the vertical and horizontal blanking signals.

Computer Vision. Perception by a computer, based on visual sensory input, in which a symbolic description is developed of a scene depicted in an image. It is often a knowledge-based, expectation-guided process that uses models to interpret sensory data.

Concurve. Sometimes used to refer to boundary representation consisting of a chain of straight lines and arcs.

Condenser. Lens used to collect and redirect light for purposes of illumination.

Congruencing. Process by which two images of a multi-image set are transformed so that the size and shape of any object on one image is the same as the size and shape of that object on the other image. In other words, when two images are congruenced, their geometries are the same, and they coincide exactly.

Connectivity Analysis. Procedure (algorithm) that analyzes the relationships of pixels within an image to define separate blobs; e.g., sets of pixels that are of the same intensity and are connected. A figure F is connected if there is a path between any two spatial coordinates or resolution cells contained in the domain of *F*.

Continuous Image. Image not broken up into its discrete parts. A photograph is a continuous image.

Continuous Motion. Reflects condition where object cannot be stopped during the inspection process.

Contrast. Range of difference between light and dark values in an image. Usually expressed as contrast ratio, contrast modulation, or contrast difference.

Contrast Difference. Difference between the higher density object or background and the lower density object or background.

Contrast Enhancement. Any image processing operation that improves the contrast of an image.

Contrast Modulation. Difference between the darker object or background gray shade and the lighter object or background gray shade divided by the sum of the object gray shade and the background gray shade.

Contrast Ratio. Ratio between the higher object transmittance or background transmittance to the lower object transmittance or background transmittance.

Contrast Stretching (Shrinking). Image enhancement operation that involves multiplying or dividing all pixels by a constant brightness.

Contrast Transfer Function. Measure of the resolving capability of an imaging system. Shows the square-wave spatial frequency amplitude response of a system. See Modulation Transfer Function.

Convolution. Generic mathematical operation that involves calculating an output pixel based on its properties and those of its surrounding neighbors; used to accomplish different effects in image enhancement and segmentation; also superimposing an m X n operator (kernel) over an m X n pixel area (window) in the image, multiplying corresponding points together, summing the result, and repeating this operation over all possible windows in the image.

Correlation. Mathematical measure of similarity between images or subimages within an image. It involves pattern matching.

Correlation Coefficient. Normalized covariance of two random variables, i.e., their covariance divided by the product of their standard deviations. The correlation coefficient ranges from zero for uncorrelated variables to 1 for perfectly correlated variables. Frequently computed for pixels in an image to measure their relations to each other.

Correspondence Problem. In stereo imaging the requirement to match a feature in one image with the same feature in the other image. The feature may be occluded in one or the other image.

Cross Correlation. Expected value of the product of two functions. A measure of their similarity.

CTF. Acronym for contrast transfer function.

Cylindrical Lens. Single-axis spherical lens that results in only one dimension of an object being in focus.

Dark Current. Current that flows or signal amplitude generated in a photosensor when it is placed in total darkness.

Dark-Field Ilumination. Technique where the illumination is supplied at a grazing angle to the surface. Ordinarily, only a negligible amount of light is reflected into the camera. Specular reflections occur off any abrupt surface irregularities and are detected in the image.

Data Flow Machine (Computer). Computer based on decentralized program control relying on data-driven architecture. With no predetermined order of events, each data word is given a label or tag (data word and label together are called *token)* describing the operation that it is to undergo. Design is such that tokens are systematically generated, and those of the same operation locate and match up with each other to execute an operation.

Decision Theoretic. Pattern recognition technique that compares an N-dimensional feature vector with reference feature vectors. Decision-making is based on a similarity rule between measured features of the image or object under test and stored model features.

Density. Measure of the light transmitting or reflecting properties of an area. It is expressed by the logarithm of the ratio of incident to transmitted or reflected light flux.

Depth of Field. In-focus range of an imaging system. It is the distance from behind an object to in front of the object within which objects appear to be in focus.

Depth of Focus. Range of lens to image plane distance for which the image formed by the lens appears to be in focus.

Design Rules Checking. Basing a vision inspection decision on the geometric constraints of the original design being satisfied.

Detection. A unit is said to be detected if the decision rule is able to assign it as belonging only to sorne given subset A of categories from the set C of categories. To detect a unit does not imply that the decision rule is able to identify the unit as specifically belonging to one particular category.

Detectivity. Ability to repeatably sense the location of an edge in space.

Dichroic Mirror. Semitransparent mirror that selectively reflects some wavelengths more than others and so transmits selectively.

Diffraction. Bending of light around an obstacle.

Diffraction Grating. Substrate with a series of very closely spaced lines etched in its surface. The surface may be transparent or reflective. Light falling on the grating is dispersed into a series of spectra.

Diffraction Limited. Optical system of such quality that its performance is limited only by the effects of diffraction.

Diffraction Pattern. Pattern produced by the bending of light into regions that would be shadows if rectilinear propagation prevailed.

Diffuse. Process where incident light is redirected over a range of angles (scattered) while being reflected from or transmitted through a material.

Diffuse Reflection. Characteristic of light that leads to redirection over a range of angles from a surface on which it is incident, such as from a matte surface.

Diffuse Transmission. Characteristic of light that penetrates an object, scatters, and emerges diffusely on the other side.

Digital Image, Digitized Image, Digital Picture Function of Image. Image in digital format obtained by partitioning the area of the image into a finite two-dimensional array of small, uniformly shaped, mutually exclusive regions called resolution cells and assigning a "representative" gray shade to each such spatial region. A digital image may be abstractly thought of as a function whose domain is the finite two-dimensional set of resolution cells and whose range is the set of gray shades.

Digital Imaging. Conversion of a video picture into pixels by means of an A/D converter where each pixel's level can be stored in a computer.

Digitalization. Process of converting an analog video image into digital brightness values that are assigned to each pixel in the digitized image.

Digital Subtraction. Process by which two images are subtracted in a computer so that information common to both images is removed; e.g., given two images of the same anatomical areas, one with dye in the blood vessels and one without, the resultant subtraction will only show the dyed blood vessels.

Digital-to-Analog (D/A) Converter. Hardware device that converts a digital signal into a voltage or current proportional to the digital input.

Digitization. See Digitalization.

Digitizer. Device to sample and quantize an incoming video signal, convert it to a digital value, and store it in memory. See also Frame Grabber.

Digitizing. See Digitalization.

Dilation. Technique used in applying mathematical morphology to image analysis. Geometric operation of forming a new image based on the union of all translations of one image by the position of each of the pixels within the second image.

Discrete Tonal Feature. On a continuous or digital image a connected set of spatial coordinates or resolution cells all of which have the same or almost the same gray shade.

Discrimination. Degree to which a vision system is capable of sensing differences.

Distortion. Undesired change in the shape of an image or waveform from the original object or signal.

Dyadic Operator. Operator that represents an operation on two and only two operands. The dyadic operators are AND, equivalence, exclusion, exclusive, OR, inclusion, NAND, NOR, and OR.

Dynamic Range. Ratio of the maximum acceptable signal level to the minimum acceptable signal level.

Dynamic Threshold. Threshold that varies with time; it is controlled either by application requirements or by local or global image parameters.

Edge. Parts of an image characterized by rapid changes in intensity value that represent borderlines between distinct regions.

Edge-Based Stereo. Stereographic technique based on matching edges in two or more views of the same scene taken from different positions.

Edge Detection. Process of finding edges in a scene by employing local operators that respond to the first or second derivative of the gray scale intensity in the neighborhood of each pixel. An edge is detected when these derivatives exceed a given magnitude.

Edge Enhancement. Image-processing method to strengthen high spatial frequencies in the image.

Edge Following. Segmentation algorithm for isolating a region in an image by following its edge.

Edge Operators. Templates for finding edges.

Edge Pixels. Pixels that lie along edges in a scene.

Elongation. A shape factor (blob feature) measurement. Equal to the length of the major axis divided by the length of the minor axis. Squares and circles have a minimum of 1; elongated objects have a value greater than 1.

Erosion. Technique used in applying mathematical morphology to image analysis; geometric operation of forming a new image based on the union of all translations of one image where that image is completely contained in the second image; see Shrinking.

Euler Number. Number of objects in a binary image minus the number of holes.

Extended Gaussian Image (EGI). Mathematical representation of the surface orientations of an object can be pictured as a distribution of material over the surface of the Gaussian sphere.

Feature, Feature Pattern, Feature n-tuple, Pattern Feature. An n-tuple or vector with (a small number of) components that are functions of the initial measurement pattern variables or some subsequent measurement of the n-tuples. Feature n-tuples or vectors are designed to contain a high amount of information relative to the discrimination between units of the types of categories in the given category set. Sometimes the features are predetermined, and at other times they are determined at the time the pattern discrimination problem is being solved. In image pattern recognition, features often contain information relative to the gray shade, texture, shape, or context.

Feature Extraction. Process in which an initial measurement pattern or some subsequent measurement pattern is transformed to a new pattern feature. Sometimes feature extraction is called property extraction. The pattern is used in three distinct senses: (1) as measurement pattern, (2) as feature pattern, and (3) as the dependency pattern or patterns of relationships among the components of any measurement n-tuple or feature n-tuple derived from units of a particular category and that are unique to those n-tuples; i.e., they are dependencies that do not occur in any other category.

Feature Selection. Process by which the features to be used in the pattern recognition problem are determined. Also called property selection.

Gradient Space. Coordinate system (p, q), where p and q are the rates of change in depth of the surface of an object in the scene along the x and y directions (the coordinates in the image plane).

Gradient Vector. Orientation and magnitude of the rate of change in intensity at a point in the image.

Gray Level (Gray Scale, Gray Shade, Gray Tone). Quantized measurement of image irradiance (brightness) or other pixel property; description of contents of an image derived by conversion of analog video data from a sensor into proportional digital numbers. The number is proportional to the integrated output, reflectance, or transmittance of a small area, usually called a resolution cell or pixel, centered on the position (x, y). The gray shade can be expressed in any one of the following ways: (1) transmittance, (2) reflectance, (3) a coordinate of the

ICS color coordinate system, (4) a coordinate of the tristimulus value color coordinate system, (5) brightness, (6) radiance, (7) luminance, (8) density, (9) voltage, or (10) current.

Gray Scale Modification. Image enhancement operations that involve altering the gray scale value of a pixel (see Contrast Stretching and Brightness Sliding).

Gray Scale Projection. Profile of an image or portion of an image obtained by adding the gray scale values in a particular direction, i.e., horizontal projections and vertical projections.

Gray Scale Vision. Analysis of an image based on shade-of-gray content.

Guidance. Deriving properties in an image to describe position.

Haar Transform. See Nonlinear Filter.

Hadamard Transform. Specific type of linear transformation used for image representation based on a matrix that is a square array of + 1's and - 1's whose rows and columns are orthogonal.

Halo. Appearance of black border around exceptionally bright objects in an image.

Halogen. Gas such as iodine that is placed inside an incandescent lamp to get the evaporated filament off the bulb and redeposit it back onto the filament.

Heuristics. "Rules of thumb," knowledge, or other techniques used to help guide a problem solution.

Heuristic Search Techniques. Graph-searching methods that use heuristic knowledge about the domain to help focus the search. They operate by generating and testing intermediate states along potential solution paths.

Hierarchical Approach. Approach to vision based on a series of ordered processing levels in which the degree of abstraction increases as the system proceeds from the image level to the interpretation level.

Higher Level. Interpretative processing stages such as those involving object recognition and scene description, as opposed to the lower levels corresponding to the image-processing and description stages.

Highlight. Region of maximum brightness of an image.

Histogram. Frequency count of the occurrence of each intensity (gray level) or other characteristic in an image.

Horizontal Blanking. Blanking of a picture during a period of horizontal retrace.

Horizontal Retrace. Return of a scanning (beam) from one side to the other of the scanning raster after the completion of one scan line.

Horizontal Sync. Circuit to retrace the horizontal scan line to the beginning of the next line.

Hough Transform. Global parallel method for finding straight or curved lines, in which all points on a particular curve map into a single location in the transform space.

Hue. Dominant wavelength of light representing the color of an object. It is the redness, blueness, greenness, etc., of an object.

Hueckel Operator. Method for finding edges in an image by fitting an intensity surface to the neighborhood of each pixel and selecting surface gradients above a chosen threshold value.

Iconic. Image-like.

Identification. Determination of the identity of an object by reading symbols on the object. A unit is said to be recognized, identified, classified, categorized, or sorted if the decision rule is able to assign it to some category from the set of given categories. In military applications, there is a definite distinction between recognition and identification. Here, for a unit to be recognized, the decision rule must be able to assign it to a type of category, the type having included within it many subcategories. For a unit to be identified, the decision rule must be able to assign it not only to a type of category but also to the subcategory of the category type. For example, a small area ground patch may be recognized as containing trees, which may be specifically identified as apple trees.

IEEE 488 Bus. Data transmission bus that provides communication between devices.

IHS. Acronym for intensity, hue, and saturation; a form of representing a color image. Also called luminance, hue, and saturation.

Illuminance. Luminous flux incident on a surface; luminous incidence.

Illumination. Application of light to an object.

Image. Optically formed projection of a scene into a plane. Usually represented as an array of brightness values.

Image Algebra. Image-processing algorithms whose variables are images and whose operations are logical and/or arithmetic combinations of the images.

Knowledge Base. Artificial intelligence data bases that are not merely files of uniform content but are collections of facts, inferences, and procedures corresponding to the types of information needed for problem solution.

Lag. In a television pickup tube, the persistence of electrical charge image for two or more scan periods after excitation (light) is taken away.

Laplacian Operator. Sum of the second partial derivatives of the image intensity in the x and y directions. The Laplacian operator is used to find edge elements by finding points where the Laplacian is zero.

Laser. Acronym for light amplification by stimulated emission of radiation. Device that produces a coherent monochromatic beam of light.

Laser Scanner. Imaging device that scans a spot of laser light across the image area and detects the reflected or transmitted light with a photodetector.

Learning. Process whereby a vision system develops a set of criteria so that it can match an observed object with a known object.

LED. Acronym for light-emitting diode.

Lens. Transparent optical component consisting of one or more pieces of optical glass (elements) with curved surfaces (usually spherical) that converge or diverge transmitted light rays.

Lens Speed. Ability of a lens to transmit light; represented as the ratio of the focal length to the diameter of the lens. See f-Number.

Level Slice. See Threshold.

LHS. Acronym for luminance, hue, and saturation.

Light. Electromagnetic radiation. Visible light, light detectable by the eye, has wavelengths in the range of 400-750 nm.

Light-Emitting Diode. Semiconductor diode that converts a portion of the electrical energy in the junction to light.

Light Sectioning. See Structured Light.

Light Striping. See Structured Light.

Line. Thin connected set of points contrasting with neighbors on both sides. Line representations are extracted from edges.

Line Detectors. Oriented operators for finding lines in an image.

Line Following. Accepting a new element according to how close it is to the linear continuation of the current line being tracked.

Line Scan. One-dimensional image sensor.

Linear Array. Solid-state video detector consisting of a single row of light-sensitive semiconductor devices. This is used in linear array cameras.

Linear Filters. See Spatial Filters. Can be (1) low pass, (2) high pass, or (3) edge detectors. Low-pass filters accentuate low-frequency information in an image and result in smoothing the image. High-pass filters accentuate high-frequency data such as edges. Edge detector filters accentuate high-frequency data such as edges and eliminate low-frequency data. A linear spatial filter is a spatial filter for which the gray shade assignment at coordinates *(x, y)* in the transformed image is made by some weighted average (linear combination) of gray shades located in a particular spatial pattern around coordinates *(x, y)* of the domain image. The linear spatial filter is often used to change the spatial frequency characteristics of the image. For example, a linear spatial filter that emphasizes high spatial frequencies will tend to sharpen the edges in an image. A linear spatial filter that emphasizes the low spatial frequencies will tend to blur the image and reduce salt-and-pepper noise.

Linear-Matched Filtering. Image-processing technique that convolves image windows with a template of the desired feature and seeks the maximum value.

Linking Procedure. In edge or boundary feature extraction, process of fitting straight-line segments or polynomials to edge points.

Local. Image operator that produces values depending on the values of the pixels near where the operator is being applied.

Local Feature Analysis. Type of structural pattern recognition algorithm that represents the geometric relationships of "local" objects by a graph structure. The graph obtained from the observed image is matched to a stored model.

Local Iterative Modification. See Neighborhood Processing.

Local Windowing. Selection of one area of the scene. This allows special attention to be given that area; e.g., permitting different threshold values to be assigned to various areas of a scene.

Location. Process of determining quantitatively the position of objects within a scene; coordinates of an image or object relative to an observer. An image region's location is defined by the *X, Y* coordinates of its centroid. A three-dimensional object's location is defined by the *X, Y,* and Z coordinates of some point, such as its center of gravity.

Logical Operation. Execution of a single computer instruction.

Log Spiral. Nonlinear coordinate transform resulting in logarithmic polar mapping of a scene.

Lookup Table (LUT). Memory that sets the input and output values for gray scale thresholding, windowing, inversion, and other display or analysis functions. Input values are for the storage of video input pixels into image memory, and output values are for the display of stored pixels on the monitor. Also called translation table.

Low-Level Features. Pixel-based features such as texture, regions, edges, lines, comers, etc.

Luminance. Luminous intensity (photometric brightness) of any surface in a given direction per unit of projected area of the surface as viewed from that direction and measured in foot-lamberts.

Machine Vision. Use of sensors (e.g., TV, X-ray, UV, laser scan, and ultrasonic) to receive signals representative of the image of a real scene, coupled with computer systems or other signal-processing devices to interpret the signals received for image content. The purposes of these systems are to obtain data and/or control machines or processes. Also, the application of computer vision techniques to manufacturing problems for the purpose of control: machine, robot, process, or quality control.

Magnification. Relationship of the length of a line in the object plane to the length of the same line in the image plane. It may be expressed as image magnification (image size per object size) or its inverse, object magnification.

Masking. Operation where regions of an image are set to a constant value, usually white or black; also the process of creating an outline around a standard image and then comparing this outline with test images to determine how closely they match.

Matched Filtering. Template-matching operation done by using the magnitude of the cross-correlation function to measure the degree of matching.

Mathematical Morphology. Mathematics of shape analysis.

Matrix Array. Camera image sensor that produces a matrix or two-dimensional image.

Measurement. Checking that parts are the right size, that they conform to specified dimensional tolerances.

Median Filtering. Method of local smoothing by replacing each pixel with the median of the gray levels of neighboring pixels.

Mensuration. Measurement; geometry applied to the computation of lengths, areas, or volumes from given dimensions or angles.

Menu. List of the programs that can be used within a package or list of all the functions that can be performed by a program.

Model-Based Vision System. Utilizes a given model to derive a desired description of the original scene from an image.

Modulation. Process, or results of the process, whereby some characteristic of one signal is varied in accordance with another signal. The modulated signal is called the carrier. The carrier may be modulated in three fundamental ways: by varying the amplitude, called amplitude modulation; by varying the frequency, called frequency modulation; and by varying the phase, called phase modulation.

Modulation Transfer Function. Measure of the resolving capability of an imaging system. Shows the sine wave spatial frequency amplitude response of a system. Formally defined as the magnitude of the Fourier transform of the line spread function of the imaging system or component.

Moire. In TV, the spurious pattern in the reproduced picture resulting from interference beats between two sets of periodic structures in the image.

Monochromatic. Property of light composed of only one wavelength.

Monochromator. Device for isolating narrow portions of the spectrum by dispersing light into its component wavelengths.

Monochrome. Black and white with shades of gray. Excludes color.

Morphology. Mathematics of shape analysis.

MOS. Acronym for metal-oxide semiconductor. Method for constructing semiconductor devices where the controlling element (the metal) is separated from the controlled element (the semiconductor) by a thin insulating oxide layer.

Motion Sensing. Ability of a vision system to (1) form an image of an object in motion and (2) determine the direction and speed of that motion.

MTF. Acronym for modulation transfer function.

Multiplexer. In TV, a specialized optical device that makes it possible to use a single TV camera in conjunction with one or more motion picture projectors and/or slide projectors in a film chain. The camera and projectors are in a fixed relationship, and prisms or special (dichroic) mirrors are used to provide smooth and instantaneous nonmechanical transition from one program source to the other.

Nearest Neighbor Classification. Statistical technique used to match features of an observed object with corresponding features of a learned object.

Nearest Neighbor Decision Rule. Treats the units assigned by a decision rule independently and assigns a unit u of unknown identification and with pattern measurements or features d to category c, where d is that pattern closest to d by some given metric or distance function.

Needle Diagram. Figure that depicts lines perpendicular to the surface of an object at each point. The shape of the object can be represented by the length and direction of the perpendicular lines.

Neighborhood Processing. Changing the value of each pixel based on the values of a set of nearby pixels.

Neutral Density Filter. Optical device that reduces the intensity of light without changing the spectral distribution of light energy.

Noise. Irrelevant or meaningless data resulting from various causes unrelated to the source of data being measured or inspected; random undesirable video signals.

Noise Equivalent Power (Noise Equivalent Exposure). Radiant flux necessary to give an output signal equal to the detector noise.

Noncomposite Video. Video signal containing all information except sync.

Nonlinear Filters. Class of image enhancement operators not based on linear combinations of pixel gray levels. Examples are Sobel, RobertsCross, etc.

NTSC. Acronym for National Television Systems Committee. Committee that worked with the Federal Communication Commission in formulating standards for present-day U.S. color TV system. See PAL.

N-Tuple, Measurement Pattern, Pattern, Measurement Vector. Ordered n-tuple of measurements obtained of a unit under observation. Each component of the n-tuple is a measurement of a particular quality, property, feature, or characteristic of the unit. In image pattern recognition, the units are usually picture elements or simple formations of picture elements, and the measurement n-tuples are the corresponding gray shades, gray shade n-tuples, or formations of gray shade n-tuples.

Object. Three-dimensional shape that is the subject of analysis by a vision system. An object forms the basis for creating an image.

Objective Lens. Optical element that receives light from an object or scene and forms the first or primary image.

Object Plane. Imaginary plane that is focused onto the specified image plane of a lens.

Object Recognition. See Pattern Recognition.

Oblique Lighting. Where a light source is placed at an angle generally to emphasize shadows.

Occlusion. Result in an image of an object in the optical path that is not part of the scene but blocks off some portion of the scene.

OCR. Acronym for optical character recognition.

Optical Axis. Straight line passing through the centers of the curved surfaces of a lens or lens system.

Optical Character Recognition. Identification of alphanumeric characters by a vision system.

Optical Comparator. Gaging device that allows visual measurement of dimensions of objects by an operator using backlighting. This device does not analyze the image or make decisions.

Optical Flow. Relative displacement of pixels from one image to the next that occur when a static scene is viewed by a moving sensor yielding a succession of images.

Optical Flow Analysis. Inferring relative range from image sequences obtained from a moving sensor by analysis of the motions of corresponding pixels from frame to frame.

Optical Processing. Employs coherent optical interference methods to produce spatial Fourier transform of objects and then uses transforms to recognize objects.

Optical Scanner. Device used to convert a scene into an array of numbers representing the positional distribution of optical density within the picture. A device that optically scans printed or written data and generates its digital representations.

Orientation. Angle formed by the major axis of an object image or region relative to a reference axis. For an object, the direction of the major axis must be defined relative to a three-dimensional coordinate system.

PAL. Acronym for phase-alternating line system. A color TV system in which the subcarrier derived from the color burst is inverted in phase from one line to the next in order to minimize errors in hue that may occur in color transmission. See also NTSC.

Parallax. Change in perspective of an object when viewed from two slightly different positions. The object appears to shift position relative to its background, and it also appears to rotate slightly.

Parallel Pipeline. Machine vision system design in which the image enters a sequential pipeline of processors in which many different operators are applied to the image simultaneously.

Parallel Processing. Processing of pixel data in such a way that a group of pixels are analyzed at one time rather than one pixel at a time; method of computing whereby computations are done in parallel in contrast to conventional "von Neumann" machines that perform computations in sequence. See also Data Flow Machine.

Parent. Term often used in the SRI algorithm to denote the relationship between one object (the child) wholly constrained within another object (the parent).

Parsing. Process of determining the structure of a sentence (string of symbols).

Pattern Description Language. In syntactic pattern recognition, the language that provides the structural description of patterns in terms of pattern primitives and their composition.

Pattern Discrimination. Concerned with how to construct the decision rule that assigns a unit to a particular category on the basis of the measurement pattern(s) in the data sequence or on the basis of the feature pattern(s) in the data sequence.

Pattern Recognition. Concerned with, but not limited to, problems of (1) pattern discrimination, (2) pattern classification, (3) feature selection, (4) pattern identification, (5) cluster identification, (6) feature extraction, (7) preprocessing, (8) filtering, (9) enhancement, (10) pattern segmentation, or (1 1) screening and thus classification of images into predetermined categories, usually using statistical methods.

Pattern Segmentation. Problem of determining which regions or areas in an image constitute the patterns of interest, i.e., which resolution cells should be included and which excluded from the pattern measurements.

Pel. See Pixel.

Perfect Hull. Windowing technique in which the object itself becomes its own window.

Photometric Stereo. Method of interpreting shape from local surface gradients determined by operating on each of three images taken by varying the direction of illumination.

Photometry. Measurement of light visible to the human eye.

Photon. Particle of light. A quantum of electromagnetic energy moving at the speed of light.

Photopic Vision. Vision that occurs at moderate and high levels of luminance and permits distinction of colors. This is light-adapted vision. It is attributed to the retinal cones in the eye. The contrasting capability is twilight, or scotopic, vision.

Pickup Tube. Television camera image pickup tube.

Picture Element. See Pixel.

Pincushion Distortion. Effect that makes the sides of an image appear to bulge inward on all sides similar to a pincushion. Caused by an increase in effective magnification as points in the image move away from the image center.

Pipeline Architecture. Method for implementing image-processing functions in serial stages.

Pixel. Acronym for picture element. Spatial resolution element; smallest distinguishable and resolvable area in an image; individual photosite in a solid-state camera.

Point Transformation. Replacement of one pixel value for another where the new value is a function of only the previous value.

Polarization. Restriction of the vibration of the electric or magnetic field vector to one plane.

Pose Estimation. Estimation of the six degrees of freedom that specify the position and orientation of an object.

Position. Definition of an object's location in space.

Positional Repeatability. Specification that reflects how repeatably in space an object can be expected to pass (or be held in place at) an inspection station; generally expressed in inches.

Precision. Amount of spread of the distribution of measurements around some average value.

Preprocessing. Operation applied before pattern identification is performed. Preprocessing produces, for the categories of interest, pattern features that tend to be invariant under changes such as translation, rotation, scale, illumination levels, and noise. In essence, preprocessing converts the measurements patterns to a form that allows a simplification in the decision rule. Preprocessing can bring an image into registration and into congruence, remove noise, enhance images, segment target patterns, and detect, center, and normalize targets of interest.

Prewitt Operator. Edge detection operator based on two-dimensional gradient.

Primal Sketch. See Raw Primal Sketch.

Primary Colors. Three colors such that no mixture of any two of the colors can produce the third. The most common groups of primary colors are red, green, and blue or cyan, magenta, and yellow.

Principal Plane. Imaginary plane in or near a lens where the light rays appear to have bent.

Procedural Knowledge Representation. Representation of knowledge about the world by a set of procedures, small programs that know how to do specific things (how to proceed in well-specified situations).

Process Control. Operation that results in feedback for corrective action.

Processing Speed. Time required for a vision system to analyze and interpret an image. Typical vision systems can inspect from 2 to 15 parts per second.

Production Rule. Modular knowledge structure representing a single chunk of knowledge, usually in an if-then or antecedent-consequent form. Popular in expert systems.

Projection. Feature extraction technique making use of profiles of images in particular directions, e.g., horizontal or vertical projections.

Property Extraction. See Feature Extraction.

Prototype Pattern. See Signature.

Proximity Sensor. Device that senses that an object is only a short distance (e.g., a few inches or feet) away and/or measures how far away it is. Proximity sensors work on the principles of triangulation of reflected sound or intensity-induced eddy currents, magnetic fields, back pressure from airjets, and others.

Quadtree. Representation obtained by recursively splitting an image into quadrants until all pixels in a quadrant are uniform with respect to some feature (such as the gray level).

Quantizer. Instrument that does quantizing. The quantizer has three functional parts. The first part allows the determining and/or setting of the quantizing intervals, the second part is a level slicer that indicates when a signal is in any quantizing interval, and the third part takes the binary output from the level slicers and either codes it to some binary code or converts it to some analog signal representing quantizing interval centers or means.

Quantizing. Process by which each gray shade in an image of photographic, video, or digital format is assigned a new value from a given finite set of gray shade values. There are three often used methods of quantizing: (1) In equal-interval quantizing or linear quantizing, the range of gray shades from the maximum gray shade to the minimum gray shade is divided into contiguous intervals each of equal length, and each gray shade is assigned to the quantized class that corresponds to the interval within which it lies. (2) In equal-probability quantizing, the range of gray shades is divided into contiguous intervals such that after the gray shades are assigned to their quantized class, there is an equal frequency of occurrence for each quantized gray shade in the quantized digital image or photograph; equal-probability quantizing is sometimes called central stretching. (3) In minimum-variance quantizing, the range of gray shades is divided into contiguous intervals such that the weighted sum of the variance of the quantized intervals is minimized. The weights are usually chosen to be the gray shade interval probabilities, which are computed as the proportional area on the photograph or digital image that have gray shades in the given interval.

Quantum Efficiency. (1) For a radiating source the ratio of the number of photons emitted per second to the number of electrons flowing per second. (2) For detectors, the ratio of the number of electron-hole pairs generated to the number of incident photons.

Radiant Flux. Time rate of flow of radiant energy per unit area.

Radiometry. Measurement of light within the total optical spectrum.

Random Interlace. Technique for scanning that is often used in closed-circuit television systems where there is no fixed relationship between adjacent lines in successive fields.

Range Sensing. Measurement of distance to an object.

Raster. Scanning pattern employed to given substantially uniform coverage of an area. The area of a picture tube scanned by the electron beam.

Raw Primal Sketch. Image representation created by approximating of zero-crossing contours by sequences of short line segments and evaluating each for position, orientation, length, and rate at which intensity changes across the segment.

Real Time. Operation that takes place in a short time so that its use does not affect the time required to perform any other operations.

Real-Time Processing. Ability of a vision system to interpret an image in a short enough time to keep pace with manufacturing operations; processing video data at the same rate at which it is being generated.

Recognition. Match between a description derived from an image and a description obtained from a stored model. See also Identification.

Rectifying. Process by which the geometry of an image area is made planimetric. For example, if the image is taken of an equally spaced rectangular grid pattern, the rectified image will be an image of an equally spaced rectangular grid pattern. Rectification does not remove relief distortion.

Rectilinear Transmission. Characteristic of light observed when light passes through an object without diffusion.

Reflectance, Reflection Coefficient. Ratio of the energy per unit time per unit area (radiant power density) reflected by the object to the energy per unit time per unit area incident on the object. In general, reflectance is a function of the incident angle of the energy, viewing angle of the sensor, ral wavelength and bandwidth, and nature of the object.

Reflectance Map. Representation of brightness as a function of surface useful to determine shape from shading.

Reflection. Process by which incident light leaves a surface or medium from side on which it is incident; an image symmetric with the original age through the origin; geometric complement of a binary image.

Refraction. Bending of light rays as they pass from one medium into other.

Region. Contiguous area or connected set of pixels of an image that have a common characteristic, such as average gray level, color, or texture.

Region Growing. Process of initially partitioning an image into elementary regions with a common property (such as gray level) and then successively merging adjacent regions having sufficiently small differences in the selected property until only regions with large differences between them remain; similar to clustering.

Region of Interest. Area inside defined boundaries that the user wants to analyze.

Registration. Processing images to correct geometric and intensity distortions, relative translational and rotational shifts, and magnification differences between one image and another or between an image and a reference map. When registered, there is a one-to-one correspondence between a set of points in the image and in the reference.

Relaxation. Interative problem-solving approach in which initial conditions are propagated utilizing constraints until all goal conditions are adequately satisfied.

Reliability. See Accuracy. In this context, reliability is distinguished from the frequency of failure of the system.

Repeatability. Ability to reproduce a result.

Resolution. Generic term that describes how well a system, process, component or material, or image can reproduce an isolated object or separate closely spaced objects or lines. The limiting resolution, resolution limit, or spatial resolution is described in terms of the smallest dimension of the target or object that can just be discriminated or observed. Resolution may be a function of object contrast, and spatial position, as well as element shape (single point number of points in a cluster, continuum, line, etc.).

Resolution Cell. Smallest, most elementary area constituent of gray shades considered by an investigator in an image. A resolution cell is referenced by its spatial coordinates. The resolution cell or formations of resolution cells can sometimes constitute the basic unit for pattern recognition of image format data.

Resolving Power of Imaging System, Process, Component, or Material. Measure of its ability to image closely spaced objects. The most common practice in measuring resolving power is to image a resolving power target composed of lines and spaces of equal width. Resolving power is usually measured at the image plane in line pairs per millimeter, i.e., the greatest number of lines and spaces per millimeter that can just be recognized. This threshold is usually determined by using a series of targets of decreasing size and basing the measurement on the smallest one in which all lines can be counted. In measuring resolving power, the nature of the target (number of lines and their aspect ratio), its contrast, and the criteria for determining the limiting resolving power must be specified.

Responsivity. Relative sensitivity of a photodetector to different wavelengths of light.

Retained Image (Image Burn). Change produced in or on the image sensor that remains for a large number of frames after the removal of a previously stationary image and yields a spurious electrical signal corresponding to that image.

Reticle. Pattern mounted in the focal plane of a system to measure or locate a point in the image.

Retroreflector. Device used to return radiation in the direction from which it arrived.

RGB. Acronym for red, green, and blue; a three-primary-color system used for sensing and representing color images.

Roberts Cross Operator. Operator that yields the magnitude of the brightness gradient at each point as a means of edge detection in an image.

Robot Vision. Use of a vision system to provide visual feedback to an industrial robot. Based upon the vision system's interpretation of a scene, the robot may be commanded to move in a certain way.

Rotationally Insensitive Operators. Image-processing operators insensitive to the direction of a line.

RS-170. Electronic Industries Association (EIA) standard governing monochrome television studio electrical signals. Specifies maximum amplitude of 1.4 V peak to peak, including synchronization pulses. Broadcast standard.

RS-232. EIA standard reflecting properties of serial communication link.

RS-232-C, RS-422, RS-423, RS-449. Standard electrical interfaces for connecting peripheral devices to computers. EIA standard RS-449, together with EIA standards RS-422 and RS-423, are intended to gradually replace the widely used EIA standard RS-232-C as the specification for the interface between data terminal equipment (DTE) and data circuit terminating equipment (DCE) employing serial binary data interchange. Designed to be compatible with equipment using RS-232-C, RS-449 takes advantage of recent advances in IC design, reduces crosstalk between interchange circuits, permits greater distance between equipment, and permits higher data signaling rates (up to 2 million bits per second). RS-449 specifies functional and mechanical aspects of the interface, such as the use of two connectors having 37 pins and 9 pins instead of a single 25-pin connector. RS-422 specifies the electrical aspects for wideband communication over balanced lines at data rates up to 10 million bits per second. RS-423 does the same for unbalanced lines at data rates up to 100,000 bits per second.

RS-330. EIA standard governing closed-circuit television electrical signals. Specifies maximum amplitude of 1.0 V peak to peak, including synchronization pulses.

Run Length Encoding. Data compression technique in which an image is raster scanned and only the lengths of "runs" of consecutive pixels with the same color are stored.

Sampling. Mechanism of converting a continuous image (e.g., a photograph) into an image composed of discrete points.

Saturation. Degree to which a color is free of white.

Scan. To move a sensing point around an image.

Scan Line. One scanned line of an image.

Scattering. Process by which light passing through a medium or reflecting off a surface is redirected throughout a range of angles. See Diffuse.

Scene. Three-dimensional environment from which an image is generated.

Scene Analysis. Process of seeking information about a three-dimensional scene from information derived from a two-dimensional image.

Seam Track. Noncontact sensing method of providing feedback on the width of a gap to control a process, typically, arc welding. (a) "Through the arc": Real-time feedback in advance of the welding process; reflects influence of the welding process itself on the gap. (b) "A priori": Gap measurement before welding, the system generates gap data and the path to be followed by the robot.

Seam Tracking. Mechanical probe with feedback mechanism to sense and allow for changes in a given taught path or a vision system to look at a given path (or set points) and determine if it has changed its location with respect to the robot, or a voltage within the welding arc that is read on each side of the welding arc when that arc is oscillated. The difference (or changes) in voltage sends a signal to the robot to change its path accordingly.

Search Area. Area in which the vision system will look for a part. This area is defined by how much the part is expected to move.

SECAM. See NTSC.

Segmentation. Process of separating objects of interest (each with uniform attributes) from the rest of the scene or background; partitioning an image into various clusters.

Semantic Network. Knowledge representation for describing the properties and relations of objects, events, concepts, situations, or actions by a directed graph consisting of nodes and labeled edges (arcs connecting nodes).

Semantic Primitives. Basic conceptual units in which concepts, ideas, or events can be represented.

Sensitivity. Factor expressing the incident illumination upon the active region of an image sensor required to produce a specified signal at the output.

Sentence. String of symbols constructed according to a particular grammar.

Sequential. Scanning system of TV scanning in which each line of the raster is scanned progressively.

Shading. Fidelity of gray level quantizing over the area of the image; method using levels of the gray scale to differentiate a specific blob from neighboring black objects and the white background.

Shape from Contour. Inference of information about the surface in a scene from the shapes of edges.

Shape from Shading. Analysis of clues derived from changes in gray level to infer three-dimensional properties of objects.

Shape from Shape. See Shape from Contour.

Shape from Texture. Analysis of texture variations to infer three-dimensional properties of objects.

Shift Register. Electronic circuit consisting of a series of storage locations (registers). During each clock cycle the information in each location is moved (shifted) into the adjacent location.

Shrinking. Image-processing technique that has the effect of reducing patterns on a binary image by successively peeling their boundaries for purposes of segmentation or simplifying the scene content thinning; see Erosion.

Sibling. Term often used in the SRI algorithm to denote the relationship between two children of the same parent. See Child.

Signal-to-Noise Ratio. Ratio of the peak value of an output signal to the amplitude of the noise affecting that signal (usually expressed in decibels).

Signature. Observable or characteristic measurement or feature pattern derived from units of a particular category. A category is said to have a signature only if the characteristic pattern is highly representative of the n-tuples obtained from units of that category. Sometimes a signature is called a prototype pattern.

SIMD. Acronym for single-instruction/multiple-data computer-processing architecture. See Array Processor.

Simple Decision Rule. Decision rule that assigns a unit to a category solely on the basis of the measurements or features associated with the unit. Hence, the units are treated independently, and the decision rule may be thought of as a function that assigns one and only one category to each pattern in measurement space or to each feature in feature space.

SISD. Acronym for single-instruction/single-data computer-processing architecture.

Slant Transform. See Nonlinear Filters.

Slow Scan. System of scanning in which the time needed to read one line has been increased.

Snap. Term frequently used in machine vision to indicate the loading of a camera image into a buffer.

Snap Shot. Scan disruption of beam for a period of one frame or longer while the layer of the imaging device integrates the incident light.

Sobel Operator. Operator that yields the magnitude of the brightness gradient as a means of edge detection in an image.

Software. Term used to describe all programs and instructions whether in machine, assembly, or high-level language.

Solid-State Camera. Camera that uses a solid-state integrated circuit to convert light to an electrical signal.

Sort (Sorting). Determination of which of a number of unknown objects or patterns is present; analogous to identification.

Span. Allowance of gray level acceptance in thresholding techniques; usually adjustable from 0 to 100% of black to white.

Spatial. Directional characteristics of light in space.

Spatial Filter. Class of image enhancement operators that create an output image based on the spatial frequency content of the input image. See Linear Filters and Nonlinear Filters.

Spatial Frequency. Reciprocal of line spacing in an object or scene.

Spatial Noise. Unwanted artifacts in an image.

Spectral. Distribution of light by wavelength within an electromagnetic spectrum.

Spectral Analysis. Interpreting image points in terms of their response to various light frequencies (colors).

Spectrum. Range of frequencies or wavelengths.

Specular Reflection. Characteristic of light that leads to highly directional redirection from a surface on which it is incident.

Spherical Aberration. Degradation of an image due to the shape of the lens elements.

Spread Function of Image System. Process, component, or material describing the resulting spatial distribution of gray shade when the input to the system is some well-defined object much smaller than the width of the spread function. If the input to the system is a line, the spread function is called the line spread function. If the input to the system is a point, the spread function is called the point spread function.

Square-Wave Response. In image pickup tubes, the ratio of the peak-to-peak signal amplitude given by a test pattern consisting of alternate black and white bars of equal widths to the difference in signal between large-area blacks and large-area whites having the same illuminations as the black and white bars in the test pattern.

SRI Vision Module. Object recognition, inspection, orientation, and location research vision system developed at SRI International; based on converting the scene into a binary image and extracting the calculated needed vision parameters in real time as the scene is sequentially scanned line by line.

Stability. Measure of the amount of change of the shape or size of an image with time (due to electronic signal change, not part change).

Stadimetry. Determination of distance based upon the apparent size of an object in the camera's field of view.

Stereo. Approach to image analysis that uses a pair of images of the same scene taken from different locations where depth information can be derived from the difference in locations of the same feature in each of the images.

Stereopsis. Measurement of distance by use of stereoimages.

Stereoscopic Approach. Use of triangulation between two or more views obtained from different positions to determine range or depth.

String. One-dimensional set of symbols.

Strobe Lamp. Lamp that generates a short burst of high-intensity light through gas discharge.

Structural Pattern Recognition. Pattern recognition technique that represents scenes as strings, trees, or graphs of symbols according to geometric relationships between segmented objects. The set of symbols is compared to a known set of symbols.

Structured Element. Area of a specific shape used in conjunction with morphological operations.

Structured Light. Projected light configurations used to directly determine shape and/or range from the observed configuration that the projected line, circle, grid, etc., makes as it intersects the object.

Subpixel Resolution. Any technique that results in a measurement with a resolution less than one pixel.

Subtraction. Image-creating operation that creates a new image by subtracting corresponding pixels of other images.

Surface. Visible outside portion of a solid object.

Symbolic Description. Noniconic scene descriptions such as graph representations.

Sync. Abbreviated form of synchronization. The timing pulse that drives the TV scanning system.

Sync Generator. Electronic circuit that produces the sync pulses.

Synchronous Processing. Digital signal processing synchronized with a master clock such as a pixel clock.

Syntactic. Relationship that can be described by a set of grammatical rules.

Syntactic Analysis. Process used to determine whether a set of symbols (e.g., a sentence) is syntactically (grammatically) correct with respect to the specified grammar.

Syntactic Pattern Recognition. Scene that is represented by a set of symbols is "recognized" through a formal set of rules similar to the role played by grammar in the English language.

Systolic Array. Matrix of simple identical processor elements with a nearest neighbor interconnection pattern that can perform local convolutions for edge detection, image enhancement, spatial filtering, and differential imaging; also, computer architecture that accesses data only once and proceeds to operate on those data in parallel.

Target. In image pickup tubes, a structure employing a storage surface that is scanned by an electron beam to generate an output signal corresponding to a charge density pattern stored thereon. The charge density pattern is usually created by light from an illuminated scene. One type of category used in the pattern recognition of image data. It usually occupies some relatively small area on the image and has a unique or characteristic set of attributes. It has a high, a priori interest to the investigator.

Target Identification, Target Recognition. Process by which targets contained within image data are identified by means of a decision rule.

Telephoto Lens. Compound lens constructed so that its overall length, from rear focal point to front element, is less than its effective focal length.

Template. Prototype model that can be used directly to match to image characteristics for object recognition or inspection.

Template Match. Operation that can be used to find out how well two images match one another. The degree of matching is often determined by cross-correlating the two images or by evaluating the sum of the squared corresponding gray shade differences. Template matching can also be used to best match a measurement pattern with a prototype pattern.

Tessellation. Pixel pattern. Most tesselations are square (square pixels); some are rectangular (rectangular pixels).

Texture. Local variation in pixel values that repeats in a regular or random way across a portion of an image or object. Texture is concerned with the spatial distribution of the gray shades and discrete tonal features. When a small area of the image has little variation of discrete tonal features, the dominant property of that area is gray shade. When a small area has wide variation of discrete tonal features, the dominant property of that area is texture. There are three things crucial in this distinction: (1) the size of the small areas, (2) the relative sizes of the discrete tonal features, and (3) the number of distinguishable discrete tonal features.

Texture Gradient. Refers to the increasing fineness of visual texture with depth observed when viewing a two-dimensional image of a three-dimensional scene containing approximately uniform textured surfaces.

Three-Dimensional Analysis. Type of vision algorithm that develops a three-dimensional model of a part and matches it to stored models.

Threshold. Intensity (specific pixel value) below which a stimulus produces no effect or response. Dynamic threshold: When threshold result depends on the location of a pixel, the characteristic pixel value and some local property. Global threshold: When threshold result depends only on comparison of a pixel value with a global constant (a single brightness value, a single color value, etc.). Local threshold: When threshold result depends on comparison of a joint local property and a characteristic pixel value with a single decision value, e.g., Threshold is based on comparing the difference between pixel value and neighborhood value with a global value.

Thresholding. Scene segmentation process based on converting a gray scale image into a binary image by reassigning pixel gray levels to only two values. Regions of an image are separated based on pixel values above and below a chosen intensity level.

Threshold Value. Decision point above or below which a decision is made, e.g., a pass or fail point.

Throughput Rate. Generally refers to the number of objects to be examined per unit of time.

Time of Flight. Technique of inferring shape from reflection of active light signal off object based on measuring elapsed time for signal to return when an impulse source is used or by modulating a CW source signal and matching the return signal to measure phase differences that are in turn interpreted as range measurements.

Token. In a data flow machine, term associated with combination of data word and label or tag.

Tolerance. Amount that establishes the range upon which to base the differentiation between good and bad products.

Top-Down Approach. Goal-directed image analysis approach in which the interpretation stage is guided in its analysis by trial or test descriptions of a scene. Sometimes referred to as "hypothesize and test."

Tracking. Processing sequences of images in real time to derive a description of the motion of one or more objects in a scene.

Transforms. Various mathematical transformations applied to image data for the purpose of analysis. See Image Transformation.

Transition. In a binary image, the point where the pixels change between light and dark.

Transition Counting. Modeling of a scene based on the number of white-to black and black-to-white pixel changes.

Translation. Movement left or right, up or down, but not rotated; geometric operation that shifts the position of an image from its original position.

Transmission. Passage of light or other signal.

Transmittance, Transmittance Coefficient. Ratio of the energy per unit time per unit area (radiant power density) transmitted through the object to the energy per unit time per unit area incident on the object. In general, transmittance is a function of the incident angle of the energy, viewing angle of the sensor, spectral wavelength and bandwidth, and the nature of the object.

Tree. Hierarchical representation of a scene.

Triangulation. Method of determining distance by trigonometry.

Two and One-Half Dimensions (2 1/2 D). Photometric stereo; see Needle Diagram.

Ultraviolet (UV). Region of the electromagnetic spectrum adjacent to the visible portion of the spectrum but with wavelengths between 100 and 400 nanometers.

Union (of Two Images). Logical operation forming a new image that is black at all points where either of the two images are black.

Vector Encoding. Method of characterizing line segments extracted from a scene, specifying each segment as a pair of image coordinates corresponding to the segment's end points.

Verification. Activity providing qualitative assurance that a fabrication or assembly process was successfully completed.

Vertex. Point on a polyhedron common to three or more sides.

Vertical Resolution. Number of horizontal lines that can be seen in the reproduced image.

Vertical Sync. Circuit to retrace the scan from bottom to top.

Video. Analog time-varying output signal from an image sensor that conveys the image data.

Video Image. Image in electronic signal format capable of being displayed on a cathode-ray tube screen. The video signal is generated from such devices as a vidicon or flying spot scanner, which convert an image from photographic form to video signal form by scanning it line by line. The video signal itself is a sequence of signals, the signal representing the line of the scanned image.

Vidicon. Image pickup tube in which a change density pattern is formed by photoconduction and stored on that surface of the photoconductor that is scanned by an electron beam.

Visible. Light that can be seen by the eye. Having wavelength between 400 and 750 nanometers.

Vision. Process of understanding the environment based on sensing the light level or reflectance property of objects.

Von Neumann Architecture. Current standard computer architecture that uses sequential processing.

Walsh-Hadamard Transform. See Hadamard Transform.

Wavelength. Reciprocal of frequency. The distance covered by one cycle or event.

Window. Selected portion of an image. Also, a limited range of gray scale values.

Windowing. Technique for reducing data-processing requirements by electronically defining only a small portion of the image to be analyzed. All other parts of the image are ignored.

Wireframe Model. Three-dimensional model, similar to a wireframe, in which the object is defined in terms of edges and vertices.

Zoom. To enlarge or reduce the size of an image. It may be done electronically or optically.

Zoom Lens. Optical system of continuously variable focal length, the focal plane remaining in a fixed position.

Appendix B

Machine Vision Application Checklist

This is a form to assist in developing ideas and requirements for machine vision applications. It contains many typical questions to help determine feasibility, benefits and cost. Permission is granted to use this form.

SECTION 1: PRODUCTION PROCESS

1. What do you make?

2. How do you make it?

3. What is the expected life of the product?

4. Is the product and the problem going to be around long enough to justify the purchase of a system?

5. Why do you need to inspect or control the process? Are there problems? Is process improvement the goal?

6. What is the current reject rate of bad parts?

7. What is the accuracy of the current inspection system?

8. If you inspect or improve control, what are the specific benefits you expect to achieve?

9. Will the system be for a new line or an old line (retrofit)?

10. Does your application involve: one object at a time?

Multiple objects? How many different objects?

What are the different part numbers?

11. Is it a batch operation? or a continuous dedicated process/line?

12. What are the changeover times and the frequency of changeovers?

13. What are the skill levels involved in changeover?

14. How is inspection and/or function to be replaced currently being performed? Is it effective?

15. Is inspection to be: On line? or Off line?

16. Must every produced item be inspected, or can you randomly sample?

17. Will new part models or variations be added to the system at a later date? Define any potential future inspections that may be required of the same machine vision systems:

18. Are product design or production process changes anticipated?

19. Where do parts come from?

20. Can rejected parts be repaired?

21. Can vision assist in the diagnosis?

22. Where do pass and fail parts go?

23. When is the machine vision system needed by?

24. How many shifts will the system be used?

25. How many lines/machine will the vision system be needed for?

26. What is the attitude of the plant floor people towards machine vision/automation?

27. What is the attitude of the plant's management toward machine vision/automation?

28. What is your attitude towards machine vision/automation?

29. Can a representative sample of parts be provided to system vendors or integrators for evaluation?

30. Can drawings be provided?

31. Can video of line be provided?

32. Can the vision suppliers observe production at your facility?

SECTION 2: BENEFITS OF INSPECTION

1. When an incorrect or flawed part escapes detection, what are the downstream effects? (quality, repair, machine downtime, etc.):

2. If a bad part is assembled, does it cause problems with the overall assembly?

3. If inspection is implemented, can any downstream testing requirements be relaxed?

4. If inspection is implemented, do you expect the yield through test to be improved?

5. At the inspection point, what is the cost (qualitative or quantitative) of a bad (faulty) part that escapes?

6. At the inspection point, what is the cost (qualitative or quantitative) of a good part that is falsely rejected?

SECTION 3 APPLICATION

1. Describe the application:

2. What distinguishes a bad part from a good part?

3. Generically, does the application involve:

Gaging? (Show a sketch or drawing, if possible. Highlight critical items.)

What are tightest tolerances? On what specific dimension?

What is the design goal for accuracy?

What is the design goal for repeatability?

Are there features that serve as references?

Describe calibration requirements:

Assembly Verification?

Dimensions of assembly:

Presence/absence:

Orientation:

What is the smallest piece to be verified?

What are the dimensions of that piece?

What is the largest piece to be verified?

What are the dimensions of that piece?

Do you also need to verify the correctness of the part?

Flaw inspection ?

Describe flaw types:

What is the smallest size flaw?

Does the flaw affect surface geometry?

Does the flaw affect surface reflectivity?

Is it more of a stain?

Is classification of flaws required?

Location analysis?

What is the design goal for accuracy?

What is the design goal for repeatability?

What is the area over which "find" is required?

Describe calibration requirements:

Pattern recognition?

What is the size of the pattern?

Describe the differences between patterns:

Is there a background pattern?

Color ?

Geometric?

Number of different patterns?

Purpose: to identify? to sort? other?

Are pattern differences geometric?

Are pattern differences photometric?

Are pattern differences color?

Is application specifically OCR? OCV

Handwritten characters? Fixed font? Variable font?

What is the height of the characters?

What is the stroke width?

What is the spacing between characters?

What is the spacing around the characters

How many characters in a string?

How many lines?

Describe background

What is the color of the characters?

What is the color of the background?

SECTION 4: PART TO BE INSPECTED

1. Describe the part(s) to be inspected (consider those conditions that can change appearance of part or background). Are drawings available?

2. What is specific material (steel, plastic, etc.)?

3. What is specific finish (texture) like?

Is the surface finish the same on all faces of the part?

Is the surface finish the same for all part numbers and/or production runs?

Describe any differences:

Describe the platings:

Coating?

Thin films (oils, mist, etc.)?

Paint? Dull? Glossy?

Specular? Highly reflective (mirror like)? Poorly reflective? Dull? Matte?

Will the reflectivity of the part change from part to part?

Over time?

4. Are there any machining marks on the part?

Does the part generally have scratches, nicks, burrs, dents, etc.?

Is there any porosity on the parts?

5. What are the object's shapes? –Flat? Curved ? Gently curved? Other?

Irregular? Grooved surface ?

Sharp radii (prismatic)? Mixed geometric properties ?

6. Is part always oriented in the same direction?

7. What is the temperature of the part?

8. What is the size of part(s)?

Smallest:

Largest:

9. Are there different colors for different models?

Does the color of the part change from part to part?

Color:

- single hue
- variations in saturation
- subtle color variations
- discrete color variations
- mixed with broad and discrete colors

10. Discuss conditions such as warpage, shrinkage, bending that could be experienced.

11. Is there any change in appearance over time due to environment? (rust inhibitors, corrosion, lubricants, dirt, perishability, etc.)

12. Are there any markings on the part?

13. Is it possible to make a refernce mark on the part, if necessary?

14. What are part appearance variables?

15. Is surface translucent?
    Describe variations in translucent optical density/degree of opaqueness:
16. Is surface transparent? Totally? Partially?
17. Is part sensitive to heat?
18. Is object sensitive to light?
    If yes, what type of light?
    Ultraviolet?
    Visible?
    Infrared?

SECTION 5: MATERIAL HANDLING?

1. Describe the material handling system (current or planned):
2. Is object subject to damage in handling?
   Describe precautions to take in handling:
3. Will inspection be done at a station that also performs other functions?
   Describe those functions:
4. What is the production rate? (How many parts per minute on average?)
   During production catch up mode or peak rates?
5. Are there any expected changes that might affect the above rates?
6. Are parts static/indexed? or moving continuously? If
   indexed:
   How long stationary?
   Total in-dwell-out time:
   What is the settling time?
   Is there any acceleration and, if so, where in the cycle?
   If parts are moving continuously, what is the speed?
   Regulation of that speed?

7. What are the maximum positional variations that can be expected?

+/-X

+/-Y

+/-Z

+/- degrees around X direction

+/- degrees around Y direction

+/- degrees around Z direction

8. How much spacing is there between parts? Is the spacing random? or constant?

Part spacing repeatability:

Do parts ever touch?

Do parts ever overlap?

9. Is there more than one stable state involved? How many?

10. If there will be multiple inspections, will the part maintain the same orientation throughout the process?

11. What is the volume envelope available for an inspection station?

12. Are there any restrictions or obstructions to viewing the product?

13. Is there a weight constraint?

14. How close can an associated electronic enclosure be located?

15. How far will the vision system controller be from other equipment that the system will be interfaced with?

16. Describe any other physical constraints surrounding the proposed installation site:

17. If conveyor - what type?

What color?

What are the appearance variables of the conveyor (specular, uniformity, over time, etc.)?

18. Is a by-pass mode required?

19. Describe action to take when reject is detected:

Describe additional action desired if several consecutive objects are rejected?

What should be the number of consecutive rejects that trigger that action?

SECTION 6: OPERATOR INTERFACE

1. Describe the operator interface that you would like. Try to keep it as simple as possible. Describe "must have" items and then "like to have"

must have items:

like to have items

2. Describe the operators who will operate the equipment - educational level, familiarity with machinery, electronics, computers, experience, etc.:

3. Describe personnel access requirements (password protection, etc.):

4.Are there specific enclosure requirements?

5. Is it desirable to have display of objects under test?

6. Is it desirable to be able to display the last reject condition?

7. Is fail-safe operation required (i.e., part considered a reject unless it passes)?

8. What are the desired program-storage requirements?

9. What are the desired data-storage requirements?

10. Power-failure program storage preservation? Yes No

Power-failure data storage preservation? Yes No

11. Describe reporting requirements:

Must reports be generated without interrupting inspections? Yes No

12. What type of data will the system be required to display to a CRT or communicated to an external device?

If statistics are required, how often will the reports be generated?

Will the reports need to be printed and/or displayed?

13. What false reject rate is acceptable?

14. What escape rate or false acceptance rate is acceptable?

SECTION 7: MACHINE INTERFACES

1. Alarms desired?

2. What other machines must this system be integrated with mechanically or electrically?

3. What event will trigger an inspection?

How will the event be detected?

How will this be communicated to the inspection system?

4. How will the results of the inspection be communicated and implemented?

5. Describe machine interfaces required/handshaking signals, etc.:

part in position sensor type

opto-isolation AC DC

voltage level signal conditioning required

RS 232 RS 422 RS 449 IEEE 488 PCI

Parallel PLC MAP

Ethernet Other interface details:

6. Describe hierarchical interfaces anticipated (to host computer, PLC, etc.)(data, programs, etc.):

SECTION 8: ENVIRONMENTAL ISSUES

1. Describe the environment in which the system must operate:

Factory

Clean room

Laboratory

Outdoor/indoor

2. Describe the air quality:

    Dust/Smoke

    Steam

    Oil

3. Ambient light (type-incandescent, fluorescent, etc.):

4. Dirt on parts:

    Lubricant on parts:

5. Wash-down requirements:

6. Corrosive atmosphere:

7. Temperature range: Operating Storage

8. Humidity range: Operating Storage

9. Radiation

    EMI

    RFI

10. Shock:

11. Vibration:

12. Hazardous environment:

13. Utilities available:

    Compressed air How clean

    Water

    Input power - Regulated Unregulated

    120V 240V Three Phase

    Vacuum

SECTION 9: SYSTEM RELIABILITY/AVAILABILITY:

1. Number of hours use per week?

2. Number of hours available for maintenance per week?

3. Describe calibration procedures required:

4. Describe challenge procedure to routinely verify performance:

5. Maintenance time between failures allowed:

6. Maintenance time to repair allowed:

7. Response time to service:

SECTION 10: OTHER ISSUES/REQUIREMENTS

1. Special paint colors:

2. Shipping:

3. Installation:

4. Warranty:

5. Spare parts:

6. Cost:

7. Documentation (instruction manuals for operator, maintenance, engineering, programming, prints, schematics, spare parts lists, software, etc.):

8. Training - Where and when?

Details (subjects to be covered):

operator:

maintenance:

programming:

For how many?

9. Software Issues - ownership of

Revisions

Will vendor be required to support?

For how long?

Who will make software changes?

Who will be responsible for cost?

Who keeps backup?

Is there a license fee?

10. Hardware Issues -

Who is responsible for changes?

Who is responsible for spares?

11. Describe any safety issues that must be considered:

SECTION 11: ACCEPTANCE TEST/BUYOFF PROCEDURE

1. How will we be sure the machine is functioning properly? (Describe performance test hurdles.)

2. Tests at vendor site:

a. Can good, bad, and/or marginal parts be provided?

b. What is the sample-size for each challenge?

c. Define acceptability criteria for each challenge

d. Define part variations or parts to be used during acceptance testing:

e. Define part position variations to be used during acceptance testing:

f. Define lighting variations to be used during acceptance testing:

g. Define environmental conditions:

3. Tests at installation site:

a. Can good, bad, and/or marginal parts be provided?

b. What is the sample-size for each challenge?

c. Define acceptability criteria for each challenge

d. Define part variations or parts to be used during acceptance testing:

e. Define part position variations to be used during acceptance testing:

f. Define lighting variations to be used during acceptance testing:

g. Define environmental conditions:

SECTION 12: OTHER RESPONSIBILITIES

Describe responsibilities for:

a. Design and build of fixtures:

b. Installation:

c. Start up:

d. Other:

Index